University Bookstore
Fall '87
Rented to _____
Student No. _____

UNIVERSITY BOOKSTORE
SPRING '87
RENTED TO Mark Gammel
STUDENT NO. 430-41-1332

Principles and Practices
of Heavy Construction

Principles and Practices

of Heavy Construction

THIRD EDITION

Ronald C. Smith

(RETIRED), STRUCTURES DEPARTMENT, SOUTHERN ALBERTA INSTITUTE OF TECHNOLOGY

Cameron K. Andres, P. Eng.

STRUCTURES DEPARTMENT, SOUTHERN ALBERTA INSTITUTE OF TECHNOLOGY

PRENTICE-HALL, Englewood Cliffs, N.J. 07632

Library of Congress Cataloging-in-Publication Data

SMITH, RONALD C.
   Principles and practices of heavy construction.

   Includes index.
   1. Building.   I. Andres, Cameron K.   II. Title.
TH145.S58 1986b   690   85-16689
ISBN   0-13-701939-4

Editorial/production supervision and
   interior design: Tom Aloisi
Cover design: Joseph Curcio
Manufacturing buyer: John Hall
Page layout: Diane Koromhas

Printed in the United States of America

10 9 8 7 6 5 4 3 2 1

ISBN: 0-13-701939-4   025

Prentice-Hall International, Inc., *London*
Prentice-Hall of Australia Pty. Limited, *Sydney*
Editora Prentice-Hall do Brasil, Ltda., *Rio de Janeiro*
Prentice-Hall Canada Inc., *Toronto*
Prentice-Hall Hispanoamericana, S.A., *Mexico*
Prentice-Hall of India Private Limited, *New Delhi*
Prentice-Hall of Japan, Inc., *Tokyo*
Prentice-Hall of Southeast Asia Pte. Ltd., *Singapore*
Whitehall Books Limited, *Wellington, New Zealand*

Contents

Preface

The remains of many ancient edifices are historic proof of the devotion and dexterity of the master builders of the past. Today's construction industry has progressed to a level sufficient to be considered by many the backbone of modern civilization. Indeed, those persons participating in the construction industry can be justifiably proud of their contribution to society.

The builders of today have many more concerns to resolve than did their forebears. Not only must buildings achieve an aesthetic value, satisfy all regulatory codes, be constructed within budget and time parameters, but they must also provide a comfortable and pleasant atmosphere for the occupants without the necessity of undue maintenance.

The third edition of *Principles and Practices of Heavy Construction* attempts to acquaint the reader with the latest methods, materials, and equipment used within the industry and to familiarize the reader with those concepts of the construction industry that have stood the test of time. The traditional materials—reinforced concrete, masonry, steel, and timber—are thoroughly examined; in addition, recent developments in the construction industry are introduced.

The text considers all aspects of the construction procedure as it relates to commercial construction and progresses through the stages of (1) site investigation and building layout, (2) site excavation and foundations, (3) the structural frame (the materials used and the construction methods applied), and (4) roofing, insulation, and finishing. Illustrations, diagrams, and tables are utilized extensively throughout the book to aid the reader in visualizing the concepts presented. It is believed that those interested in the various aspects of the construction industry will find the text practical and informative.

Ronald C. Smith and Cameron K. Andres
Calgary, Alberta, Canada

Approximate
Conversion Factors
Imperial to Metric (SI)

Area

$$1 \text{ ft}^2 = 0.093 \text{ m}^2$$
$$1 \text{ in.}^2 = 645.16 \text{ mm}^2$$
$$1 \text{ yd}^2 = 0.836 \text{ m}^2$$

Density

$$1 \text{ lb/ft}^3 = 16.018 \text{ kg/m}^3$$

Force

$$1 \text{ lb} = 4.448 \text{ N (newtons)}$$
$$1000 \text{ lb (1 kip)} = 4.448 \text{ kN (kilonewtons)}$$

Length

$$1 \text{ ft} = 304.8 \text{ mm}$$
$$1 \text{ in.} = 25.4 \text{ mm}$$
$$1 \text{ mile} = 1.609 \text{ km (kilometers)}$$

Mass

$$1 \text{ lb} = 0.454 \text{ kg (kilograms)}$$

Mass per unit area

$$1 \text{ lb/ft}^2 = 4.882 \text{ kg/m}^2$$

Mass per unit length

$$1 \text{ lb/ft} = 1.488 \text{ kg/m}$$

Principles and Practices
of Heavy Construction

Site Investigation

Each construction project begins with the selection of a building site. To ensure proper planning and design of the structure, as much information must be obtained above and below the surface of the proposed building site as is practical. Geotechnical experts—individuals specializing in soil sampling and testing—are retained by the project managers to establish the parameters that will be used in the design of the building foundations.

✳ The amount of testing that is done on the site depends on a number of conditions—the size and complexity of the structure, the type of soil encountered, proximity of the proposed structure to existing buildings, and the level of the groundwater table are the more important items that must be considered. The information is then passed on to the structural designers, who must decide on the type and size of foundations that will be used.

Many projects exceed their budgets and their completion dates because of unforeseen problems during the excavation and construction of their foundations. To ensure that these problems are kept to a minimum, a thorough site investigation is a wise investment.

PRELIMINARY INVESTIGATION

Soil Bearing Capacity

Investigation of the site early in the planning stage is essential for a number of reasons. One reason is the neces-sity of determining the load-bearing capacity of the soil. Heavy buildings require soils of high bearing capacity, rock in some cases, to carry their loads. The depth at which this bearing is available will determine the feasibility of locating the building on the proposed site. If the project is feasible from this standpoint, this investigation will help to estimate the cost and influence the design of the foundations.

The proposed building plans will indicate the distribution of building loads and the proposed size of footings under various parts of the building. A site investigation will show whether the load-bearing capacity of the soil is constant over the entire area or whether special footing plans must be made to overcome an unequal bearing problem.

Topography — DRAINAGE - CONTOUR! TEMP(CONST) PERMANENT

The topography of the area will also be studied during this investigation. Is the site relatively level? Will it be necessary to remove large quantities of earth or, alternatively, to do considerable filling? Is there rock to be removed, and if so, how much?

Kinds of Soil

It is important to know the kinds of soil that will be encountered during excavation. Are they loose—requiring shoring—or are they relatively firm? What type of machinery will be best for excavating this soil? Must the excavated earth by removed from the site or is there room for storage on the property?

Groundwater Level

Another reason for preliminary site investigation is the necessity of determining the level of the water table and the possible presence of underground streams. If groundwater is close to the surface, what are the possibilities of lowering the water table in order to excavate? The presence of an underground stream could present a serious problem if uncovered during excavating operations. If water is there, a decision must be made as to the best method of removing and disposing of it. One means of assessing the possibility of an underground stream is to study available data on investigations of the surrounding area.

Proximity of Other Buildings

How close are other buildings to the construction site? If the excavation is close to an existing building, some means must be found of protecting it from being undermined. If blasting is required, surrounding buildings must be protected from flying debris and from shock of blasting.

Area of Site

The area of the site must be known. How much of the total area will actually be occupied by the building? Is there room left for adequate storage of building materials? If space is at a premium, a careful schedule will have to be worked out for the ordering and delivery of materials, so that the right ones will be on hand at the right time. Too many materials stored in a limited area can only result in confusion and work delays.

Subsurface Exploration

Subsurface exploration for the purpose of determining the type or types of soil to be encountered is a matter of prime importance in preparing for the construction of a heavy building. Neglecting to investigate the subsurface properly can result in foundation faults that may be very expensive to remedy (Fig. 1-1).

There are a number of methods available for examining subsurface conditions, and the one adopted in any particular case will depend on circumstances. It may be that more than one method will have to be used to make the investigation complete. These methods include the use of (1) sounding rods, (2) augers, (3) test pits, (4) wash borings, (5) dry sample borings, (6) rock drillings, and (7) geophysical instruments.

(1) A sounding rod is not used to bring up soil for

FIGURE 1-1 Failure in building wall due to differential settlement of foundation.

examination but rather to determine the depth below the surface at which rock appears and whether the soil resistance is increasing or decreasing as the test proceeds. This method cannot be used in soil containing rocks of any significant size since the rod would be deflected from its vertical course.

The rod is of solid steel, from $\frac{5}{8}$ to $1\frac{1}{2}$ in. in diameter, about 5 ft long. It has a pointed and enlarged head (see Fig. 1-2) for easy penetration and reduced friction. The length is increased by adding sections by means of threaded couplings. The top end is fitted with a flat driving cap. The rod may be driven either by hand or by a mechanical driver.

(2) An auger is useful for bringing up samples from relatively shallow depths in soils which are cohesive enough to be retained in the tool while it is being raised to the surface. It consists of a cylinder, usually 2 in. in diameter, with cutting lips on the lower end (see Fig. 1-3).

FIGURE 1-2 Sounding rod. (*Courtesy Soiltest Inc.*)

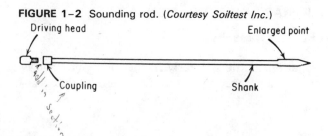

Driving head   Enlarged point

Coupling   Shank

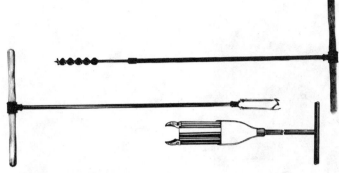

FIGURE 1–3 Hand-operating auger.

It is connected by ordinary couplings to a series of pipe sections. As the auger is turned, layers of earth are peeled off and forced up into the auger cylinder. When the cylinder is full, the auger is brought to the surface, emptied, cleaned, and returned. Power augers, like the one shown in Fig. 1–4, bring up what are known as *disturbed auger borings.*

⟡ (3) The use of test pits is probably the best method of examining subsurface soil because it is possible, by this method, to examine the layers of earth exactly as they exist. In addition, soil moisture conditions are evident, and load tests can be made at any desired depth. This method is relatively expensive, and the depth to which examination can be carried out is limited. Excavating is usually done by hand, but if a required depth cannot readily be reached by hand, mechanical digging equipment may be used.

FIGURE 1–4 Power auger. (*Courtesy Mobile Drilling Inc.*)

⟡ (4) The wash boring method requires the use of water, and, as a result, the borings are in the form of mud. Any given sample may be a mixture of two or more layers of soil. A particular stratum may thus not be detected at all (see Fig. 1–5).

FIGURE 1–5 Wash boring rig on construction site.

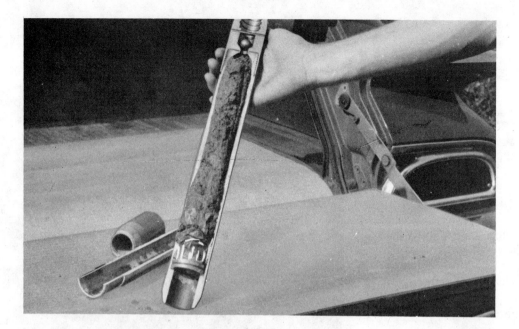

FIGURE 1-6 Split-spoon sampler. (*Courtesy Soiltest Inc.*)

FIGURE 1-7 Drill bits: (a) diamond drill; (b) rotary bit; (c) cross-chopping bit. (*Courtesy Soiltest Inc.*)

(a)

(b)

(c)

The equipment for wash boring consists of an outer casing, inside of which is a hollow drill rod with a cross-chopping bit. Boring is done by raising, lowering, and turning the rod while water is forced down the rod and out through ports in the sides of the bit. Loose material is forced up between the drill rod and the outer casing by water pressure and collected through an opening near the top of the casing. This is known as a *wet* sample.

(5) A *dry* sample may be taken by removing the drill rod and inserting a rod which has a sampling spoon on the end. The spoon, which is similar to a short length of pipe split vertically, is driven into the cuttings at the bottom of the casing and lifted for inspection (see Fig. 1-6).

(6) A number of systems are employed for drilling rock, among them *diamond drilling, shot drilling,* and *churn drilling.*

A diamond drill consists of a diamond-studded bit, like the one shown in Fig. 1-7, attached to a core barrel. The barrel is in turn attached to a drill rod mounted in a rig similar to the one shown in Fig. 1-8. The drill rod is rotated, and water is forced down through the hollow rod to cool the bit and carry the drill cuttings to the surface. The bit cuts a circular groove, and the core is forced up into the core barrel. When it is of the desired length, the core is broken off by a special device and brought to the surface for inspection. Diamond drilling may be done either vertically or at an angle.

Shot drilling is similar to diamond drilling except for the bit. It consists of a circular, hollow, hard steel bit with a slot around the bottom edge to allow circulation of shot. A flow of chilled steel shot is fed through the drill rod to the bit, and as the bit turns, the shot cuts

FIGURE 1-8 Test drilling rig. (*Courtesy Canadian Industries Ltd.*)

The seismic refraction theory is based on the fact that shock waves travel at particular and well-defined velocities through materials of various densities. The denser the material, the greater the speed. The velocities may range from as low as 600 ft/sec in light, dry top soil to 20,000 ft/sec in unseasoned granite. If the speed of the shock wave is known, the type, hardness, and depth of the stratum responsible for the refracted wave can be accurately determined.

The *electrical resistivity measurement* method of analysis depends on the ability of earth materials and formations to conduct electrical current, which follows relatively good conductors and avoids poor ones. Conductivity of the material depends on its electrolytic properties. *Conductivity* and *resistivity* vary according to the *presence and quantity of fluids, mineral salt content of the fluids, volume of pore spaces, pore size and distribution, degree of saturation,* and a number of other factors.

Of course, the use of this type of equipment does not eliminate the need for test boring. Drilling and sampling are necessary for foundation investigations where accurate information on the bearing capacity of a soil is

FIGURE 1-9 Automatic engineering seismograph. (*Courtesy Soiltest Inc.*)

the rock and forms a core, which is forced up into the core barrel. Cores as large as 72 in. in diameter can be taken with a shot bit.

Churn drilling consists of operating a hard steel chopping bit attached to a drill stem inside a casing. A cable, to which is fastened a set of weights, raises and drops the bit so that it strikes the bottom, chipping the rock. Water is forced down the side of the casing and the resulting slurry is removed from the hole with a bucket.

(7) There are two basic types of geophysical instruments used for shallow (100 ft or less) subsurface investigation and exploration. They are refraction seismographs and earth electrical resistivity units. The techniques involved in the use of these instruments have been known and used successfully for many years for deep exploration common to the oil and mineral industries.

Through advances in electronic technology, the heavy, bulky equipment once required for this deep exploration has been reduced to small, compact units which can be carried quite readily by a two-man crew.

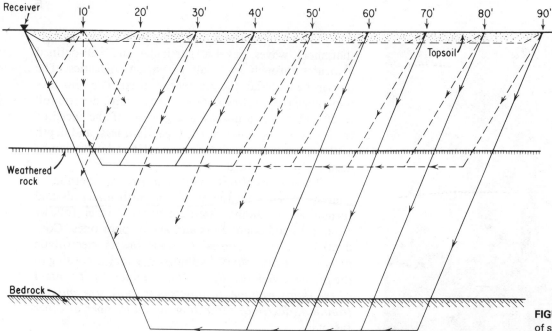

FIGURE 1-10 Schematic diagram of seismic-wave refraction principle.

(a)

(b)

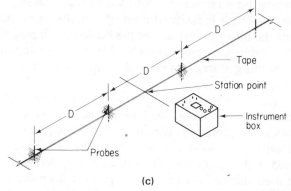

(c)

FIGURE 1-11 Earth resistivity measurement instrument: (a) instrument in position for readings; (b) diagram of set in position for reading (*Courtesy Soiltest Inc.*); (c) paths of current flow and equipotential surfaces in vertical plane.

required and where samples are needed for laboratory analysis. But the use of electronic equipment may materially reduce the amount of drilling necessary and may help in the intelligent selection of drilling sites.

Surface Testing

When investigating a site and when replacing and compacting soil, it is frequently required that tests be taken to determine *soil conditions* for foundations or to ensure that *compaction* and *moisture-control* specifications are being met. There are several methods used for making tests such as these, including a *vane tester* for making shear tests, a *penetrometer* for making penetration resistance, bearing capacity, and compaction tests, and *moisture testers* for determining the moisture content of a soil.

Penetration Tests

One instrument used for determining the *bearing capacity* of subgrades and for *compaction control* is a penetrometer. One which is in common use is the *proving ring penetrometer,* shown in Fig. 1–12, a *cone-type* penetrometer, also used as a quick method of determining the *penetration resistance* of soils in shallow exploration work.

FIGURE 1–12 Proving ring penetrometer. (*Courtesy Soiltest Inc.*)

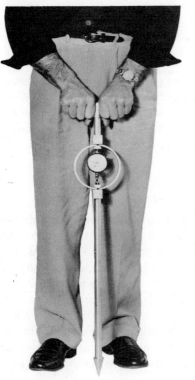

FIGURE 1–13 Hand vane shear tester. (*Courtesy Soiltest Inc.*)

To determine the penetration resistance at a given soil surface, the penetrometer is pushed firmly into the soil at a uniform rate until the top of the penetration cone is reached. The proving ring dial is then read and the corresponding penetration load determined from a calibration chart supplied with the instrument. There is also correlation between penetration, bearing capacity, and degree of compaction.

Shear Tests

It is often necessary to make tests of soil conditions for foundation work in clay soils, and one of these tests is on the *shear strength* of clay. A small tester used for making such tests is a hand vane tester, shown in Fig. 1–13. It is designed for making tests on the shear strength of clays and may be used on the surface, in excavations, or, with extension rods, down a borehole. It consists of a *four-bladed vane,* connected by a *rod* to a *torsion head* containing a *helical spring.*

To make a test, the vane is pushed into the soil and the torsion head is rotated at the rate of 1 rpm. When the clay shears, the load on the torsion spring is released and a pointer registers the maximum deflection of the spring. The shear strength of the soil is then read from a calibration curve.

Moisture Tests

There are a number of reasons for taking moisture tests on soils during construction, and several methods of doing so. One method used involves the use of a tester which operates on the principle of a calcium carbide reagent being introduced into the free moisture in the sample. This forms a gas, the amount of which depends on the amount of free moisture in contact with the reagant. The gas is confined in a sealed chamber, and, by measuring the gas pressure, the amount of free moisture in the sample is determined.

Nuclear Moisture-Density Tests

Nuclear instruments are now in use for testing materials, and among them is a lightweight nuclear meter with a dual gauge, with which both density and moisture tests may be made* (see Fig. 1-14).

With this tester, testing may be done by either the *backscatter* or *direct transmission* method. In using the backscatter method, the tester is simply set on the surface and the handle turned down to a horizontal position to activate the radioactive source. The instrument then measures the amount of radiation that is scattered back from the material. The radiation reflection is inversely proportional to the density of the material. Either the density or the moisture button is pushed to obtain the reading desired.

In direct transmission reading, a template and drive pin are used to prepare the test hole (see Fig. 1-14). The instrument is then placed over the hole, turned on, and the probe is advanced into the hole by 2-in. increments. The dial readings are then converted to density figures by the use of a chart. Density readings are taken by measuring the absorption of gamma rays in the test material. The radioactive source is located at the end of the direct transmission rod, and when it is lowered to a predetermined depth in the test hole already made, the gamma radiation travels through the material and is detected by the instrument on the surface. The denser the material, the more gamma radiation is absorbed and the less the radiation that will reach the detector.

In most jurisdictions, a license is required to own or use such a tester, and, according to the supplier, if the operating instructions are fully complied with, the instrument is safe to use.

There are two important things to remember with regard to the use of testing equipment such as described in the foregoing sections. One is that it must be in excellent working order and properly calibrated in order to produce satisfactory results. Instruments which give incorrect readings may lead to serious mistakes at later stages of the operation. It is therefore imperative that all instruments be thoroughly checked and calibration-tested before being used.

*The information in this section is provided by the manufacturer of this meter, Soiltest Inc.

FIGURE 1-14 Nuclear moisture-density tester. (*Courtesy Soiltest Inc.*)

The other important point is that the operator of such equipment must understand the operation of the instrument he is using so that the results may be accepted with any degree of certainty. Proper training of instrument operators is essential.

Nature of Soil

To make use of the samples taken during subsurface exploration, it is necessary to understand something of the nature of soil, types of soil, and how they react under various circumstances. This topic, in all its facets, is a complete study in itself, and it is intended here to deal with the subject only insofar as it pertains to construction.

What is meant by *soil?* It is the word used by engineers and architects to denote all the fragmented material found in the earth's crust. Included within the meaning of the word is material ranging all the way from individual rocks of various sizes, through sand and gravel, to fine-grained clays. It should be noted that while particles of sand and gravel are visible to the naked eye, particles of some fine-grained clays cannot even be distinguished when viewed through low-powered microscopes.

All soils are thus made up of large or small particles derived from solid rock. They consist of one or more of the minerals which make up solid rock. These particles have been transported from their original location by various means. For example, there are notable deposits of *eolian soil* in western North America which were deposited by wind. There are also numerous deposits of *glacial till,* a mixture of sand, gravel, silt, and clay, moved and deposited by glaciers. Other soils have been deposited by the action of water, while others—known as *residual soils*—consist of rock particles which have not been moved from their original location but are products of the deterioration of solid rock.

Soils are generally described and defined by the size of the particles of which they are composed. Soil types, as determined by particle size, are as follows:

- *Cobbles and boulders:* larger than 3 in. in diameter.
- *Gravel:* smaller than 3 in. and larger than #4 sieve (approximately ¼ in.).
- *Sand:* particles smaller than #4 sieve and larger than #200 sieve (40,000 openings per square inch).
- *Silts:* particles smaller than 0.02 mm and larger than 0.002 mm in diameter.
- *Clays:* particles smaller than 0.002 mm in diameter.

For purposes of establishing the abilities of these soils to safely carry a load, they are classified as *cohesionless soils, cohesive soils, miscellaneous soils,* and *rock.* Table 1-1 lists the allowable bearing values, in pounds per square foot, for the various types of soil indicated.

Cohesionless soils include sand and gravel, soils in which the particles have little or no tendency to stick together under pressure. Cohesive soils include dense silt, medium dense silt, hard clay, stiff clay, firm clay, and soft clay. The particles of these soils tend to stick together, particularly with the addition of water. Miscellaneous soils include glacial till and conglomerate. The latter is a mixture of sand, gravel, and clay, with the clay acting as a cement to hold the particles together. Rock is subdivided into *massive, foilated, sedimentary,* and *soft* or *shattered.* Massive rocks are very hard, have no visible bedding planes or laminations, and have widely spaced, nearly vertical or horizontal joints. They are comparable to the best concrete. Foliated rocks are also hard but have sloping joints which preclude equal compressive strength in all directions. They are comparable to sound structured concrete. Sedimentary rocks include hard shales, sandstones, limestones, and silt-stones, with softer

TABLE 1-1 Bearing strengths of soils

Type of soil	Allowable bearing strength (psf)
Cohesionless soils	
Dense sand, dense sand and gravel	6,000
Cohesive soils	
Dense silt	3,000
Medium dense silt	2,000
Hard clay	6,000
Stiff clay	4,000
Firm clay	2,000
Soft clay	1,000
Miscellaneous soils	
Dense till	10,000
Cemented sand and gravel	20,000
Rock	
Massive	100,000
Foliated	80,000
Sedimentary	40,000
Soft or shattered	20,000

components. Rocks in this category may be likened to good brick masonry. Soft or shattered rocks include those that are soft or broken but not displaced from their natural beds. They do not become plastic when wet and are comparable to poor brick masonry.

One of the problems caused by soil in building operations is related to backfilling around foundation walls and in service trenches. Once soil has been removed from its original location it tends to increase in bulk, because pressure is no longer keeping the particles closely packed. When this soil is returned to the excavation it is still in this bulked state, and, unless some special measures are taken during backfilling, time and weather will bring about a return to the original volume, causing shrinkage in the hole. This can be overcome by compacting the soil while it is being replaced.

Tests have shown that mixing a certain amount of water with the soil enables the particles to be packed more closely together so that all the voids that existed between the particles during the bulked state have been filled. The addition of too much water will simply tend to separate the particles and so increase the overall volume.

The practical implications are important. If soil can be replaced and packed with this optimum amount of water, no shrinkage will take place later. The optimum moisture condition is usually close to the condition of the soil as it was removed. Thus, it is clear that soil removed from an excavation should be protected against drying out as much as possible, if it is to be used again as backfill. However, if the moisture content is changed, it should be allowed to dry or water should be added, as the case may be, before it is returned. Soil should be replaced in thin layers, not more than 6 in. thick, in order to get it into place at its maximum density. Each layer must be treated by some type of compaction machinery. If this process is followed carefully, it is possible to backfill trenches, even very large ones, so that the soil approximates its original condition, and no appreciable settling of the surface occurs.

Frost Penetration

Buildings that are located in cold climates or that are subjected to artificially produced temperatures (ice arenas, cold storage plants, ice plants, etc.) can experience differential movements in their foundations as a result of frost action.

In many instances, when the ground freezes, there are no outwardly visible changes in the soil. If the soil contains moisture, the freezing action will bind the soil particles and alter the strength of the soil. In some cases the soil can experience an increase in volume, causing unwanted pressure on the foundations.

As the mean air temperature drops, the surface of the ground will freeze. With the lower temperature of approaching winter, the freezing plane slowly penetrates the soil. In a fine-grained and moist soil, a peculiar phenomenon occurs. At the freezing plane, the water in the soil turns to ice. This is in effect a drying action, and water in the unfrozen soil beneath moves toward the freezing plane in the same way as water will move from moist to dry soil. This water, upon reaching the freezing plane, is able to flow through and around the soil particles and to join the ice crystals above, thus adding to the growth of a layer of pure ice. Pressure develops so that the ice and soil above it are lifted.

Heaving pressures range over wide limits and depend mainly on the type of soil and its moisture content. A saturated soil will develop the maximum heaving pressure; as the moisture content drops, heaving pressure drops also, and is reduced to zero in a soil with low moisture content. The type of soil is influential, with finer-grained soils developing more pressure than coarser-grained ones. Thus, clay soils develop higher pressure than silts, and silts higher pressure than fine sands. In general, it can be said that coarse sands and gravels do not heave.

The three basic requirements for frost heaving are (1) a freezing plane in the soil, (2) a fine-grained soil through which moisture can move, and (3) a supply of water. If any of these factors can be controlled, frost heaving can be prevented. In a site investigation for a building project, it may be necessary to determine whether ice lenses will form in the soil. Since it is seldom economically possible to control soil temperature, frost heaving is usually prevented by replacing the fine-grained soil with coarse, granular material. Soil moisture can also be controlled by careful attention to drainage, thus greatly reducing the extent of frost heaving.

SECONDARY INVESTIGATION

Secondary does not necessarily refer to importance, but to chronological sequence. The primary investigations discussed in the first part of this chapter are carried out before plans and specifications are drawn—that is, they are made by the planners. But once the building has been decided upon and the tenders let, then the prospective contractor must carry out some investigations of his own.

Access to the Site

The contractor will want to know what roads or streets give access to the site and if it is feasible to move the necessary equipment over them. It may be necessary to build a private road.

If a navigable waterway is accessible, is it practical and economical to use it for the delivery of equipment and material to the site? Is air transportation available? It may be that fast transportation of urgently needed men and supplies will be economical in the long run.

In any case, it is most often advantageous to give some thought and study to the most practical and economical routes by which equipment and materials are to be moved to the job site.

Availability of Services

One of the most important considerations is the availability of electrical power at the job site. Practically no modern construction job can be carried out without it. Is power close at hand, or must it be brought in over long distances? Will the power line be supplied by the power company, or will this be the responsibility of the construction company?

Gas is another very important commodity at a construction site, for both heating and cooking. If there is gas in the area, the location and size of the main must be known in order to bring gas to the site.

Water must be available when a construction job begins. The builder must therefore ascertain whether or not there is a water main in the vicinity. If not, some other method of supplying water must be found. This may be accomplished by means of a well, pipeline, or tank trucks.

Local Building Bylaws

Local and national building codes and zoning bylaws are formulated to set safety standards, to stipulate proper methods of construction, and to designate proper use and occupancy of buildings coming under their jurisdiction. Restrictions stipulate the type of construction required in various fire zones. Other rules govern the distance which a building must be set back from the street, the maximum height of buildings in various localities, the number of stories allowed before there must be a setback in the building, location of fire escapes, etc. Other regulations provide safe working standards for workers; specify types of scaffolds, ladders, etc., which must be used and limit the loads allowed on streets and roads. All these rules and bylaws must be known to the builder so that he may govern his operations accordingly.

Local Labor Supply

An important consideration when planning a construction job is the availability of labor, both skilled and unskilled. The contractor will usually provide his own staff of technical workers (engineers, estimators, and superintendents) but may wish to rely on local supply for the remainder. In some areas unskilled labor will be in adequate supply, while skilled workers, such as carpenters, bricklayers, electricians, and plumbers, will have to be brought in. In other instances, the entire labor force my have to be imported, while in still other situations an adequate supply of labor of all kinds will be locally available. Importing help may mean that the contractor will have to feed and house these workers during construction at his own expense.

Site Conditions

The prospective builder will of course want to know what sort of soil will be encountered during excavation and what problems he may have with underground water. He should be able to obtain this information from the planner, but if it is not available, subsurface exploration will have to be carried out as previously described.

Local Weather Conditions

Climate plays an important part when planning a construction job. Is the area subject to a great deal of rain, is it unusually dry, or just average? The answer to these questions will have an effect on how equipment and materials are to be stored, what types of equipment will work best under those conditions, and whether special construction schedules will have to be developed.

Is the job in an area where a long winter season is likely to be encountered? If so, this will affect the types of equipment to be used, the heating and lighting requirements, and the proper protection that must be provided in order to carry on winter work.

Unforeseen inclement weather may disrupt construction schedules to the extent that completion dates cannot be met. This could result in penalties which may significantly increase the overall cost of the project.

In short, the more a builder learns about the conditions to be met during the operation, the more likely he is to complete the job with a minimum of trouble. He will be able to arrive at a realistic cost figure, to plan a reasonably smooth working schedule, and to complete the building in a manner satisfactory to both himself and the owner.

REVIEW QUESTIONS

1. Outline two basic differences between *preliminary* and *secondary* site investigation.

2. Explain clearly: (a) the principle on which a refraction seismograph instrument works, and (b) the principle involved in the operation of an earth resistivity meter.

3. Give a brief explanation of each of the following terms: (a) *complementary* usage of soil testing methods, (b) bulking of soil, (c) residual soil, and (d) frost heaving in soil.

4. Give an explanation for the statement ''As a general rule, the resistivity readings for sand and gravel will be high and uniform, while those for bedrock will be high and erratic.''

5. Explain: (a) what is meant by the *shear strength* of clay, (b) what is meant by *degree of compaction,* (c) why is it important to know the degree of compaction in certain backfilling operations, and (d) why moisture control is important in compacting soil.

Site Layout

Once the building site has been established, it is then necessary to locate the building precisely within the boundaries of the site. It is also necessary to indicate to what depth excavation must be carried out. When the site is not level, excavation depths will vary, and these variations must be shown.

Included among the drawings for a building is the site plan (see Fig. 2-1), which shows the property lines, available utilities, location of trees, etc., and the location of the building. Approaches and slopes of finished grades are usually shown. In addition, a *bench mark* or *datum point* is very often indicated on the plan. A bench mark is a point of known elevation, established by registered survey and marked by a brass plate on a post at or near ground level or by a brass plug set into a building, bridge, or other permanent structure, near ground level. A datum point is a reference point, usually given an arbitrary elevation, e.g., 100.00 ft or 0.0 ft, such as a manhole cover or some permanent surface at or near ground level on an adjacent building or other structure.

LEVELING INSTRUMENTS

Leveling instruments are used to run the lines, lay out the angles, and ascertain the various differences in elevation required during the construction of a building and its foundations. These are precision instruments incorporating a telescope, leveling devices, horizontal and vertical cross hairs, and finely calibrated scales for measuring horizontal and vertical angles.

Three basic types of leveling instruments are used in the building construction industry: the *builders' level,* the *level transit,* and the *transit.* A variety of styles in each type of instrument can be found, depending on the manufacturer, but the basic purpose and operation of the instruments in one type are similar.

Builder's Level

The builder's level, two varieties of which are shown in Figs. 2-2 and 2-3, is made to operate in a horizontal plane only, and its use is limited to those operations which can be done in that plane, such as *measuring horizontal angles* and *differences in elevation.* All will have the same basic parts, namely, a leveling mechanism, a level vial with bubble, a telescope with an objective lens, an eyepiece and a focusing knob, a horizontal circle with vernier scale, a mechanism for fine adjustments of horizontal movement, cross-hair focusing, as well as a number of others which may or may not be common to all, depending on the make (See Fig. 2-2). Figure 2-4 is an illustration of an automatic level.

Level Transit

Besides turning in a horizontal plane a level transit will also tilt through a vertical arc, up to 45° from the hori-

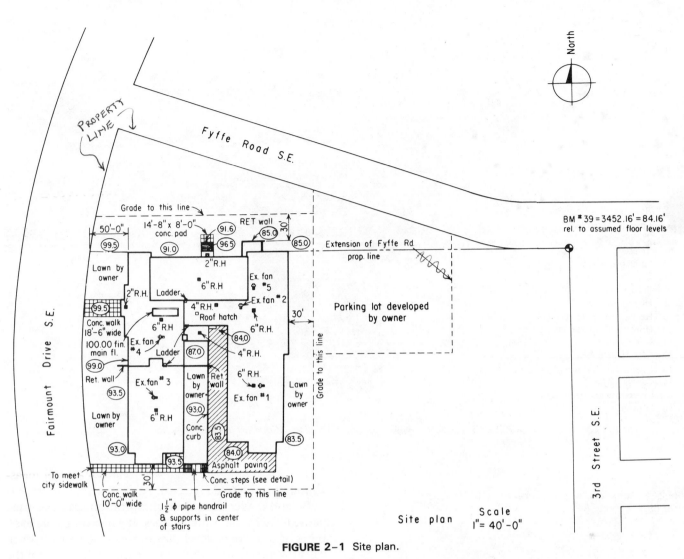

FIGURE 2-1 Site plan.

FIGURE 2-2 Builders' level, three-point leveling. (*Courtesy Wild of Canada, Inc.*)

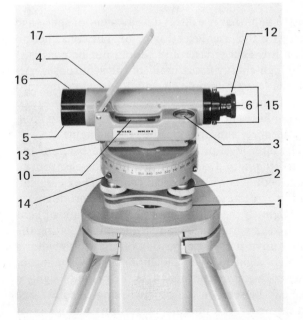
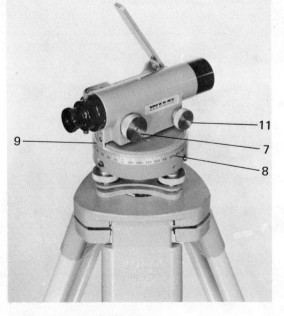

1. Base plate	7. Focusing knob	13. Tubular level adjusting screw
2. Leveling screw	8. Horizontal circle	14. Adjusting screw for leveling screw
3. Circular level	9. Horizontal circle vernier	15. Cross-hair adjusting screws
4. Telescope	10. Tubular level	16. Sunshade
5. Objective lens	11. Horizontal motion tangent screw	17. Bubble viewing mirror
6.	12.	

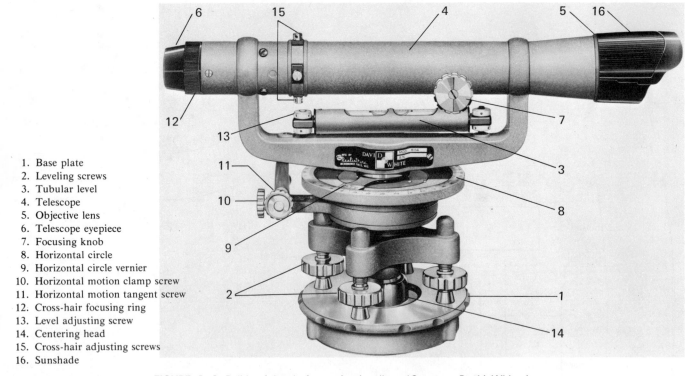

1. Base plate
2. Leveling screws
3. Tubular level
4. Telescope
5. Objective lens
6. Telescope eyepiece
7. Focusing knob
8. Horizontal circle
9. Horizontal circle vernier
10. Horizontal motion clamp screw
11. Horizontal motion tangent screw
12. Cross-hair focusing ring
13. Level adjusting screw
14. Centering head
15. Cross-hair adjusting screws
16. Sunshade

FIGURE 2-3 Builders' level, four-point leveling. (*Courtesy David White Instruments.*)

FIGURE 2-4 Automatic level. (*Courtesy Hickerson Instrument Co.*)

zontal position (see Fig. 2–5). As a result, it is more versatile and makes possible such operations as measuring vertical angles and running lines. It will have all the basic parts of a builders' level plus those involved with vertical movement.

Laser Level

One instrument that is a direct development of space-age technology is the laser level (Fig. 2–6). The construction laser is a low-energy laser and should not be confused with the high-energy lasers used for other purposes. Although the laser is completely safe to use, looking into the laser beam itself should be avoided since damage to the eye could result—an effect not dissimilar to looking directly into the sun or any other bright light.

The laser level produces a controlled beam of red light 5 to 10 mm in diameter by exciting the electrons in a helium-neon gas with electrical energy. The beam can be made to rotate about the instrument, producing a plane of light. If the instrument is level, the plane of light can be used as a reference plane for leveling. Setting the laser level up in a strategic location on the job site allows the plane of light to cover the entire site. When an object such as a survey rod intersects the plane of light, a red line can be seen on the object. Using this line, one person can accurately obtain an elevation reading; a second person is not required on the instrument.

1. Base plate
2. Leveling screws
3. Tubular level
4. Telescope
5. Objective lens
6. Telescope eyepiece
7. Focusing knob
8. Horizontal circle
9. Horizontal circle vernier
10. Horizontal motion clamp screw
11. Horizontal motion tangent screw
12. Cross-hair focusing ring
13. Level adjusting screw
14. Centering head
15. Cross-hair adjusting screws
16. Sunshade
17. Vertical motion clamp screw
18. Vertical motion tangent screw
19. Locking lever
20. Vertical arc
21. Vertical vernier

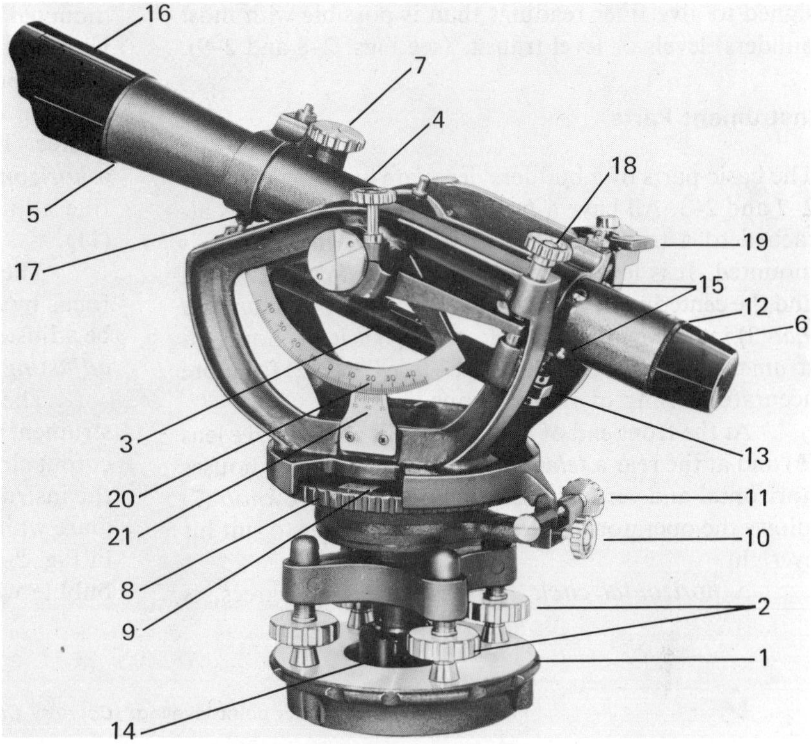

FIGURE 2–5 Level transit. (*Courtesy David White Instruments.*)

FIGURE 2–6 Laser level.

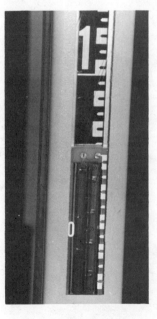

FIGURE 2–7 Laser-sensitive target on survey rod.

Levels can be taken using a normal surveying rod or with the use of a special rod that has a traveling laser-sensitive target (Fig. 2–7). The target moves up and down the rod until it intersects the rotating beam of light. Once the plane of light strikes the target, the target locks on the light beam allowing an accurate rod reading to be taken.

Transit

The basic difference between a level transit and a transit is that the telescope of the latter may be rotated 360° in a vertical plane. The instrument is designed primarily for measuring angles—both horizontal and vertical—and consequently the circle graduations and scales are designed to give finer readings than is possible with most builders' levels or level transits (see Figs. 2–8 and 2–9).

Instrument Parts

The basic parts of a builders' level are indicated in Figs. 2–2 and 2–3. All have a *base plate* (1), by which it is attached to a tripod and upon which the instrument is mounted. It is leveled by means of *leveling screws* (2), and the centering of a bubble in a *circular* or *tubular level vial* (3) indicates when the telescope (4) is level. Some instruments have a second level tube (see Fig. 2–2) for more accurate leveling of the telescope.

At the front end of the telescope is an *objective* lens (5) and at the rear a *telescope eyepiece* (6), which houses horizontal and vertical cross hairs. A *focusing knob* (7) allows the operator to adjust the object focus to suit his eyesight.

A *horizontal circle* (8), graduated in degrees, is mounted below the telescope and the telescope and frame, with an attached *vernier scale* (9) containing an index pointer or mark, revolve on it, making possible the measurement of horizontal angles, usually to within 5 ' of a degree. The instrument can be held in any position by a *horizontal motion clamp screw* (10) and brought into fine adjustment by a *horizontal motion tangent screw* (11).

The cross hairs can be brought into sharp, clear focus by a *cross-hair focusing ring* (12), while they may be adjusted with relation to one another by the *cross-hair adjusting screws* (15).

The *centering head* (14) (see Fig. 2–3) allows the instrument to be moved laterally within the confines of the cutout circle in the base plate. This facilitates positioning the instrument over a *point*. A sunshade (16) cuts down glare while sighting, and the *bubble-viewing mirror* (17) in Fig. 2–2 allows the operator to see the telescope level bubble while sighting through the eyepiece.

FIGURE 2–8 Transit four-point leveling. (*Courtesy David White Instruments.*)

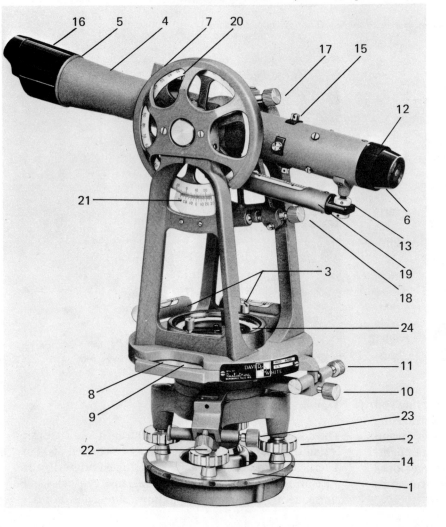

1. Base plate
2. Leveling screw
3. Tubular base levels
4. Telescope
5. Objective lens
6. Telescope eyepiece
7. Focusing knob
8. Horizontal circle
9. Horizontal circle vernier
10. Horizontal motion clamp screw
11. Horizontal motion tangent screw
12. Cross-hair focusing ring
13. Telescope level adjusting screw
14. Centering head
15. Cross-hair adjusting screws
16. Sunshade
17. Vertical motion clamp screw
18. Vertical motion tangent screw
19. Telescope level
20. Vertical circle
21. Vertical vernier
22. Horizontal circle clamp screw
23. Horizontal circle tangent screw
24. Compass

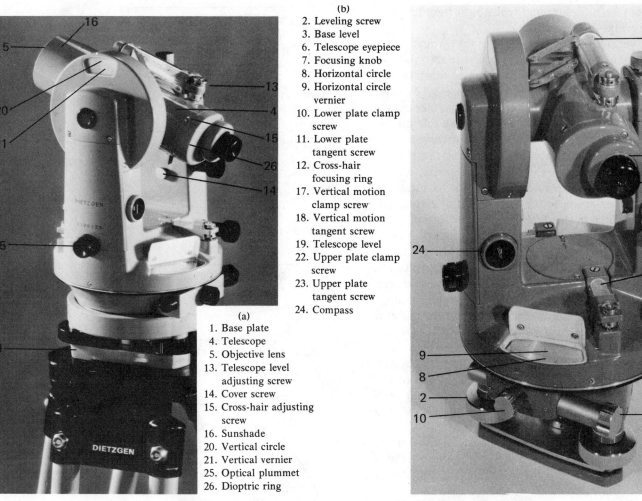

(b)

2. Leveling screw
3. Base level
6. Telescope eyepiece
7. Focusing knob
8. Horizontal circle
9. Horizontal circle vernier
10. Lower plate clamp screw
11. Lower plate tangent screw
12. Cross-hair focusing ring
17. Vertical motion clamp screw
18. Vertical motion tangent screw
19. Telescope level
22. Upper plate clamp screw
23. Upper plate tangent screw
24. Compass

(a)

1. Base plate
4. Telescope
5. Objective lens
13. Telescope level adjusting screw
14. Cover screw
15. Cross-hair adjusting screw
16. Sunshade
20. Vertical circle
21. Vertical vernier
25. Optical plummet
26. Dioptric ring

FIGURE 2–9 Transit three-point leveling. (*Courtesy Eugene Dietzen Co.*)

The level transit, shown in Fig. 2–5 has all the parts of the builders' level. In addition, it has a *locking lever* (19) to hold the telescope in a horizontal position and, when it is disengaged, a *vertical motion clamp screw* (17) to hold the telescope in any desired position. A *vertical motion tangent screw* (18) will allow fine vertical motion adjustments. A *vertical arc* (20), marked off in degrees, and a *vertical vernier* (21), containing an index mark and a minute scale, provide the means for reading the angles of *depression* or *elevation* (below or above the horizontal), usually to within 5′ of a degree.

A transit (see Figs. 2–8 and 2–9) has base levels (3) used for leveling the whole instrument and a *telescope level* (19) for fine leveling of the telescope itself.

In place of a vertical arc, a transit has a full *vertical circle* (20), in graduations of ⅓° or ½° (20′ or 30′ divisions) and a *vernier* (21) which will allow a reading to within 20″ or 30″ of a degree. The horizontal circle and its vernier will be similarly graduated.

Measuring Angles with Builders' Level or Level Transit

With many levels and level transits, angles can be measured to within one-twelfth (5′) of a degree.

The horizontal circle and the vertical arc are marked off in degrees and numbered every 10° (see Fig. 2–10). The vernier is marked into 12 divisions of 5′ each, the 12 representing 1° (60′). Those 12 spaces are equal to 11 spaces (11°) on the circle or arc. The vernier has zero

FIGURE 2–10 Horizontal circle scale.

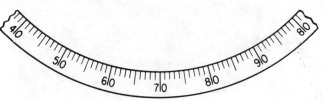

(marked 60) in the center and reads both right and left to 30′ (see Fig. 2-11).

The vernier makes it possible to take a reading to within 5′ (one-twelfth of a degree). If the zero (60) on the vernier coincides *exactly* with a degree mark on the circle, then the reading will be $x°0′$. Figure 2-12 shows a reading of 76°0′. But if the zero on the vernier comes *between* two degree marks on the circle, we rely on the vernier to read the fraction of a degree involved.

Remember, each space on the vernier is one-twelfth smaller than a space on the circle. When zero on the vernier is just enough past a degree mark to allow the first line on the vernier to the right or left of zero to coincide with a line on the circle, the vernier zero must be one-twelfth of a degree (5′) past the degree mark. Therefore, the reading would be $x°5′$. Figure 2-13 illustrates a reading of 63°5′. If the second line on the vernier coincides with a degree mark, it means that the vernier zero is two-twelfths of a degree (10′) past the degree reading. In other words, in order to read the number of minutes by which the vernier zero is past a degree mark, find a line on the vernier which coincides with a line on the circle. If none is found between 0 and 30, in the direction being read, go back to the 30 at the opposite end of the vernier and read toward 60 again. Remember that each space on the vernier, to the right or left, equals 5′. Figure 2-14 shows a reading of 45°50′.

The horizontal circle is marked off in quadrants, with the degrees reading from 0 to 90, then down to 0 again, repeated in the two other quadrants (see Fig. 2-15).

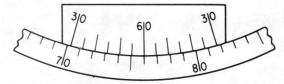

FIGURE 2-12 Reading of 76°0′.

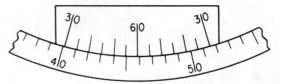

FIGURE 2-13 Reading of 63°5′.

FIGURE 2-14 Reading of 45°50′.

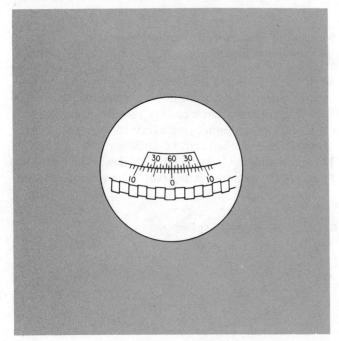

FIGURE 2-11 Vernier scale.

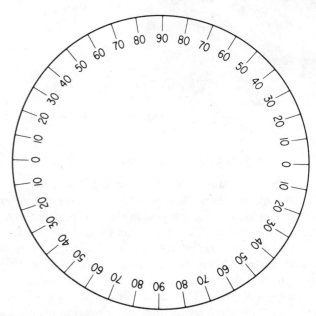

FIGURE 2-15 Horizontal circle quadrants.

Measuring Angles with a Transit

Because a transit's primary purpose is to measure angles, horizontal or vertical, the circles on a transit are usually more finely graduated than those on a level or level transit. The circle may be numbered by one of the systems shown in Fig. 2-16 or by a variation of either one. Each degree will be divided into two or three divisions (30′ or

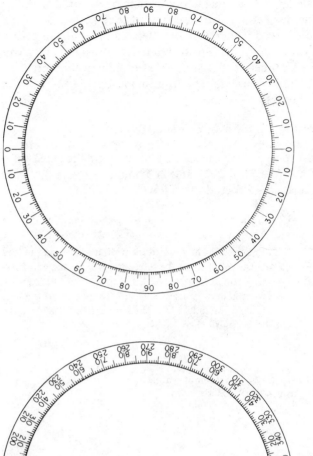

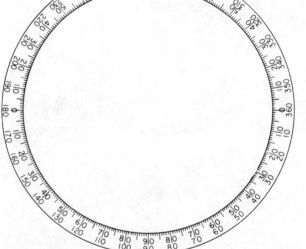

FIGURE 2-16 Transit circles.

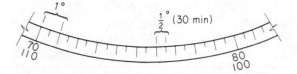

Section of transit circle, graduated into 30 min. intervals

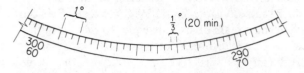

FIGURE 2-17 Graduations on transit circles.

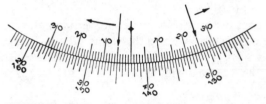

FIGURE 2-18 Vernier reading to 1′.

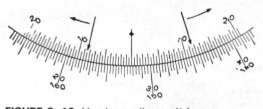

FIGURE 2-19 Vernier reading to ½′.

FIGURE 2-20 Vernier reading to 20″.

20′ intervals), as indicated in Fig. 2-17. Similarly, the vernier will be graduated into 1′, 30″, or 20″ intervals. (See Figs. 2-18, 2-19 and 2-20.)

It is important to remember that, with scales which have a double row of numbers, the circle scales and the vernier must be read in the direction of increasing numbers on the circle.

In Fig. 2-18, the circle scale is divided into ½° or 30′ divisions and the vernier into 30 divisions on each side of the index mark. Following the same vernier principle described in the foregoing section, 30 vernier scale divisions are equal to 29 circle scale divisions and each

vernier scale division will be equal to one-thirtieth of a circle scale division or 1′.

If the angle is turned in a clockwise direction, the scale of increasing numbers will be the lower one, so the vernier index will be seen to be past the 142°30′ graduation on the circle. Then, reading clockwise on the vernier, it will be seen that the fifth vernier mark (5′ mark) coincides with a scale graduation. The correct reading then will be 142°30′ + 5′ = 142°35′.

If the angle is turned in a counterclockwise direction, the reading will be taken in the direction of increasing numbers in the upper row on the circle. In this case, it will be seen that the vernier index is past the 37° mark on the circle scale. Now, reading the vernier in the same direction, it is found that the 25′ vernier mark coincides with a circle scale graduation. The correct reading will be 37° + 25′ = 37°25′.

In Fig. 2–19, each degree on the circle scale is marked off into thirds (20′ divisions) and the vernier is graduated into 40 divisions on each side of the index. Forty vernier scale divisions equal 39 circle scale divisions, so each vernier scale division is equal to one-fortieth of a circle scale division—one-fortieth of 20′ = ½′ or 30″.

If the angle is turned counterclockwise, the vernier index is past 27°40′. Reading counterclockwise on the vernier, it can be seen that the 11½′ vernier mark coincides with a circle scale graduation. Thus, the reading will be 27°40′ + 11′30″ = 27°51′30″.

If the angle is turned clockwise, the vernier index is past 152°. Reading clockwise on the vernier, the 8½′ vernier mark is seen to coincide with a circle graduation. Therefore, the reading will be 152° + 8′30″ = 152° 8′30″.

In Fig. 2–20, the circle scale is graduated into 20′ divisions and the vernier scale into 60 divisions, beginning at one end of the scale. Since 60 vernier scale divisions are equal to 59 circle scale divisions, each vernier

scale division is equal to one-sixtieth of 20′ or 20″. Notice that this vernier scale has an index at each end, so the scale is read from *either end*, depending on the direction in which the angle is being read. The left-hand index will be used when reading an angle turned in a counterclockwise direction, while the right-hand one will be used in the opposite situation.

Setting Up a Level or Level Transit

To use a leveling instrument, it must be mounted on a *tripod* (see Figs. 2–21 and 2–22) and properly adjusted before use. Proceed as follows:

1. Check the tripod to see that all screws, nuts, and bolts are properly tightened (see Figs. 2–21 and 2–22).
2. Set the tripod in the desired location and spread the legs so that the eyepiece of the telescope will be in a comfortable position for sighting. A spacing of about 3 ft for the legs is generally satisfactory. The spacing of the legs should be adjusted so that the tripod head is as level as can be accomplished by eye.

FIGURE 2–22 Instrument mounted on tripod. (*Courtesy Wild of Canada, Ltd.*)

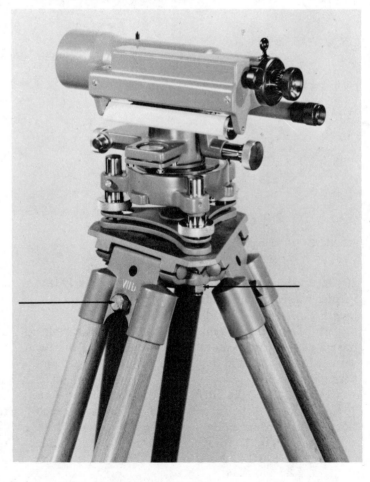

FIGURE 2–21 Instrument tripod. (*Courtesy David White Instruments.*)

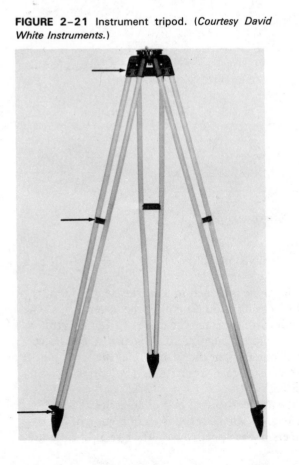

3. Push the legs firmly into the ground. On a paved surface, be sure the points will hold securely.

4. Tighten the wing nuts at the top of the tripod legs into a reasonably firm position.

5. Remove the cap from the tripod head and place it in the instrument box. Lift the instrument from the box by the frame, *not by the telescope,* and set it on the tripod head. Make sure the horizontal clamp screw is loose.

6. Holding the level by the upper structure, screw the leveling head firmly into place.

Leveling the Instrument

Leveling is the most important operation in preparing to use the instrument. It is good practice to adopt the following procedure for four-point leveling:

1. Make sure that the locking levers holding the telescope in a horizontal position are locked, if a level transit is being used.

2. Loosen two adjacent leveling screws, as shown in Fig. 2-23 (a).

3. Turn the telescope so that it is parallel to two opposing screws—Fig. 2-23(b).

4. Bring the bubble to the center of the level tube (see Fig. 2-24) by loosening one screw and tightening the opposite one. Try to turn them about the same amount simultaneously. Figure 2-25 shows how to hold and turn the screws for adjustment. The pressure of the screws against the leveling head should be light and only finger tight.

5. Move the telescope 90° over the other pair of screws and repeat step 4.

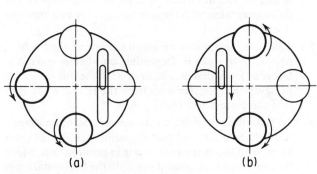

FIGURE 2-23 Turning level screws.

FIGURE 2-24 Level bubble centered.

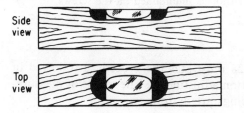

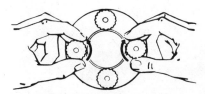

FIGURE 2-25 Adjusting level screws.

6. Turn back over the first pair of screws and center the bubble again.

7. Turn back to the second pair of screws and check the levelness.

8. Turn the telescope 180° over the same pair of screws as in step 6. The bubble should now be centered in this or any other position of the telescope.

Three-Point Leveling

For instruments with only a circular level vial, leveling is a simple operation. The instrument is leveled by adjusting the leveling screws in pairs until the *circular level bubble* is brought to the exact center of the circular vial.

For instruments which have a second, tubular telescope level (see Fig. 2-2), the procedure is as follows:

1. Level the instrument by using the leveling screws and the circular level bubble.

2. Turn the instrument until the telescope level is parallel to one pair of leveling screws.

3. Turn that pair of screws by the same amount, in opposite directions, until the telescope level bubble is accurately centered.

4. Rotate the instrument 180°. If the bubble does not remain centered, correct half of the deviation by adjusting the level screws and the other half by adjusting the level-adjusting screw (see 14, Fig. 2-2).

5. Turn the instrument 90° so that one end of the telescope lies over the third level screw and the other end between the first two. If the bubble is not centered, center it by adjusting the third level screw.

Some instruments with a second level have a *split bubble* system in that level (see Fig. 2-26). The telescope bubble is viewed through a viewing microscope, and if the telescope is not level, the bubble will appear to be split, as in Fig. 2-27(a). When the two halves are made to coincide, as in Fig. 2-27(b), the bubble is centered. The procedure is as follows:

1. Center the lower spirit level bubble by adjusting the level screws, as above.

2. Bring the two halves of the split bubble into coincidence by adjustment of the *tilting screw* (see 3, Fig. 2-26).

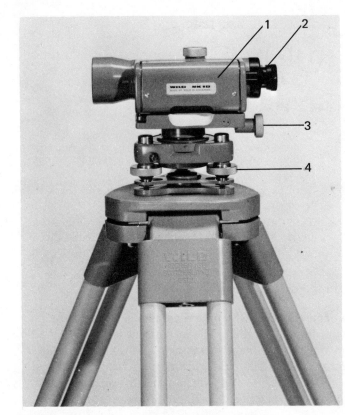

FIGURE 2–26 Level with split-bubble system: 1, level cover plate; 2, bubble-viewing eyepiece; 3, tilting screw; 4, level screw. (*Courtesy Wild of Canada, Ltd.*)

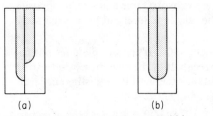

(a)   (b)

FIGURE 2–27 Split telescope bubble.

Sighting Adjustments

The instrument is now ready for sighting adjustments. These may vary slightly for different individuals.

1. Aim the telescope at the object and sight it first along the top of the tube.
2. Then look through the eyepiece and adjust the focus by turning the focusing knob until the object is clear.
3. Now turn the eyepiece cap or focusing ring until the cross hairs appear sharp and black.

4. Adjust the telescope until the object is centered as closely as possible. *Do not put your hands on the tripod.* Tighten the horizontal clamp screw.
5. Center the object exactly by bringing the vertical cross hair into final position with the horizontal motion tangent screw.

Collimation Adjustments

The horizontal cross hair is on the *line of sight*, which should be on a true horizontal plane for any direction in which the telescope is pointed. The vertical distance from the ground on which the tripod stands or from a stake from which measurements are being taken to that line of sight is called the *height of instrument* (H.I.).

Collimation of a leveling instrument means checking and adjusting to make sure that the line of sight through the telescope *is* parallel to a true horizontal plane. To check and make this adjustment, proceed as follows:

1. Drive two stakes, A and B, into the ground about 100 ft apart.
2. Set the instrument up close enough to stake A so that the telescope eyepiece will be within 4 in. of a level rod held on the stake. Level the instrument very carefully.
3. Sight through the *objective end* of the telescope, and, using a pencil held on the rod, determine the H.I. at stake A.
4. Move the rod to stake B, and, using a rod target, determine the distance from the top of stake B to the line of sight.
5. Move the instrument to a point beside stake B, and repeat steps 3 and 4 to find an H.I. and a rod reading on stake A.
6. Calculate the true difference in elevation between stakes A and B. The difference is equal to the average of the differences taken with the instrument at the two locations A and B.
7. Calculate the amount by which the line of sight is high or low, from stake B. Depending on whether stake A is higher or lower than B, add or subtract the true difference in elevation from step 6 to/from the H.I. at stake B to find the correct rod reading at A.
8. To make the correction, set the correct reading on a target rod held at stake A. Make the adjustment to the cross hairs, with the instrument sitting in position at B. Move the cross-hair ring up or down until the horizontal cross hair exactly coincides with the correct rod reading set on the target. The cross-hair ring is moved up or down by the adjustment of the upper and lower cross-hair ring adjustment screws. Turn each by the same amount so that the same tension is maintained on the screws.

Example

Figure 2–28 illustrates the procedure followed in making collimation adjustments. The results of measurements and readings are as follows:

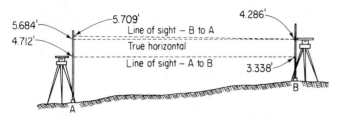

FIGURE 2-28 Example of instrument collimation.

1. H.I. at stake A   = 4.712 ft
2. Rod reading at B, from A   = 3.338 ft
3. Apparent difference in elevation   = 1.374 ft
4. H.I. at stake B   = 4.286 ft
5. Rod reading at A, from B   = 5.709 ft
6. Apparent difference in elevation   = 1.423 ft
7. True difference in elevation between A and B =

$$\frac{(1.374 + 1.423)}{2} = 1.398 \text{ ft}$$

8. Correct rod reading at A = 4.286 + 1.398 = 5.684 ft
9. The line of sight was high by 5.709 − 5.684 = 0.025 ft

Leveling Rods and Their Uses

One of the important uses of a leveling instrument is to find the difference in *grade* or *elevation* between two points. To do this, it is necessary to use a *leveling rod*, and, to use a leveling rod, one must be able to read it properly.

A section of a leveling rod is shown in Fig. 2-29. Some rods are equipped with a *target* (see Fig. 2-29) which can be moved up or down the rod until the horizontal target line corresponds to the instrument's line of sight (horizontal cross hair). The target may be mounted on a bracket which is held in place by friction, or, on some rods, the target is equipped with some type of clamp, one of which is illustrated in Fig. 2-30.

Rods may be graduated in feet and inches (see Fig. 2-29), or they may be graduated in feet and decimal fractions of a foot (see Fig. 2-30). In the latter case, the target on such a rod may have a vernier scale on it which will allow a rod reading to be made to three decimal places—thousandths of a foot.

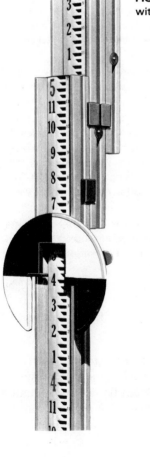

FIGURE 2-29 Section of leveling rod with target.

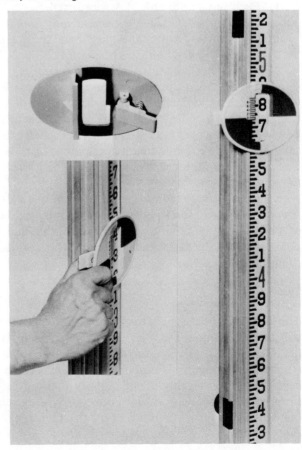

FIGURE 2-30 Section of leveling rod with vernier. (*Courtesy C. L. Berger & Sons.*)

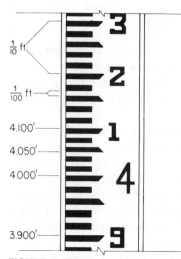

FIGURE 2–31 Graduations on leveling rod.

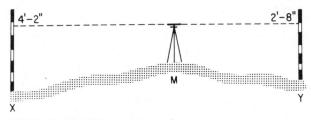

FIGURE 2–32 Difference in grade.

How to Read a Rod with Vernier Target

In the first place, the rod is graduated in 1-ft intervals, marked in large numbers (see Fig. 2–30). Next, each 1-ft division is graduated into 10 equal parts, each one one-tenth of a foot, numbered 1 through 9 (see Fig. 2–31). Next, each one-tenth of a foot is graduated into 10 equal parts, each one equal to one-hundredth of a foot (see Fig. 2–31). Each one-hundredth of a foot is represented by either a black or a white band—consequently, one reads to either the top or bottom of a black band to ascertain the correct reading in hundredths of a foot (see Fig. 2–31).

But the horizontal line on the target may lie somewhere between the top and bottom of a black or white band, and, to obtain a final reading to three decimal places, one must consult the vernier scale on the target. The scale is graduated into 10 equal parts, each one-tenth smaller than the one-hundredth graduations on the rod. The index (0) on the scale coincides with the horizontal line on the target.

To get a final reading, find the line on the vernier scale which exactly corresponds to the top or bottom edge of a black band on the rod. The number of that line will represent the third decimal place in the reading—thousandths of a foot.

Elevations

The difference between the rod readings at two locations will be the difference in *elevation* or grade. This is illustrated in Fig. 2–32. To read this difference, set the instrument in a convenient location from which both points can be seen, preferably about midway between them. Level the instrument as explained previously. Have someone hold a rod vertically at one of the two points, or *stations*, and sight on it. Record the reading at which the

horizontal cross hair cuts the graduations on the rod. Have the rod held over the second station and swing the instrument around without otherwise disturbing it and take another reading on the rod. The difference between the two readings is the amount that one station is above or below the other. For example, in Fig. 2–32, the reading at station *X* is 4 ft 2 in. and at station *Y*, 2 ft 8 in. The difference between these two, 1 ft 6 in., is the amount by which station *Y* is higher than station *X*. In other words, there is a difference in grade between the two points of 18 in. The elevation of a whole series of stations can be compared to a given point by the same method.

Another use of a leveling instrument is to find the elevation at a particular point or points, knowing the elevation of some designated point, a *bench mark*. Proceed as follows:

1. Set up the instrument at a convenient point between the bench mark and the unknown elevation, where the rod (held on the bench mark) will be in sight. Level the instrument.

2. Take a sight on the rod and record the reading. This is called a *backsight*. That reading, *added* to the bench mark elevation, is the H.I.

3. Have the rod moved to a convenient location between the instrument and the unknown elevation. Loosen the clamp screw and swivel the instrument around so that a reading can be taken on the rod at its new location. This is a *foresight*. Record that reading and *subtract* it from the H.I. The result is the elevation of the point on which the rod rests—station 1.

4. Now move the instrument to a new position between station 1 and the unknown elevation and take a backsight. Add that backsight reading to the elevation of station 1 and you have a new H.I.

5. Have the rod moved to a new station and take a foresight. From it establish the elevation of station 2.

6. This procedure is repeated until the final station reaches the unknown elevation.

Notice that a backsight was added to a known elevation to obtain the H.I. It is thought of as a *plus* quantity. A foresight is subtracted from the H.I. to obtain a station elevation. It is thought of as a *minus* quantity. In reality, this continuous process of adding and subtracting is not necessary. The backsights are merely all recorded as being *plus*, and their sum is a *plus* quantity. The foresights are all recorded as *minus*, and their sum is a *minus* quantity. The difference between these two totals is a net *plus* or *minus* quantity, depending on which is larger. The difference, added to or subtracted from the original bench mark, will give the elevation of the point in question.

For example, the elevation of a point A is required, from a bench mark recorded as 2642.62 ft, far enough away that three stations are required between the bench mark and the point A. The recorded readings were as follows:

	Backsight	Foresight
1	4.26 ft	3.19 ft
2	6.19 ft	8.27 ft
3	5.35 ft	6.92 ft
4	3.68 ft	7.75 ft
	+ 19.48 ft	− 26.13 ft

Net difference = 26.13 − 19.48 = − 6.65 ft.
Elevation of point A = 2642.62 − 6.65 = 2635.97 ft.

Running Straight Lines

One particular use of a level transit is for running straight lines. A straight line must be run from a point in a given direction or to another point.

1. Set the instrument up over the starting point. To do this, use the plumb bob in the instrument box. Attach its cord to the eyelet in the middle of the centering head. Adjust the position of the tripod legs until the plumb bob is as nearly centered as possible over the starting point.
2. Loosen two adjacent screws and move the centering head within its limits until the plumb bob is aligned exactly over the center of the starting point.
3. Level the instrument and unlock the locking levers.
4. Sight on the second point given or point the telescope in the given direction and lock it in that position with the horizontal clamp screw.
5. By tilting the telescope up or down, a number of positions can be sighted and marked on the ground, all of which will be in a straight line from the starting point (see Fig. 2–33).

FIGURE 2–33 Running a straight line.

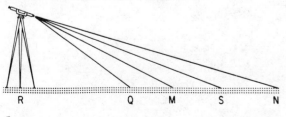

Measuring Angles

A level or level transit is readily used for laying out angles or for measuring a given angle. To lay out an angle, the *intersection* of the two legs must be known, the *direction* of one leg, and the *size* of the angle.

1. Set up the instrument over the point of the angle, using the plumb bob as previously described, and level the instrument.
2. Unlock the locking levers and sight on a point (a stake) on the given leg of the angle. Record the reading on the horizontal circle or turn the horizontal circle to a zero reading. The bottom edge of the horizontal circle is knurled, and the circle can be turned by applying light pressure with a finger.
3. Loosen the horizontal clamp screw, and turn the instrument through the required angle. It can be brought to the required angle by using the horizontal tangent screw. Tighten the clamp screw again.
4. Run a line in this direction for the required distance.

For measuring angles already laid out, the procedure is much the same. Figure 2–34 shows a plot outline for which the angles may have to be measured.

1. Set up the instrument at station 1 and center it over the station pin.
2. Level the instrument and release the locking levers.
3. Sight on station 2, centering the vertical cross hair on the station pin. Set the horizontal circle to zero or record the reading.
4. Turn the instrument and sight accurately on station 4. Record the new reading on the circle. In this case it is 120°.

FIGURE 2–34 Measuring angles.

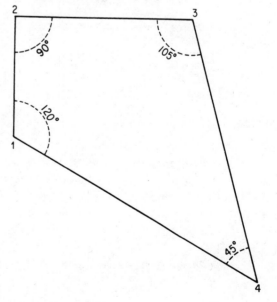

5. Now set up over station 2, sight back to station 1, and around to station 3. Read the angle—90°.

6. Repeat the procedure at stations 3 and 4.

Sights in Vertical Plane

A level transit is required for taking vertical sights, such as plumbing building walls, piers, columns, doorways, and windows. First level the instrument; then release the locking levers that hold the telescope in the level position. Swing the telescope vertically and horizontally until the line to be established is directly on the vertical cross hair. Tighten the horizontal clamp screw. If the telescope is now rotated up or down, each point cup by the vertical cross hair is in a vertical plane with the starting point (see Fig. 2–35).

Leveling the Transit

1. Set up the tripod with the legs extended to the approximate height required and the head in an approximately level position.

2. Place the instrument on the tripod head and turn it down snugly or fasten it in place with the bolt, depending on the type of instrument.

3. Place the telescope tube in the horizontal position and rotate it until it is approximately parallel with one pair of leveling screws.

For transits with four-point leveling, proceed as described for leveling other instruments.

For transits with three-point leveling,

4. Use the two leveling screws and the plate vials to make

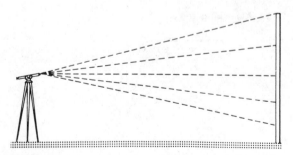

FIGURE 2–35 Establishing a vertical line.

necessary adjustment to bring the bubble to the center of the telescope vial.

5. Rotate the instrument 180° and make the necessary adjustment to center the bubble.

6. Rotate to the original position and check.

7. Rotate the telescope tube 90°, to place it directly over the third leveling screw. Adjust that screw to bring the bubble to center.

8. Rotate 180° and make the adjustment necessary.

9. Bring the instrument back to its original position for a final check.

Collimation of Transit

To collimate the instrument is to make the *line of sight perpendicular to the horizontal axis* of the instrument. If the instrument is not collimated, a straight line cannot be extended by direct means. The recommended procedure is as follows:

1. Set the instrument up where clear sights of approximately 200 ft (60 m) are available on both sides and carefully level the instrument.

2. Sight the instrument accurately on a point on stake A (see Fig. 2–36). Lock the horizontal motion.

3. *Plunge* the instrument (rotate the telescope on its horizontal axis so that the positions of the telescope and level tube are reversed).

4. Sight on stake B and set a point on it to coincide with the vertical cross hair.

5. Leaving the telescope in the reversed position, loosen the horizontal clamp screw and rotate the instrument until the vertical cross hair coincides with the point on stake A. Use the horizontal tangent screw for the final alignment.

6. Leave the horizontal motion clamped and plunge the telescope again (bring it back to its original position).

7. Sight in the direction of stake B. If the vertical cross hair coincides with the point previously set on stake B, no adjustment is necessary.

8. If they do not coincide, set stake C and a point on it which *does* coincide with the vertical cross hair.

9. Since double-reversing is involved, the apparent error is four times the actual error. Therefore, set stake D and mark a point on it which is one-fourth of the distance from stake C to stake B.

FIGURE 2–36 Collimating transit.

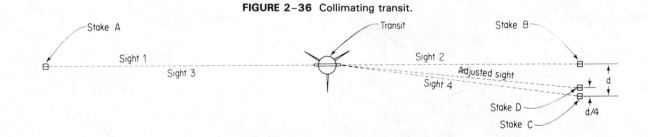

10. Shift the cross-hair ring laterally by adjusting the capstan screws at the ends of the horizontal cross hair until the vertical cross hair coincides with the point on stake D.

For further adjustments to this or any other instrument, consult the manual prepared by the manufacturer and shipped with the instrument (see Fig. 2-37).

LAYOUT BEFORE EXCAVATION

The boundaries of the lot on which a building is to be constructed should be established by markers, called *monuments*, set by a registered surveyor. The site plan shows the positions of the boundary lines, relative to streets, roads, etc. (see Fig. 2-38). The contractor locates the boundary markers and, by running lines between them, can establish the boundaries. Now, from the distances given on the site plan and using a tape, the building can be laid out by driving stakes at the corners and on the column center lines (see Fig. 2-39). The accuracy of the staking may be checked by measuring the diagonals of the rectangles concerned. Each pair will be exactly equal if the stakes are set in their correct position.

These corner stakes will be lost during excavating operations, and it is therefore necessary to record their position beforehand. This is done by means of *batter boards*, on which are marked points that are on the various building and column lines. These batter boards are horizontal bars fastened to posts, which are set up singly or in pairs around the important points in the building (see Fig. 2-40). The basic purpose of the stak-

FIGURE 2-37 Transit in use over point.

ing outlined above is to enable the workers to locate batter boards correctly. They must be set far enough back from the stakes to ensure that excavating will not disturb them.

FIGURE 2-38 Site plan.

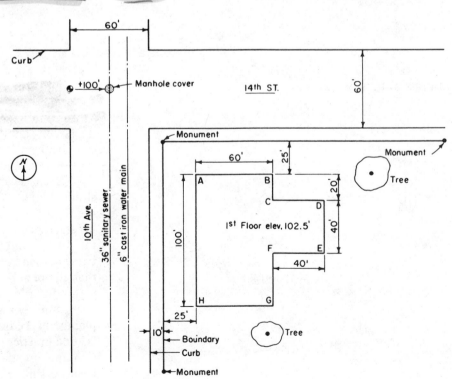

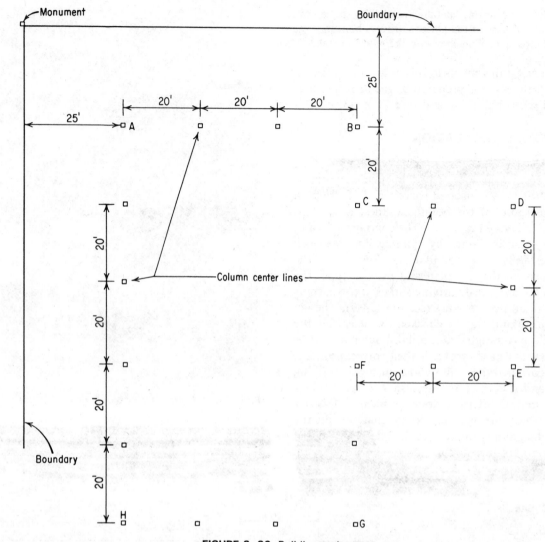

FIGURE 2–39 Building stakes.

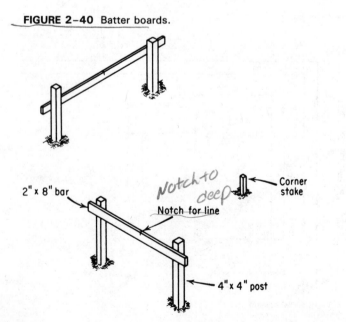

FIGURE 2–40 Batter boards.

Lines running between designated points on two opposite batter boards represent a building line (outside of wall frame, center line of column footings, etc.). Two intersecting lines represent a building corner, and a plumb bob dropped at that intersection will pinpoint a corner on the excavation floor.

Because wire lines are often used to withstand the strain of pulling them taut, strong batter boards and posts are required. The posts may be timbers (4 in. by 4 in. or larger), small steel beam sections, or pipe driven firmly into the ground and braced if necessary. The horizontal bars are then bolted or welded to the posts, making sure that all are at the same level. That level should coincide with a floor level of the building, if possible, or at least bear some definite relation to it. The elevation should be plainly marked on one or more of the batter boards. It is good practice to check these elevations during excavation and foundation work to make sure that there has been no shift or displacement.

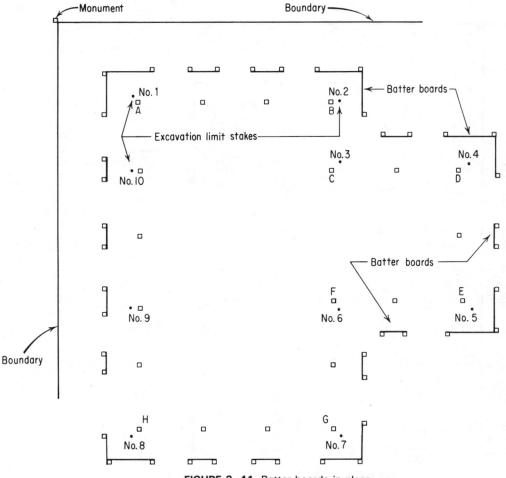

FIGURE 2-41 Batter boards in place.

The limits of the excavation now have to be staked out. The excavation will usually extend 2 ft or more beyond the boundaries of the building itself—to the payline—in order to allow room to work outside the forms. Excavation limit stakes are accordingly driven at the required distance from the corner stakes, as shown in Fig. 2-41.

PRELIMINARY LAYOUT PROBLEMS

Setting Batter Boards to a Definite Level

For the building indicated in Fig. 2-38, it has been decided to set the batter boards to the first floor elevation, shown on the plan as 102.5 ft. Proceed as follows:

1. Set up the leveling instrument so that sights can be taken on the bench mark and the batter boards and level it.

2. Take a backsight on the bench mark. Suppose that the rod reading is 4.86 ft. The H.I. is then 104.86 ft.

3. The difference between the H.I. and the batter board elevation is 104.86 − 102.5 = 2.36 ft. Set the target on the rod at 2.36 ft.

4. Now place the rod alongside a post of the batter board at A, Fig. 2-41, and move it up or down until the cross hair coincides with the target. Mark the height of the bottom of the rod on the post.

5. Fasten one end of a batter board to the post with its top edge at that mark. Level the batter board and fasten to the second post.

6. Set the remainder of the batter boards at exactly the same level, using the instrument to check their height and levelness.

Projecting Building Lines onto Batter Boards

1. Measure accurately (from the boundary intersection) the distances to stations A, B, C, D, E, and F. These distances will be 25, 45, 65, 85, 115, and 125 ft. respectively (Fig. 2-42).

2. Set up the instrument over station A, level it, and backsight on the boundary intersection marker. Now set the horizontal circle to 0.

3. Turn the telescope to 90° and tighten the horizontal clamp screw. Sight across the intervening batter boards at stakes 1 and 2 and mark carefully on each batter board where

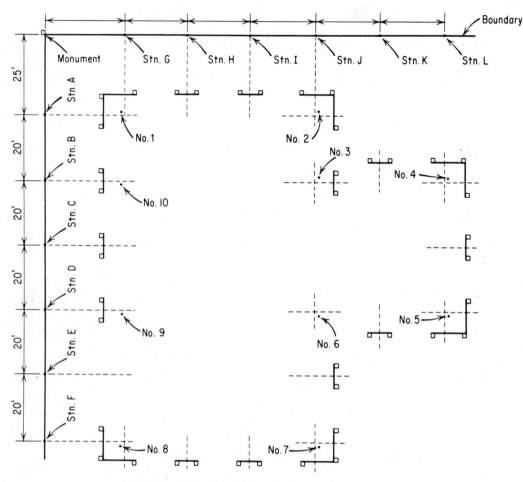

FIGURE 2-42 Running lines on batter boards.

the vertical cross hairs intersect the top edge of the batter board.

4. Set up over station B and mark the building line on batter boards at stakes 10 and 4.

5. Repeat the procedure at stations C, D, E, and F.

6. From the boundary intersection, measure the distances to stations G, H, I, J, K, and L. They will be 25, 45, 65, 85, 115, and 125 ft, respectively.

7. From these stations mark the remainder of the batter boards.

Marking the Depth of Cut on Excavation Limit Stakes

Figure 2–43 shows the excavation limit stakes for the building in Fig. 2–38. The foundation plan indicates that the top of column footings are to have an elevation of +89.5 ft. The excavation is to be taken 6 in. below that level.

1. Set up the instrument at a convenient location from which it is possible to sight on the bench mark and the limit stakes.

2. Find the differences in elevation between the bench mark and each of the limit stakes and record these differences. Suppose that they are

Stake no.	Elevation (ft)	Difference (ft)
1	101.5	+ 1.5
2	99.2	− 0.8
3	99.0	− 1.0
4	98.0	− 2.0
5	96.5	− 3.5
6	97.5	− 2.5
7	95.0	− 5.0
8	99.0	− 1.0
9	100.0	0.0
10	100.8	+ 0.8

3. Calculate the difference in elevation between the bottom of the excavation and the bench mark. In this case the difference will be $100 - 89 = 11$ ft.

4. Calculate the cut at each stake and indicate it on the stake (see Fig. 2–44). These will be

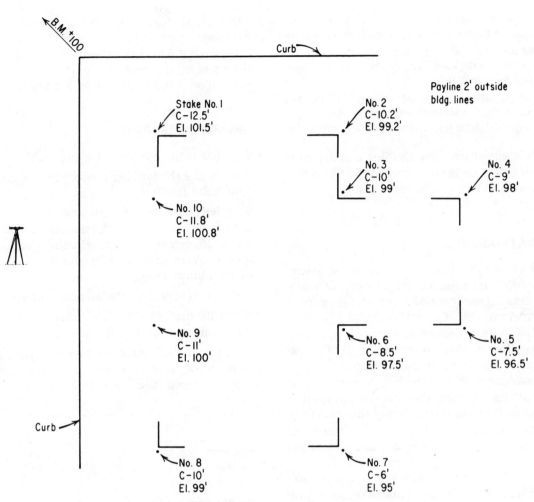

FIGURE 2-43 Excavation limit stakes.

FIGURE 2-44 Excavation stake marked for cut.

No. 1	Cut 11 + 1.5 = 12.5 ft
No. 2	Cut 11 − 0.8 = 10.2 ft
No. 3	Cut 11 − 1.0 = 10.0 ft
No. 4	Cut 11 − 2.0 = 9.0 ft
No. 5	Cut 11 − 3.5 = 7.5 ft
No. 6	Cut 11 − 2.5 = 8.5 ft
No. 7	Cut 11 − 5.0 = 6.0 ft
No. 8	Cut 11 − 1.0 = 10.0 ft
No. 9	Cut 11 − 0.0 = 11.0 ft
No. 10	Cut 11 + 0.8 = 11.8 ft

Running Grades

It is sometimes necessary to run *grade lines* at a construction site—lines which have a uniform drop throughout their length. This is usually expressed as a given number of feet of drop per 100 ft of length. For example, a 1% grade means that there is 1 ft of uniform drop per 100 ft of length.

One method of obtaining a uniform drop in any given distance is as follows:

1. Drive a stake to the required height at the start of the proposed line, set up the instrument over this stake, and level it.

2. Measure (with the rod or a tape) the height of the telescope above the stake; suppose, in this example, that it is 3 ft. 8 in.

3. Align the telescope in the direction of the line; this may be done by sighting on the rod held at the far end of the line.

4. Measure 100 ft along the line and set another stake.

5. Raise the target the distance the drop is to be, e.g., 6 in. This will make the target read 4 ft 2 in. in this example.

6. Drive the stake until the target is on the cross hair when the rod is held on the stake. The line between the stakes now drops 6 in. in 100 ft.

7. Set the target back to its original reading, 3 ft 8 in., and incline the telescope downward until the cross hair is again on the target. Tighten the vertical motion clamp screw.

Stakes set anywhere along this line and driven in until the cross hair is on the target when the rod is held on the stake will now be on the same grade.

CARE OF INSTRUMENTS

Builder's levels and transit levels are precision instruments and require particular care in handling. When the instrument is not attached to the tripod, see that it is properly placed in its carrying case. One end of the case is marked **eyepiece end,** so that the instrument can always be fitted in properly. Be sure that it is not subjected to shocks during packing, unpacking, or transportation.

If the instrument gets wet, let it dry—preferably in a relatively dust-free environment. If it has become dusty, clean the dust from the lens with the brush provided. Never rub the optical surfaces. Have the instrument checked and cleaned periodically by a reliable instrument man.

When the instrument is attached to its tripod, carry both as an incorporated unit. Grasp the tripod about two-thirds of the way toward the top, closing the legs, and tip it slightly so that it rests comfortably against the

shoulder. Make sure that the instrument is not jarred or bumped.

Do not stand the instrument on a surface into which the metal leg tips cannot bite. If necessary, use a wooden tripod stand such as that illustrated in Fig. 2–45.

REVIEW QUESTIONS

1. What is the purpose of a site plan?

2. Describe the leveling procedures for both three and four point leveling.

3. Briefly describe the purpose of each of the following parts to be found in a level; **(a)** horizontal motion tangent screw, **(b)** centering head, **(c)** vernier scale, **(d)** vertical motion clamp screw, and **(e)** adjustment screw at the end of the bubble tube.

4. What is meant by the *elevation* of an object?

5. Define the following: **(a)** backsight, **(b)** H.I., **(c)** turning point, and **(d)** station.

6. A bench mark is recorded as 1656.121 '. What is the elevation of point A, measured from this bench mark, if the recorded readings were as follows:

	T.P.1	T.P.2	T.P.3	T.P.4	T.P.5	T.P.6
Backsight	2.56'	6.63'	5.16'	4.20'	3.92'	6.08'
Foresight	3.12'	2.18'	6.31'	7.65'	1.58'	2.35'

7. We wish to determine the height of a chimney. The instrument is set up 100 ft from the base and a level sight is to be 4.6 ft above ground level. When sighting on the top of the chimney from the same point, the angle of inclination is found to be 31°23'. What is the height of the chimney?

8. The elevations of the corner stakes at the site of an excavation are found to be **(a)** 102.6 ', **(b)** 99.7 ', **(c)** 94.8 ', and **(d)** 95.5 ' relative to a datum point of +100.0'; calculate the amount of *cut* to be shown on each stake if the bottom of the excavation is to be 11 ft below datum.

9. Give the reading indicated on the section of scale shown below if the angle was turned clockwise.

FIGURE 2–45 Tripod stand.

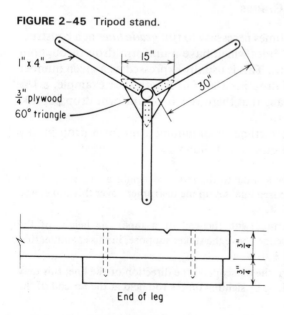

1" x 4"

15"

30"

¾" plywood
60° triangle

¾"

¾"

End of leg

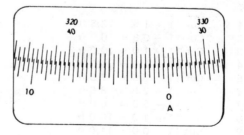

320
40

330
30

10

0
A

Excavations and Excavating Equipment

The type of foundation to be used for any particular building will have been determined at the planning stage. The type chosen will depend to a large extent on the soil or subsoil encountered at the site. Soil or subsoil in its natural state is often sufficiently stable to support the foundations of light buildings. Foundations for heavy buildings, on the other hand, will be brought to a level with sufficient bearing strength, or to bedrock (unless tests show that bedrock occurs at too great a depth to be reached by conventional foundation walls). In that event, deep foundations will be required, and these will be discussed in a subsequent chapter.

To build a normal foundation, an excavation is required, and this operation will usually be carried out by some type of power excavation equipment.

EFFECTS OF SOIL TYPES ON EXCAVATING EQUIPMENT

The type of soil at the site will be the major factor in determining the type of excavating equipment needed for the job. If the soil is loose or noncohesive—dry sand or gravel, for instance—excavating can usually be started with a clamshell bucket and completed with a payloader—a shovel mounted on a tractor—or it may be done completely by payloader or bulldozer. The use of these two machines requires access into and out of the excavation (see Fig. 3–46). In restricted locations where such access is not possible, other methods must be used.

More cohesive soils require more power, and in such cases a power shovel or pull shovel (backhoe) will be required. Again, a power shovel requires access to and from the excavation, while a pull shovel can dig below its own level—within limits. A bulldozer may also be used for excavating cohesive soils. For very large excavations in soils of either of the above types, a scraper may be employed.

Soils that are too soft or wet to support machinery traveling over them may be excavated by means of a dragline. This machine will operate from one or two stable spots at the side of the excavation.

When narrow trenches or individual holes are required—for footings, or piers—a pull shovel or a trencher is the normal equipment to use.

Boulders and broken rock will probably be handled best by a power shovel, though a clamshll bucket could be used if the pieces are small enough. All these machines will be described in some detail in subsequent paragraphs.

REMOVING GROUNDWATER

One serious problem encountered regularly during digging operations is the presence of water in the excavation. This may be caused by rain, melting snow, an underground stream, or the fact that the *water table* in the area is high. The water table is the normal level of groundwater, and if that level is close to the surface, excavating will allow the groundwater to seep and collect in the excavation.

There are two principal methods of getting rid of water. One is to use a pump (or pumps) to empty the ex-

FIGURE 3–1 Pump draining excavation. (*Courtesy Homelite Pumps.*)

FIGURE 3–2 Pump removing seepage water. (*Courtesy Homelite Pumps.*)

cavation or to keep it dry. This method is particularly useful in getting rid of collected rain water and runoff or to remove the water from an uncovered underground stream. Figure 3–1 illustrates the removal of collected rain water from a pier excavation by pump. Pumps are also used to get rid of seepage water. A sump, into which the water collects and from which the pump can then remove it, is required. Figure 3–2 shows a pump removing seepage water from behind a cofferdam.

The other method is to lower the water table in the area in which the excavating will take place. This is done by a *dewatering system*—a system which involves sinking a series of *well-points* around the area and extracting the water by suction pump. Figure 3–3 shows an excavation surrounded by a dewatering system to keep the working area dry. Figure 3–4 illustrates how a typical well-point system would be laid out, while Fig. 3–5 shows a section through such a system.

A system consists of a series of well-points (see Fig. 3–6), each with a riser pipe and a swing joint connecting to a common header or manifold. In Fig. 3–7, a profile of a well-point system shows the setup. Notice the normal water level and the predrained water level, brought about by the water extraction.

The header pipe is exhausted by a combination pumping unit, which is a centrifugal pump continuously primed by a positive displacement vacuum pump. Atmospheric pressure forces water through the ground to the well-point screens, into the well-points, up the riser and swings, and through the header to the pump. Any air entering the system is separated in the float chamber

FIGURE 3–3 Excavation surrounded by well-point system. (*Courtesy Moretrench Corp.*)

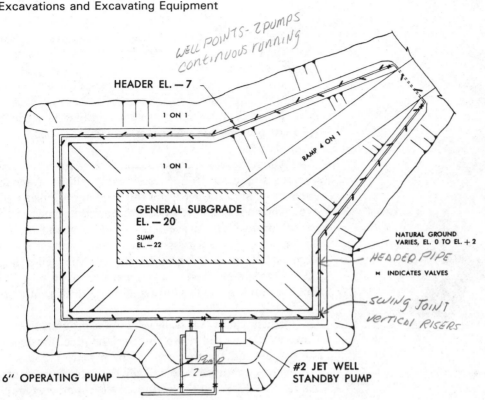

WELL POINTS- 2 pumps continuous running

HEADER EL. —7

1 ON 1

RAMP 4 ON 1

1 ON 1

GENERAL SUBGRADE
EL. —20

SUMP
EL. —22

NATURAL GROUND
VARIES, EL. 0 TO EL. +2

HEADER PIPE

⋈ INDICATES VALVES

SWING JOINT
vertical RISERS

6" OPERATING PUMP

Pump 2

#2 JET WELL
STANDBY PUMP

FIGURE 3-4 Typical well-point layout plan. (*Courtesy Moretrench Corp.*)

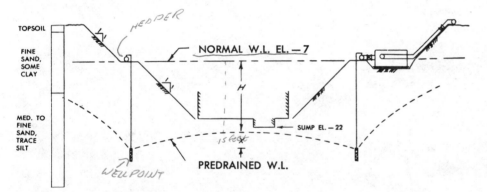

TOPSOIL

HEADER

FINE
SAND,
SOME
CLAY

NORMAL W.L. EL. —7

H

MED. TO
FINE
SAND,
TRACE
SILT

SUMP EL. —22

15 feet

PREDRAINED W.L.

WELLPOINT

FIGURE 3-5 Typical well-point layout section. (*Courtesy Moretrench Corp.*)

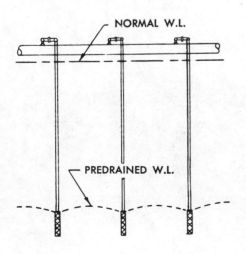

NORMAL W.L.

PREDRAINED W.L.

FIGURE 3-7 Profile of well-point
system. (*Courtesy Moretrench Corp.*)

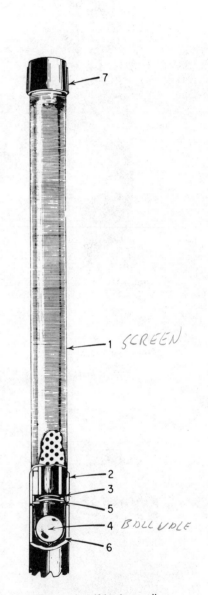

7

1 *SCREEN*

2
3
5
4 *BALL VALVE*
6

FIGURE 3-6 Two inch self-jetting well-
point. (*Courtesy Moretrench Corp.*)

FIGURE 3-8 Dewatering pump. (*Courtesy Moretrench Corp.*)

and passed to the vacuum pump. The water passes to the centrifugal pump, which discharges it to a drainage system (see Fig. 3–8).

In coarse soils, easily drained, well-points are driven directly into the soil. In fine, dense soils, they are generally set in a column of sand, as shown in Fig. 3–9, which helps to prevent the well-point screen from becoming clogged with silt.

The well-point is a hollow, perforated tube, covered with a stainless steel filter screen which allows water to enter the well-point but keeps out sand and silt (see Fig. 3–6). The bottom end is serrated to enable it to penetrate the soil more easily. In the bottom end is a chamber containing a ball valve with a wooden center, so that it will float. When water is being drawn from the well-point, the ball floats up against a ring valve at the top of the chamber and prevents sand from being drawn into the system.

When placing the well-point in the ground, a jet of water can be forced down through the pipe, loosening the soil around the tip and making penetration easier. During this operation, the ball valve is forced down into the retainer basket at the bottom of the chamber. Figure 3–10 shows a well-point being jetted into place.

FIGURE 3-9 Properly sanded well-point. (*Courtesy Moretrench Corp.*)

FIGURE 3-10 Jetting a well-point. (*Courtesy Moretrench Corp.*)

FIGURE 3–11 Drill installing perforated drains to facilitate slide control. (*Courtesy Mobile Drilling Co.*)

Another type of dewatering operation is illustrated in Fig. 3–11. Here perforated drains are being installed into the bottom of a bank to facilitate slide control.

GENERAL AND SPECIAL EXCAVATIONS

Excavating operations are of two types, *general* and *special*. General excavations include all the work, other than rock excavation, which can be carried out by mechanical equipment. For example, excavating done with draglines, shovels, or scrapers; loading with payloader or clamshell; and hauling by truck all come under the heading of general excavation.

Special excavation includes the work that must be done by blasting, by special machines, by hand, or by a combination of hand and machine.

The contractor should be fully aware of the types of excavation with which he must deal in order to estimate the cost properly and to ensure that he has the right kind of equipment on hand to do the best job.

ESTIMATING AMOUNT OF MATERIAL
TO BE REMOVED

Before an excavation is begun, its cost must be estimated. This estimate should be based on the cubic yards of material to be dug and hauled away. Plans show the area occupied by the building and the depth to which the excavation must be carried. Since there must usually be

room to work around the outside of forms in an excavation, a line is drawn at a reasonable distance (usually 2 ft) outside the building line and the excavating contractor is paid for the material excavated to this line—the *payline*.

When soil is taken from its compacted position and broken up, as in excavating, it increases in bulk. This is called the *soil swell*. Therefore, the volume of material to be removed from the site must be increased by a percentage, depending on the type of soil. Most soils will swell between 20 and 50%. Table 3–1 shows the percentages of bulking encountered with some common soil types.

TABLE 3–1 Soil swell percentage

Soil type	Percentage of swell from compact state
Silt	20
Clay	25
Sand and gravel	50
Loam	25
Stone	50

If the soil is cohesive enough, or if some means of supporting the sides of the excavation are to be used, the compact volume to be removed is simply a matter of length × width × depth. But for loose, sliding soils, the excavation must have sloping sides, usually with a 1:1 slope. This fact must be taken into account when estimating the volume of material to be removed. A formula that may be used in calculating the volume in such cases is

$$\frac{(A + B + 6C)H}{8}$$

where

A = top area

B = bottom area

C = midsection area (average of top and bottom areas)

H = vertical height

CHOOSING EXCAVATION EQUIPMENT

A number of factors must be taken into consideration when deciding what type or types of excavating machinery will do the job most efficiently. The first point to consider is the volume of material to be removed. This will influence the size of the machines to be used. The depth of the excavation will determine the height to which material must be lifted to get it into trucks.

The disposal of the excavated material may influence the type of machinery used. If it can be deposited on a spoil bank at the site, one machine may do both digging and depositing. The distance from excavation to spoil bank is also a factor. One the other hand, if the material must be removed from the site, hauling units and special loading equipment may be required.

The type of soil to be excavated will influence the kind of equipment selected. Some machines work well in loose, dry soils but not in wet or highly compacted earth.

The time allowed for excavation will also affect the type, size, and amount of equipment to be used.

With such a wide range of equipment available for earth excavating and moving, the contractor must have some method of evaluating each type and size of machine.

Many factors have a bearing on the type and size of machine that will be considered. The major considerations are:

1. Type of material to be excavated and hauled.
2. Site conditions
3. Distance of haul
4. Time allowed for job completion
5. Contract price

With these requirements in mind the contractor now must select his equipment to complete the work within the required time span and yet realize the projected profit.

To estimate the hourly production of a piece of equipment, the following formula may be used:

$$P = \frac{E \times I \times H}{C}$$

where

P = production, cu yd/hr (in-bank)

E = machine efficiency, min/hr

I = shrinkage factor for loose material

H = heaped capacity of machine, cu yd

C = cycle time of the machine, min

The production or volume of material that a piece of equipment can move is based on the volume occupied by the material in its natural state or, as it is commonly referred to, the *in-bank* condition of the material. Most materials, when disturbed, increase in volume—some as much as 50% (see Table 3–1). To allow for this increase

TABLE 3–2 Hourly shovel handling capacity[a] shovel dipper sizes (cu yd)

Class of material	⅜	½	¾	1	1¼	1½	1¾	2	2½
Moist loam or sandy clay	85	115	165	205	250	285	320	355	405
Sand and gravel	80	110	155	200	230	270	300	330	390
Good common earth	70	95	135	175	210	240	270	300	350
Clay, hard, tough	50	75	110	145	180	210	235	265	310
Rock, well-blasted	40	60	95	125	155	180	205	230	275
Common, with rock and roots	30	50	80	105	130	155	180	200	245
Clay, wet and sticky	25	40	70	95	120	145	165	185	230
Rock, poorly blasted	15	25	50	75	95	115	140	160	195

[a]Conditions:

1. Cu yd bank measurement per hour.
2. Suitable depth of cut for maximum effect.
3. Continuous loading with full dipper.
4. 90° swing, grade-level loading.
5. All materials loaded into hauling units.

Source: Reproduced by permission of Power Crane and Shovel Association.

in volume, the shrinkage factor I is applied to the heaped capacity H of the earth mover to reduce the load to the in-bank condition. In other words, the in-bank volume is equal to the machine heaped capacity multiplied by the shrinkage factor. In equation form,

$$\text{in-bank machine capacity} = H \times I$$

The shrinkage factor I can be calculated by the formula

$$I = 1 - \frac{\text{percent swell}}{100}$$

Example:

Calculate the shrinkage factor for clay based on the values in Table 3–1.

Solution:

From Table 3–1, the percent swell for clay is 20%.

$$I = 1 - \frac{20}{100}$$

$$= 0.8$$

As no machine is 100% efficient, the efficiency of the machine must also be included in the calculations. Average efficiencies for various types of equipment are as follows:

- *Crawler tractor equipment:* 50 min/hr
- *Rubber-tired hauling units:* 45 min/hr
- *Large rubber-tired loaders and dozers:* 45 min/hr
- *Small rubber-tired loaders:* 50 min/hr

That is, for every hour on the job, a piece of equipment is between 75 and 85% efficient. Some equipment can be as much as 90% efficient, depending on the material being handled.

The *cycle time* of a piece of equipment is based on the time required to obtain its load, move it to its dumping point, and return to the loading point. The cycle time is based on the sum of two factors: travel time factors and fixed time factors.

Travel time factors relate to the movement of the machine (i.e., usable speeds and pulls, load resistance, travel resistance, and machine weight). Fixed time factors are based on the characteristics built into the machine (i.e., spotting, loading, turning, dumping, and reversing). Therefore, it can be said that total cycle time is equal to cycle travel time plus cycle fixed time, or

$$C = CT + CF$$

The travel time (CT) in minutes can be calculated as

$$CT = \frac{D}{S \times 88}$$

where

D = distance traveled, feet

S = speed, miles per hour

88 = distance moved, feet per minute, when traveling 1 mph

The fixed time (CF), of course, depends on the piece of equipment being considered. All major suppliers have charts and data for the various pieces of equipment which enable the contractor to establish these factors. Fixed cycle times can vary from 0.25 min to over 5 min, depending on the circumstances.

Example:

A crawler-mounted power shovel having a 1½-cu yd bucket is to load well-blasted rock into trucks. If the machine efficiency is 50 min/hr, establish the total cycle time. Use Table 3–2 (page 38) for production data.

Solution:

From Table 3–1, the percent increase in volume for rock is 50%. The resulting shrinkage factor is

$$I = 1 - \frac{50}{100} = 0.5$$

Using the production formula,

$$P = \frac{E \times I \times H}{C}$$

Rewrite in terms of C:

$$C = \frac{E \times I \times H}{P}$$

From Table 3–2, for a shovel with a 2-cu yd bucket, $P = 230$ cu yd/hr. With $I = 0.5$, $H = 2$ cu yd, and $E = 50$ min/hr,

$$C = \frac{50 \times 0.5 \times 2}{230}$$

$$= 0.22 \text{ min}$$

Such a short cycle time indicates that ideal conditions exist for the power shovel (see conditions for Table 3–2) and the total cycle time calculated is based solely on the cycle fixed time. If the shovel was required to travel some distance between loading and dumping, the cycle time would increase correspondingly.

Example:

A rubber-tired scraper with a heaped capacity of 30 cu yd has a cycle fixed time of 1.0 min (loading, dumping, and turning). If the average speed of the scraper is 3.5 mph and it must travel 0.75 mile round trip, calculate the anticipated production of the machine if it is hauling clay with a swell factor of 25%. Assume a machine efficiency of 50 min/hr.

Solution:

$$\text{Production } P = \frac{E \times I \times H}{C}$$

Total cycle time = cycle travel time + cycle fixed time

or

$$C = CT + CF$$

$$\text{Cycle travel time } CT = \frac{D}{S \times 88}$$

$$= \frac{0.75 \times 5280}{3.5 \times 88}$$

$$= 12.9 \text{ min}$$

Cycle fixed time $CF = 1.0$ min

Total cycle time $\quad C = 12.9 + 1.0 = 13.9$ min

Shrinkage factor

$$I = 1 - \frac{\text{percent swell}}{100}$$

$$= 1 - \frac{25}{100}$$

$$= 0.75$$

Given $E = 50$ min/hr and $h = 30$ cu yd

$$P = \frac{50 \times 0.75 \times 30}{13.9}$$

$$= 81 \text{ cu yd/hr (in-bank)}$$

Crane

Before discussing machines used specifically for excavating, it would perhaps be well to look at a machine that has many uses on a building site, including adaptations for digging. This is the crane, a machine designed primarily for lifting but adapted to many other uses. Few building jobs are without a crane, and Figs. 3–12 to 3–14 illustrate some typical uses of these machines.

FIGURE 3–12 Mobile crane with gooseneck jib. (*Courtesy American Hoist & Derrick Co.*)

FIGURE 3–13 Mobile cranes installing bridge girders.

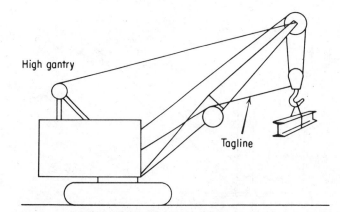

FIGURE 3-16 Crane with gantry (right, top). (*Courtesy Power Crane & Shovel Association.*)

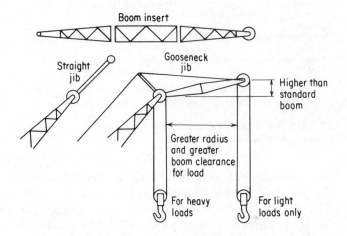

FIGURE 3-17 Boom extensions (right, bottom). (*Courtesy Power Crane & Shovel Association.*)

FIGURE 3-14 Mobile crane with concrete bucket.

A crane consists basically of a power unit mounted on crawler tracks or wheels, with a boom, and control cables for raising and lowering the load and the boom (see Fig. 3-15). A *gantry* is sometimes added (see Fig. 3-16) to provide better boom support. For greater reach, an extension may be added to the boom. This may be in the form of a *boom insert,* a section added between the upper and lower ends, or a *jib,* an extension to the end of the boom. Figure 3-17 illustrates these boom extensions. Notice that in Fig. 3-12, the crane boom has both an insert and a gooseneck jib.

FIGURE 3-15 Diagram of crane. (*Courtesy Power Crane & Shovel Association.*)

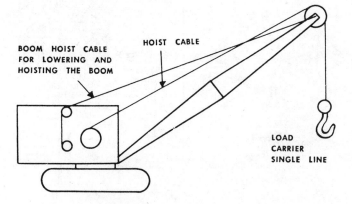

The load capacity of a crane depends on:

1. The stability of the footing. The size of the crawlers is a factor because a longer or wider mounting increases the stability of any footing.
2. The strength of the boom. This is one of the major governing factors in establishing *load ratings,* and these should never be exceeded. Ratings vary with boom length, and any extension of the boom reduces the rating. Lowering the boom also increases the radius and thus reduces the rated capacity.
3. The counterweight. This is added to the after-end of the machine, and manufacturers' specifications provide standard and maximum counterweights and give crane ratings for both. Counterweights may be increased to a specified maximum, but the operating radius *must not* exceed that given by the manufacturer.

The working range of a crane is limited horizontally for maximum lift only by the boom length. The reach below footing level is limited only by the length of the

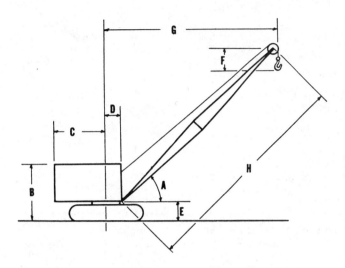

A Boom Angle
B Maximum Clearance Height of Cab
C Maximum Radius of Tail Swing
D Center of Rotation to Boom Foot Pin
E Height from Ground to Boom Foot Pin
F Distance from Center of Boom Point Sheave to Bottom of Hook
G Clearance Radius of Boom
H Length of Boom

Working radius, not shown in above diagram, is of course, of primary importance in crane selection.

FIGURE 3-18 Crane clearance diagram. (*Courtesy Power Crane & Shovel Association.*)

EXCAVATING MACHINES

Bulldozer

The term *bulldozer* is often used to mean any tractor, tracked or wheeled, on which is mounted a blade. Specifically, when the blade is mounted perpendicularly to the line of travel, the machine is a *bulldozer,* while if the blade is at an angle to the direction of travel, it is an *angledozer* (see Fig. 3-20).

Bulldozers and angledozers have limited use for excavating by themselves. They are able to loosen and remove soil from its original position but can only dispose of this soil by pushing it beyond the limits of the excavation. For this reason, a bulldozer is useful for starting an excavation, for stripping valuable topsoil from the excavation site, or for use where earth must be excavated from one part of the site and deposited as fill in another. As an excavation deepens, a bulldozer's usefulness is reduced to loosening soil and readying it for removal by other means.

Loader

A *loader* is one machine in common use to pick up excavated material. It consists of a crawler or wheeled tractor with a shovel or bucket mounted in front. A loader can excavate loose soils, but its chief use is loading excavated material into trucks for removal. Figure 3-21 shows a wheeled tractor being used as both a loader and a dozer, while Fig. 3-22 shows a crawler tractor with bucket being used only as a loader.

hoist cable. Figure 3-18 points out the various crane clearances involved in crane operation. A wide variety of tools which may be used with a crane boom is illustrated in Fig. 3-19. Three of these are excavating tools and will be discussed later.

FIGURE 3-19 Crane tools. (*Courtesy Power Crane & Shovel Association.*)

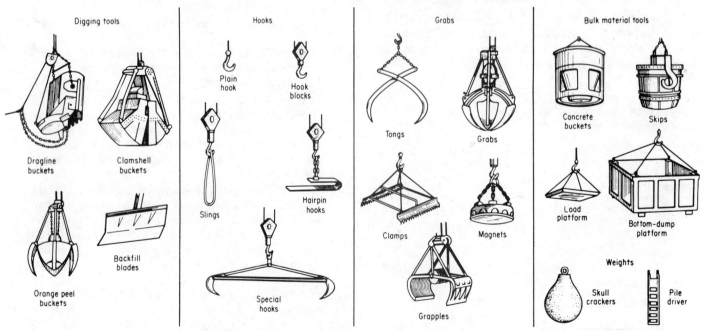

Digging tools — Dragline buckets, Clamshell buckets, Orange peel buckets, Backfill blades

Hooks — Plain hook, Hook blocks, Slings, Hairpin hooks, Special hooks

Grabs — Tongs, Grabs, Clamps, Magnets, Grapples

Bulk material tools — Concrete buckets, Skips, Load platform, Bottom-dump platform, Weights, Skull crackers, Pile driver

FIGURE 3–20 Crawler bulldozer. (*Courtesy J. I. Case Co.*)

(a)

(b)

FIGURE 3–21 Wheeled tractor: (a) as a dozer; (b) as a loader. (*Courtesy Allis-Chalmers Co.*)

FIGURE 3–22 Crawler loader. (*Courtesy J. I. Case Co.*)

Scraper

Scrapers have established an important position in the earth-moving field, particularly where large volumes must

be excavated and hauled away. A scraper is a combination machine, in that it loads, hauls, and discharges material. It may be powered by a crawler tractor or by one or two rubber-tired wheeled tractors. Some very large machines have an individual power unit driving each wheel. The crawler tractor scraper has good traction and does not usually require assistance in loading but is a slow traveler. The rubber-tired scraper, on the other hand, often needs assistance in loading if the earth is very compact but can travel at speeds in excess of 40 mph (see Fig. 3-23).

FIGURE 3-23 Wheeled scraper. (*Courtesy Euclid Division of General Motors Corp.*)

FIGURE 3-24 Power shovel. (*Courtesy Insley Equipment Co.*)

Power Shovel

A *power shovel* is used primarily to excavate earth and load it into trucks or deposit it on a spoil bank at the site. It may be mounted on crawler tracks or at the rear of a truck. A crawler-mounted unit has slow travel speeds, but wide tracks permit it to operate on soft soil. Figure 3-24 shows a crawler-mounted power shovel in operation.

A power shovel is designed to operate against a face or bank which it displaces as it moves forward. Its design permits power to be applied so as to force the dipper into the bank, and it is capable of digging into the hardest types of materials. Figure 3-25 illustrates its action and the basic parts of the digging apparatus. This type of machine can be used in a number of ways, some of which are depicted in Fig. 3-26.

Since the shipper shaft acts as a fulcrum, both dipper and dipper stick can be raised or lowered by the hoist line as well as pushed out or retracted by the crowd mechanism. These motions can be performed simultaneously and permit a flexibility of operation. The limits of this flexibility are commonly referred to as the *working ranges*. These are indicated in Fig. 3-27.

FIGURE 3-25 Shovel-digging action. (*Courtesy Power Crane & Shovel Association.*)

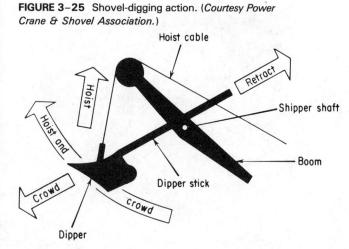

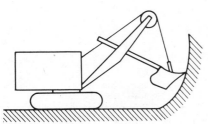

Embankment digging (above grade)

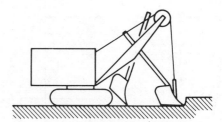

Digging along horizontal plane

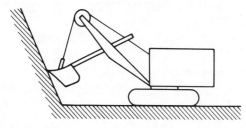

Digging slopes

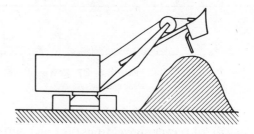

Dumping onto spoil bank

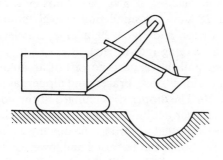

Digging below grade

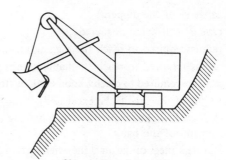

Side casting of materials

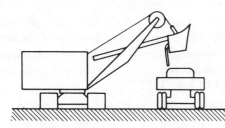

Loading into hauling units at ground level

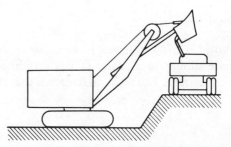

Loading into hauling units on top of bank

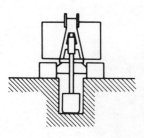

Handling shallow trenches

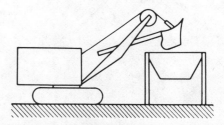

Dumping onto belt, conveyor, hopper or grizzly

FIGURE 3-26 Typical power shovel operations. (*Courtesy Power Crane & Shovel Association.*)

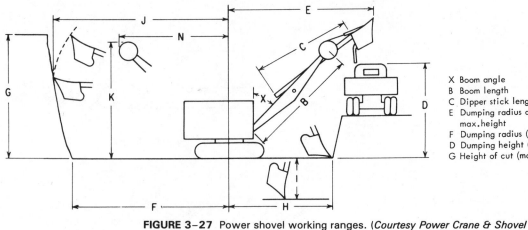

X Boom angle
B Boom length
C Dipper stick length
E Dumping radius at
     max. height
F Dumping radius (max.)
D Dumping height (max.)
G Height of cut (max.)

J Digging radius (max.)
H Floor level radius
I Max. digging depth
     below ground level
K Clearance ht. of boom
     point sheave
N Clearance radius of
     boom point sheave

FIGURE 3–27 Power shovel working ranges. (*Courtesy Power Crane & Shovel Association.*)

The size of power shovel to use for any particular job depends on a number of factors:

1. The capacity of the dipper.
2. The type of soil being extracted.
3. The radius required to reach the digging area.
4. The radius required to reach the hauling units.
5. The distance required to reach a spoil bank or stock pile.
6. The height of the face to be excavated.
7. The clearance height required to reach hauling units located on top of the bank.
8. Physical clearances of the machine when working in confined areas.

Pull Shovel

A *pull shovel* is often referred to as a *hoe* or *backhoe* and is quite similar to a power shovel insofar as basic parts are concerned. The main difference is that the position of the bucket is the reverse to that of the power shovel. It digs by pulling the load toward the power unit rather than pushing it away as a power shovel does. Figure 3–28 shows a small pull shovel mounted on a wheeled tractor, and Fig. 3–29, a large, crawler-mounted shovel discharging into a truck.

Basic pull shovel parts consist of an auxiliary gantry, boom, dipper stick, braces, and dipper, as illustrated in Fig. 3–30. The dipper stick, hinged at the boom point to provide lever action, acts as an arm for the dipper. The dipper is pulled inward as the digging cable is taken in, while the boom can be raised and lowered at the same time. All the motions of the machine combine to provide working ranges shown in Fig. 3–31.

The pull shovel is designed to dig below the level on which the machine rests and to deposit the material on a spoil bank or in trucks. Because of its positive action, it is also capable of digging in the hardest materials. The most common uses of the pull shovel are illustrated in Fig. 3–32.

FIGURE 3–28 Pull shovel on wheeled tractor or backhoe.

FIGURE 3-29 Crawler-mounted pull shovel. (*Courtesy Miron Co., Ltd.*)

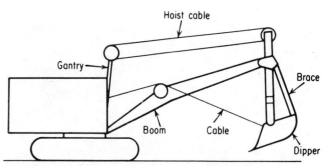

FIGURE 3-30 Basic parts of pull shovel. (*Courtesy Power Crane & Shovel Association.*)

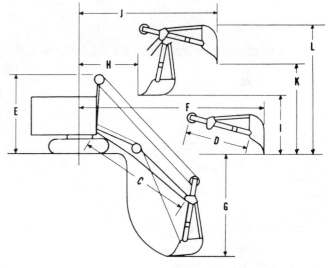

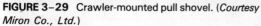

J Radius at End of Dump
I Dumping Height, Starting
K Dumping Height, Ending
L Clearance at (Max.) Dumping Height

C Boom Length
D Dipper stick Length
H Radius at Beginning of Dump
F Digging Reach (Max.)
G Digging Depth (Max.)
E Height of Hoe Gantry

FIGURE 3-31 Pull shovel working ranges. (*Courtesy Power Crane & Shovel Association.*)

FIGURE 3-32 Typical uses of pull shovel. (*Courtesy Power Crane & Shovel Association.*)

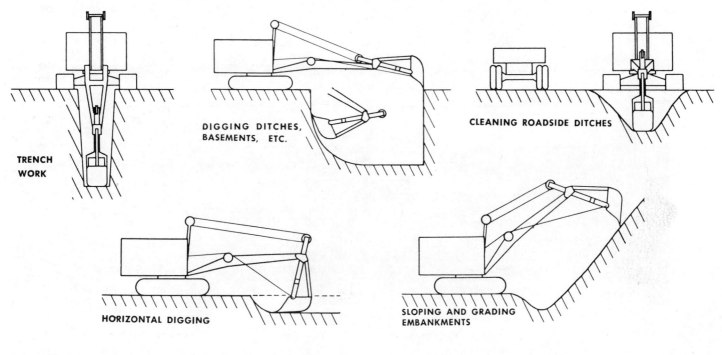

TRENCH WORK

DIGGING DITCHES, BASEMENTS, ETC.

CLEANING ROADSIDE DITCHES

HORIZONTAL DIGGING

SLOPING AND GRADING EMBANKMENTS

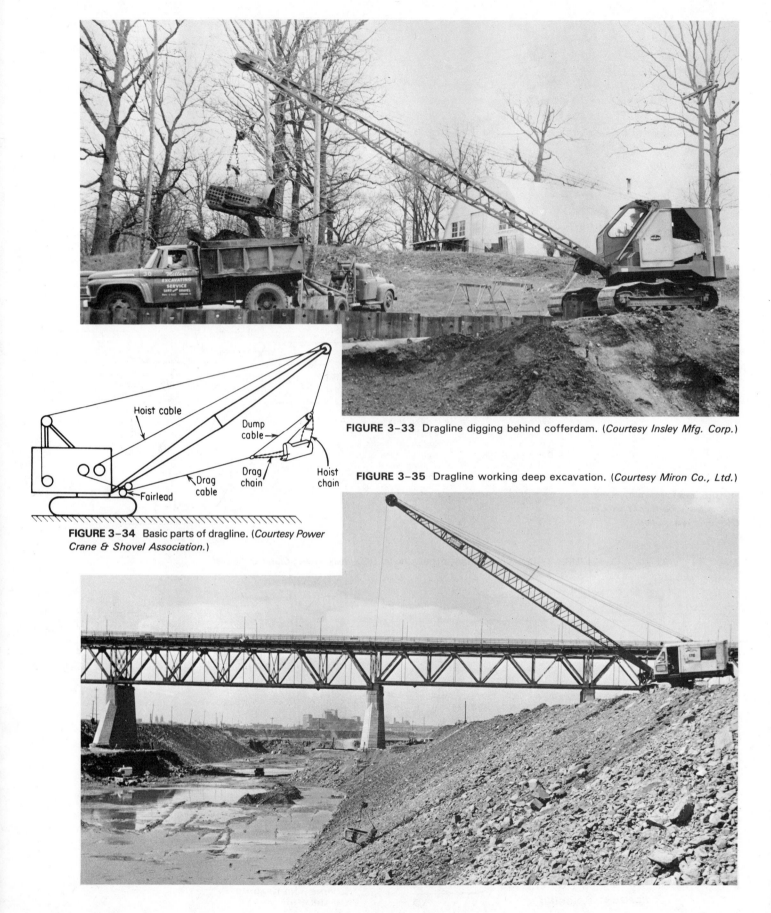

FIGURE 3-33 Dragline digging behind cofferdam. (*Courtesy Insley Mfg. Corp.*)

FIGURE 3-35 Dragline working deep excavation. (*Courtesy Miron Co., Ltd.*)

FIGURE 3-34 Basic parts of dragline. (*Courtesy Power Crane & Shovel Association.*)

Dragline

One of the attachments used on a crane boom is a *dragline,* shown in Fig. 3-33. As illustrated in Fig. 3-34, the dragline attachment consists of a dragline bucket, hoist cable, dragcable, and fairlead—used with a crane boom. The machine is operated by pulling the bucket toward the power unit, regulating the digging depth by adjusting the tension on the hoist cable.

      The dragline has a wide range of operations. It remains on firm or undisturbed ground and digs below its own level, backing away from the excavation as the material is removed (see Fig. 3-35). While it does not dig to as accurate a grade as a power shovel or a pull shovel, it has a larger working range—an advantage in many situations. Figure 3-36 outlines the dragline working range.

      In addition to its large working range, the dragline has the further advantage of being suited to digging in excavations below water level and in mud or quicksand. Figure 3-37 illustrates a few typical uses.

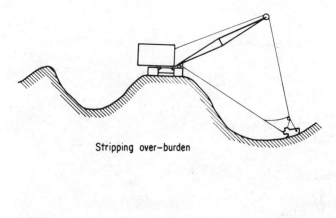
Stripping over-burden

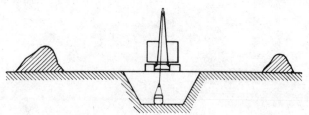
Excavating narrow ditches

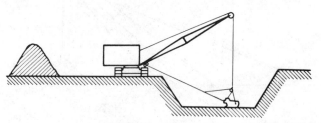
Excavating canals

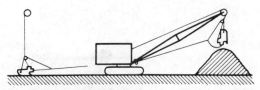
Overcasting from borrow pit

FIGURE 3-36 Dragline working range. (*Courtesy Power Crane & Shovel Association.*)

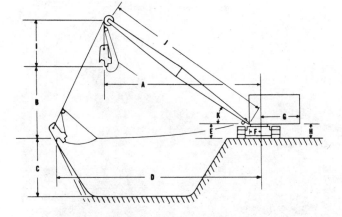

A **Dumping Radius**
B **Dumping Height**
C **Maximum Digging Depth**
D **Digging Reach (Depends on Conditions and Operator's Skill)**
E **Distance from Ground to Boom Foot Pin**
F **Distance from Center of Rotation to Boom Foot Pin**
G **Rear End Radius of Counterweight**
H **Ground Clearance**
I **Length of Bucket (Depends on Size and Make)**
J **Boom Length**
K **Boom Angle**

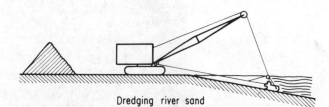
Dredging river sand

FIGURE 3-37 Typical dragline operation. (*Courtesy Power Crane & Shovel Association.*)

FIGURE 3–38 Clamshell unloading sand.

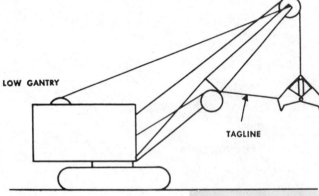

LOW GANTRY

TAGLINE

FIGURE 3–39 Clamshell bucket with tagline. (*Courtesy Power Crane & Shovel Association.*)

FIGURE 3–40 Pull shovel used for trenching. (*Courtesy Insley Mfg. Corp.*)

FIGURE 3-41 Trencher.

Clamshell

The crane boom is often used with a hinged bucket, called a clamshell (see Fig. 3–38), for vertical excavating below ground level and for handling bulk materials such as sand and gravel.

The clamshell bucket consists of two scoops hinged in the center, with holding arms connecting the headblock to the outer ends of the scoops. A closing line is reeved through blocks on the hinge and the head, so that the bucket closes when it is taken in. The bucket is opened by releasing the closing and hoisting line while holding the bucket with the holding and lowering line. Clamshells are usually equipped with a tagline to control the swing of the bucket (see Fig. 3–39).

Buckets are available in a wide variety of sizes and types. A heavy-duty type is available for digging, and lighter weight ones for general purpose work or handling light materials. Some buckets have removable teeth, used when digging in hard material, while others have special cutting lips.

Trencher

Excavating trenches (for sewer lines, pipelines, etc.) is a type of operation which usually requires specialized machines. A pull shovel may be employed, as shown in Fig. 3-40, but for speed and economy of operation, a *trencher* is used. It is a machine that digs and deposits the soil in one continuous operation. Figure 3–41 shows one type of trencher ready for operation. As earth is dug up by the wheel, it is dropped on a side delivery belt and deposited in a continuous pile alongside the trench. Backfilling the trench can be done in many ways, one of which is shown in Fig. 3-42. Bulldozers, angledozers, and

FIGURE 3–42 Backfilling trench.

loaders are also used for backfilling trenches and around foundations.

Trucks

No discussion of excavating machinery would be complete without mentioning the hauling units—usually *trucks,* with bodies made for that specific purpose. Figure 3–43 shows such a body, with a protective canopy over

FIGURE 3–43 Truck with special dump body.

the cab and a tapered back for easy dumping. In Fig. 3–44, a loader discharges into a truck with a square-end body.

Most trucks may be operated over sufficiently firm and smooth haul roads. Some units now in use are designated as *off-highway* trucks, because of their size and power.

Trucks may be classified in a great many ways, including the following:

1. According to size and type of engine
2. By the number of gears
3. As two-, four-, or six-wheel drive
4. By the number of wheels and axles
5. By the method of dumping
6. By the capacity

The truck in Fig. 3–44 is a 30-ton rear-dump, six-wheel hauler.

PROTECTION OF EXCAVATIONS

The protection of an excavation simply means taking steps to ensure that side wall cave-ins do not occur, necessitating costly work in removing fallen earth or caus-

FIGURE 3–44 Loading heavy dump truck.

FIGURE 3–45 Free-standing excavation walls. (*Courtesy Moretrench Corp.*)

ing damage to forms or injury to workers. The amount of protection required will depend on the *type of soil being excavated,* the *depth of the excavation,* the *level of the water table,* the *type of foundation* to be built, and the *available space* around the excavation.

The sides of an excavation dug in dry, compact, stable soil may need no protection at all, if it is not too deep. In the excavation shown in Fig. 3–45, for example, the sides are standing unprotected. Notice that near the top, where the soil is looser, the sides have been sloped to prevent cave-in. It is also evident that there is sufficient area for the excavation sides to be kept back from the main working area. One of the reasons for the stability of the soil in this case is that the area has been kept dry by a dewatering system.

In less stable soils, the excavation must be protected against cave-in, and this can be done in one of two ways. One is to slope the sides until the angle of repose for that particular soil is reached. In some cases this may mean a slope of 1 to 1, or 45°. This method may not be possible where space is limited or where the extra cost of

excavation is prohibitive. The alternative is to provide temporary support for the earth walls.

Many of the large buildings being built in the urban areas of the country today have large, heavy foundations which are often constructed in a deep excavation. In addition, many of them are built in congested or restricted areas (see Fig. 3–46), making very little more space available for the excavation than will be occupied by the building. The foundations of many large buildings are located in difficult earth, along waterways, in low-lying areas, or on filled land. All these factors make it necessary that large, deep excavations (see Fig. 3–47), often extending to the property limits, as, for example, the excavation shown in Fig. 5–1, be carefully planned.

Most of the methods used for supporting the earth walls around the perimeter of an excavation involve some system of sheeting and supports. The basic systems are *interlocking steel piling, steel soldier piles with horizontal sheeting,* and *concrete slurry walls.* Only in shallow excavations will ordinary *wood sheeting* (see Fig. 3–48) be used.

FIGURE 3-46 Excavation to property lines.
(*Courtesy Spencer, White & Prentis.*)

FIGURE 3-47 Deep excavation. (*Courtesy
Spencer, White & Prentis.*)

FIGURE 3-48 Excavation with timber shoring. (*Courtesy Moretrench Corp.*)

Interlocking Sheet Piling

Interlocking steel sheet piling (Fig. 3–49) is probably the most basic type of sheeting for deep excavations (see Fig. 3–50), although it is expensive unless it can be pulled and reused. It is difficult to drive in dense material and has to be carefully installed to prevent the jumping of interlocks. It is generally watertight, but, because of this, water pressure may build up behind the sheeting, which will necessitate a stronger bracing system to hold back both earth and water pressure.

FIGURE 3-49 Interlocking steel sheet piling. (*Courtesy Bethlehem Steel Co.*)

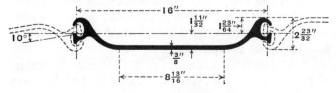

Steel Soldier Piles

The use of *soldier piles* and horizontal *timber sheeting* is an economical method for retaining excavation walls and is adaptable to many situations. The method consists of driving steel W shapes, 6 to 10 ft o.c., around the perimeter of the excavation and placing heavy wood planks, 3 to 4 in. in thickness, horizontally between the flanges (see Fig. 3–51). If the planks are spaced 1 to 2 in. apart, the buildup of water pressure behind the sheeting will be eliminated, and the earth, remaining drier, will have increased shear strength, thus reducing the tendency to slide. Such soldier piles can penetrate denser material than sheet piling and can be extended by welding, if necessary.

In some cases, the soldier piles are installed in drilled holes, which are backfilled with lean concrete after the timber has been installed. Concrete planks may replace timber ones, if necessary.

If the earth will stand unsupported for a span of about 3 ft, an alternative to using sheeting is to place piles on those centers and spray the wall surface with gunite

FIGURE 3–50 Excavation protected with steel piling. (*Courtesy Spencer, White & Prentis.*)

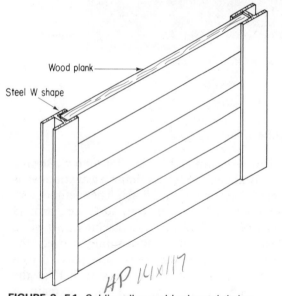

FIGURE 3–51 Soldier piles and horizontal timber.

to hold the earth in place. Wire mesh may be used in place of gunite (see Fig. 3–47), in which case the surface may be sprayed with an asphalt emulsion. In such situations, concrete piles may be substituted for steel ones.

Concrete Slurry Walls

The technique of building *slurry walls* involves the casting of concrete walls, from 18 in. to 5 ft in thickness and up

to 400 ft in depth below grade around one or more sides of an excavation. The cast-in-place, reinforced concrete wall is placed in sections or *panels,* usually not exceeding 25 ft in length and the full depth required. The completed wall may serve a dual purpose, acting as a retaining wall during the excavating phase and while the interior foundations are being built and later being used as the permanent exterior foundation walls for the building.

The wall may be built in consecutive or alternating panels, and the procedure will vary somewhat, depending on which system is used. Under the consecutive method (see Fig. 3–52), after the first panel is complete, others are constructed next to it, on either side, and, except for the first one, all the panels will be the same, with regard to reinforcing and end forming.

When the alternating panel method is used, *primary panels* are completed with sections of equal size left unexcavated between them (see Fig. 3–53). When the primary panels are finished, the *secondary* panels are excavated and concreted to form a continuous wall.

The excavation consists of a trench, with the choice of digging equipment depending on the soil involved. In soft to medium hard, loose, or cohesive soils, the popular digging tool is a clamshell bucket, similar to those illustrated in Fig. 3–54. The alignment of the trench is controlled by two concrete guide walls, about 1 ft wide and 3 to 5 ft deep, spaced the width of the trench apart and placed at grade level.

As material is excavated from the trench, bentonite slurry is immediately piped in, with the level of the slurry

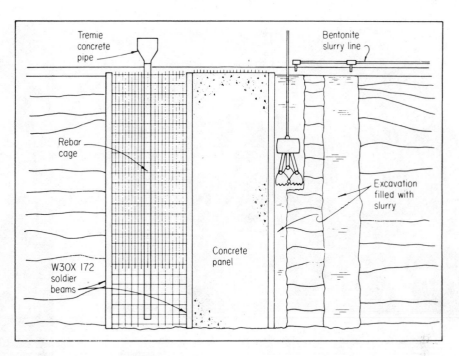

FIGURE 3-52 Consecutive slurry wall panels under construction. (*Courtesy I.C.O.S. Corp. of America.*)

being maintained close to the surface and at least 2 ft higher than the highest level of ground water, in order to produce a positive static pressure on the walls of the excavation.

Under one method of construction, when one primary panel is completely excavated, two large-diameter pipes, called *end pipes,* are placed, one at each end of the trench. A rigid, steel reinforcing cage, fabricated on the site, is lifted and placed into the slurry-filled trench.

Concrete is then placed in the panel by the tremie method, displacing the slurry, which is pumped into tanks to be reused. When the concrete has partially set, the end pipes are slowly withdrawn, leaving a semicircular concrete key at each end of the panel. In some cases, predriven steel HP shapes may be used to form a mechanical connection between panels.

Another method of providing panel end connections and reinforcing at the same time is illustrated in Fig.

FIGURE 3-53 Alternating slurry wall panels under construction.

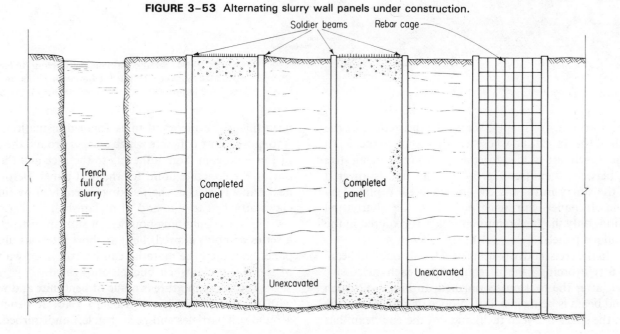

(a)

(c)

(b)

FIGURE 3–54 (a) Special clamshell bucket; (b) nine-ton clamshell bucket; (c) bucket between concrete guide walls. (*Courtesy I.C.O.S. Corp. of America.*)

3–55. Two long W shapes, whose depth is equal to the width of the trench and whose length is equal to the depth of the trench, are welded together with a reinforcing steel cage between them (see Fig. 3–56). This unit is lowered into the slurry in the trench, and concrete is placed by tremie. If panels are being placed by the alternating method, only the steel reinforcing cage is required in the secondary panels.

In the consecutive panel method, only one end beam and a reinforcing cage is required for each successive panel, after the first one is finished. In Fig. 3–57, such an end beam is being installed. Notice that on its outside face, the space between the flanges of the end beam has

been filled with pieces of thick foamed insulation. This is to prevent any concrete which escapes around the edges of the end beam from adhering to the face and thus reducing the concrete bond for the next panel. Before the next panel is placed, the packing is removed by ripping claws attached to one end of a clamshell bucket.

In soils containing boulders or similar obstructions, a series of holes is predrilled, at short intervals along a panel, and then the material can be excavated with one of the heavy clamshell buckets illustrated.

The slurry mixture is made of bentonite and water, with a specific gravity of 1.05 to 1.10. It penetrates around soil particles and gels when left undisturbed. The

FIGURE 3-55 End beams and reinforcing cage welded together. (*Courtesy I.C.O.S. Corp. of America.*)

FIGURE 3-56 Unit being lowered into trench. (*Courtesy I.C.O.S. Corp. of America.*)

FIGURE 3-57 End beam being lowered into trench. (*Courtesy I.C.O.S. Corp. of America.*)

gelled slurry maintains the soil particles in position by adhesion and creates a zone of stable soil on both sides of the trench. The lateral depth of soil so stabilized depends on the permeability of the soil and will range from several feet in loose material to very little in dense clay. The fact that the reinforcing is installed in slurry does not appreciably affect its performance, and good bond stresses with the concrete are obtained.

Slurry walls are unlikely to be competitive in price with steel sheeting or soldier piles in soils in which driving is relatively easy or in areas which may be easily dewatered and excavated. But they do provide a viable alternative where land is wet, the digging difficult, or the depth beyond an economical limit for sheeting or piles.

A new technique has recently been developed in which reinforced precast concrete walls are installed in the slurry-filled trenches. The slurry, in this case, is made up of a mixture of bentonite, cement, and additives with water. It will remain fluid until after the precast walls have been positioned but sets up later to hold the panels in place. The panels have a tongue-and-groove joint to make a continuous wall, and anchors may be cast into them, similar to standard slurry walls.

EXCAVATION WALL BRACING

No matter what system is used to build protecting walls for an excavation, they will have to be braced against earth and water pressure, and the problem is basically the same regardless of the type of wall. The important thing to realize is that bracing has to be *carefully designed and installed* to do its job properly and that failure of the braces will cause a great deal of trouble and expense, as well as the risk of injury and death to workers. To try to ensure safe practices in excavations and bracing, mandatory safety practices are outlined in the Federal Safety and Health Act. In it the angle of repose of the sides of excavations is specified in terms of the type of soil, the depth of excavation, etc.

Internal Sloping Braces

One common method of supporting excavation walls is to provide *internal sloping braces*. Two or more braces are used at each bracing point, the top end fixed to a waler or pile and the bottom end anchored into the floor of the excavation. Since they are loaded axially, members which make good columns will also make good braces. In recent years, steel pipe bracing has commonly been used, in spite of the fact that the connections at the walers are usually more difficult than for other types of bracing. Top braces are the most critical, since they are the longest and usually carry the greatest loads.

Horizontal Diagonal Bracing

One of the results of using internal bracing is that such bracing interferes with the excavating process. One way of partially overcoming this problem is to replace the top sloping braces with two or more rows of *horizontal, diagonal corner braces,* as illustrated in Fig. 3–50. In this case, the bracing is all steel pipe, the bottom ones sloping and the top ones two rows of pipe set diagonally into the corners. Such diagonal bracing should not be used where the angles are greater than 90°.

Tie Backs

In recent years the interference with excavating by internal bracing has led to the development of *tie back systems,* which provide unobstructed excavation areas (see Fig. 3–47). Basically the systems involve the use of steel ties of some kind, with one end anchored into the earth or rock *outside* the wall and the other either fastened to the outside of the wall or run through it to anchor to something on the inside (see Fig. 3–58). In the excavation shown in Fig. 3–47, the tie backs run through the soldier piles and are anchored on the inside. Tie back systems should be used in *firm ground, hardpan, glacial*

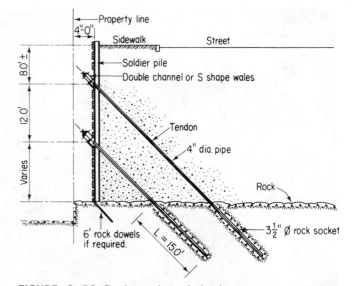

FIGURE 3–58 Rock anchor tie-back system. (*Courtesy Spencer, White & Prentis.*)

till, or *rock* but should not be used in soft clay or unconsolidated granular material. One of the disadvantages with the tie back systems is that they extend beyond the property lines, making it necessary to obtain permission to install them. This is sometimes difficult to do.

Rock Tie Backs

When this type of tie is designated, 4- to 6-in. pipe is driven through the earth overburden at a 45° angle, until rock is reached, using conventional drilling equipment. Then a 3½-in. hole, up to 15 ft long, is drilled into the rock and high tensile rod or wire strands are grouted into the hole and tested to a predetermined working stress (see Fig. 3–58). The top end of the rod or wire may be anchored to a concrete wall, passed through a hole in a soldier pile, and anchored on the inside or passed through sheet piling and anchored to a waler.

Earth Tie Backs

As the name implies, *earth tie backs* are used where there is no rock present for anchorage. They are more difficult to use because of the lower anchorage values, but they use a flatter angle of penetration, 15 to 30° from the horizontal being common. There are two types: *large diameter* and *small diameter*. Large-diameter tie backs require a tie hole about 12 in. in diameter, up to 65 ft deep, belled at the bottom (see Fig. 3–59). When the required depth is reached, rod or cable is inserted into the hole, which is then grouted to the *slip plane* (see Fig. 3–59). In some cases the bell at the end of the hole is eliminated, the hole is made longer, and then the ties depend on the friction between concrete and earth for their anchor.

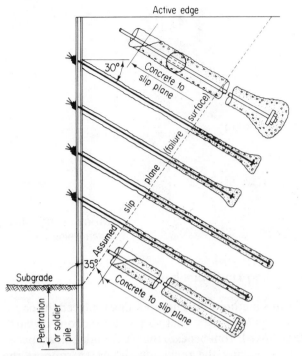

FIGURE 3-59 Belled and straight shaft earth tiebacks. (*Courtesy Spencer, White & Prentis.*)

Small-diameter tie backs are used in earth which tends to cave during drilling. This technique consists of installing 4-in.-diameter steel casing, using rotary drilling equipment. When the casing has reached the required depth, rod or cable is inserted and grouted to the slip plane, the casing being extracted during the grouting operation. This leaves the grout in contact with the earth to provide a friction anchor. In some cases, the casing may be left in the ground, and the anchor is provided by the friction between earth and steel.

PROTECTION OF ADJACENT BUILDINGS

In many instances, it is necessary to carry the excavation for a new building right up to the foundations of one or more existing buildings (see Fig. 5-1). This presents a problem, particularly if the new excavation is to be deeper than the foundations of the existing building. Part of the support for that foundation will be removed, and it is the responsibility of the builder to protect the building against movement caused by settlement during and/or after construction of the new building. Temporary support may be provided by *shoring* or *needling,* if the building is relatively light, while permanent support is provided by *underpinning*—extending the old foundation down to a new bearing level. In the case of heavy buildings, underpinning will usually be installed during an early stage of excavating, without shoring or needling.

Shoring

Shoring is the simplest method of providing temporary support for an existing building, but it is only useful for light structures. It involves only the use of shoring members and jacks (see Fig. 3-60). If they are short, the shores may be timber; otherwise they will be HP shapes or steel pipe. Screw jacks are most suitable if the pressure is to remain for an extended period of time.

There must be solid bearing for the end of the shoring member near the top of the wall being supported. One simple method is to use a waler bolted to the face of the wall, as illustrated in Fig. 3-60.

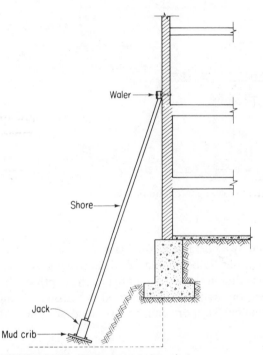

FIGURE 3-60 Wall supported to shore.

The lower end of each shore rests on a jack head. Each jack rests on a *crib* large enough to bear the load safely. Be sure to set the crib at the proper angle to ensure that jacking pressure is in direct line with the shore.

There are disadvantages to the shoring method of support which should be kept in mind. First, it is only practical for light buildings. Second, the height to which shores will reach effectively is limited, and therefore the method is not practical for tall buildings. Third, the lower ends of shores extend into the new working area and are an inconvenience to workers and machine operators.

Needling

Needling involves the use of a *needle beam* which is thrust through the wall of the building being supported. The inner end stands on solid blocking, while the outer end

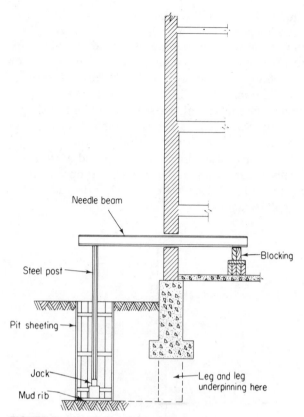

Needle beam

Blocking

Steel post

Pit sheeting

Jack

Mud rib

Leg and leg
underpinning here

FIGURE 3–61 Wall supported by needle beam.

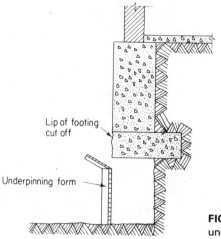

Lip of footing
cut off

Underpinning form

FIGURE 3–62 Short
underpinning form.

is supported by a post resting on a jack. Figure 3–61 illustrates how a needle beam is used. The post jacks must be set in pits dug in the working area to the depth of the new excavation. Work is then carried out around them until permanent support can be provided for the old building.

Underpinning

Underpinning is the provision of permanent support for existing buildings by extending their foundations to a new, lower level, containing the desired bearing stratum. There are a number of reasons this may be necessary. As mentioned previously, the removal of part of the supporting soil by a new, adjacent excavation may make underpinning necessary. Also the addition of new loads to an existing structure, either in the form of new machinery or by adding extra floors to the structure, may make additional support necessary. Underpinning may also be indicated if the existing building shows signs of sinking or settling due to poor bearing soil.

There are a number of ways in which underpinning may be provided, the method used depending on such factors as freedom to work, building loads involved, depth to new bearing, and type of earth encountered.

Leg-and-Leg Underpinning

Under this method, the underpinning is placed in spaced sections, or *legs,* with relatively small sections of earth being removed from beneath the old foundation at spaced intervals, leaving the remainder intact to support the building for the time being. Forms are set, and concrete legs are placed in each such section. When the concrete has hardened, the space between the top of the leg and the underside of the old foundation is dry-packed and wedged, or sometimes just wedged, to transfer the load to the new bearing member.

Then a new section is excavated and formed alongside each completed leg and new legs are placed. Adjoining legs should be tied together with *keys* or *steel dowels*. This process is repeated until the entire length of the wall has been underpinned. Forms for such underpinning are simple, as illustrated in Fig. 3–62.

Pit Underpinning

This type of underpinning is similar to leg-and-leg, the difference being in the depth to which the work is carried. When the new bearing level is a considerable distance below the old foundation—in excess of 5 or 6 ft—*spaced pits* may be dug under it—in fashion similar to the technique employed in leg-and-leg—cribbed, and filled with concrete. When the concrete has hardened, the space above is dry-packed and wedged to transfer loads to the new columns. The second series of pits is then dug alongside the completed ones and so on until the underpinning is completed. Depths of 45 to 50 ft may be reached by this method.

Jacked Cylinder Underpinning

Sectional steel cylinders are *jacked* down beneath the foundation to be underpinned until they reach a suitable baering. Then they are cleaned out and filled with concrete. A concrete cap, similar to a pile cap, is then formed and cast between the tops of the cylinders and the underside of the foundation. When the cap has hardened, the space between it and the foundation is dry-packed so that the loads are transferred to the underpinnings (see Fig. 5–1, right-hand side).

Pretest Underpinning

Any type of footing, when forced into the ground under a load, builds up a resistance beneath it which finally prevents any further penetration. Tests indicate that when this happens under a test load and then that load is released the footing will *rebound,* and if the load is reapplied, an equivalent resistance will finally be created but at a lower level than before. Tests further indicate that rebound follows each release of load and that equal resistance under subsequent reloading is found only at greater depth.

It follows that if any type of underpinning is test-loaded for adequacy and the load released before the cap or other connection is placed between it and the foundation, the underpinning will sink again when the foundation load comes on and further settlement will occur before it develops the resistance created under the test load. To overcome this, the *pretest* technique has been developed.

Sectional steel cylinders are jacked down beneath the foundation, cleaned out, and filled with concrete, as described above. When the concrete is ready, the cylinder is test-loaded to an overload capacity, usually 50% in excess of the permanent load, using two jacks, as illustrated in Fig. 3–63. While the full test pressure is maintained on the jacks, a short steel column is placed between the top of the cylinder and the underside of the foundation and the load of the foundation is permanently transferred to the underpinning cylinder by steel wedges.

ROCK EXCAVATION

Excavating sometimes includes the removal of solid rock, and this will involve blasting, a job that is normally carried out by specialists. It is important, however, that the basic principles of rock removal by blasting be understood by those involved with the construction job.

In the building construction industry, blasting is used for such purposes as rock excavation, demolition work on buildings and foundations, stump clearing, and the breaking up of boulders too large to handle (see Fig. 3–64). The techniques of blasting have been made possible through the development of gunpowder and the refinement of explosives such as dynamite.

FIGURE 3–64 Demolition blasting. (*Courtesy Canadian Industries Ltd.*)

TYPES OF EXPLOSIVES

Explosives commonly used in commercial blasting are practically all solid-solid or solid-liquid mixtures capable of rapid and violent decomposition, with resultant conversion into large volumes of gas. Decomposition of a *high* explosive, such as dynamite, takes place with extreme rapidity, while a *low* explosive, such as black blasting powder, decomposes more slowly, simulating rapid burning. As a result, high explosives are called *detonating* ex-

FIGURE 3–63 Pretest underpinning.

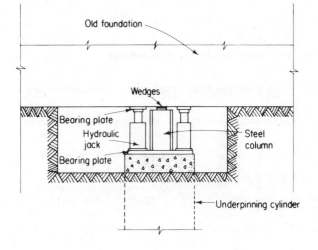

plosives and low explosives are referred to as *deflagrating* explosives.

The most widely used high explosive in the construction field is dynamite. It is basically composed of liquid nitroglycerin as a *sensitizer,* sawdust or wood pulp as a *liquid absorber,* an *oxygen supplier* such as sodium nitrate, and a small amount of *antacid* material such as zinc oxide or calcium carbonate. Several types of dynamite exist, the common ones being (1) *straight,* (2) *extra* or *ammonia,* (3) *gelatin,* and (4) *permissive* dynamite.

Dynamite is produced in various grades, depending on the percentage (by weight) of nitroglycerin the material contains. Thus, a 30% grade contains 30% nitroglycerin, while a 60% grade contains 60% nitroglycerin, etc. This does not mean that 60% dynamite is twice as strong as 30% dynamite—in fact, it is only about 1⅓ times as strong.

For practical use, dynamite is rolled into cartridges covered with wax-impregnated, 70-lb manila paper. Cartridges range from ⅞ to 8 in. in diameter and from 8 to 24 in. in length.

To produce an explosion, dynamite may be ignited in any one of three ways, depending on the type of dynamite used and the general conditions at the time of blasting. One method is to use a safety fuse connected directly to the cartridge. Another is to use a blasting cap at the end of a fuse, and the third is the use of electric blasting caps.

A safety fuse consists of black powder protected by a flexible fabric tube. Safety fuses are available in two burning rates: 1 yd in 120 sec or 1 yd in 90 sec. Fuses are generally lit with matches, and their burning time provides an opportunity to clear the blast area.

Many dynamites cannot be ignited by the burning action of a fuse and must be set off by a primary blast wave. This is done by placing a blasting cap on the end of a safety fuse and inserting the cap into a cartridge. The burning fuse ignites the cap, which detonates and in turn sets off the dynamite.

In many cases, safety fuses and detonators are replaced by electric blasting caps (see Fig. 3–65). These are small copper cylinders about 1⅛ in. long and ⅜ in. in diameter containing a small explosive charge. Two conducting wires lead into one end and are connected by a resistance bridge. When an electric current passes across this bridge, the heat generated ignites the charge, which in turn detonates the main charge. Two kinds of blasting caps known as *delay* caps are also made. They are the ordinary delay cap, which allows approximately ½ sec between groups or firing periods, and the millisecond delay cap. In the latter, various delays are available, ranging from 0.025 to 0.1 sec.

Electric blasting circuits may be wired in series, in parallel, or in series-parallel (see Fig. 3–66) (employed when a large number of caps are to be set off). Circuits are energized either by batteries or by a blasting machine. Batteries are used only when a blasting machine is not available. The latter produces the desired current and produces it every time the machine is used. Three types of blasting machines are the twist handle, rack bar, and condenser-discharge.

Excavating with explosives is carried out primarily by drilling holes in the rock, loading the holes with explosives, and firing the charge. Holes are drilled horizontally (Fig. 3–67), vertically (Fig. 3–68), or at an angle, depending on circumstances. Holes of all diameters and for different explosives are all loaded basically in the same manner.

When drilling is completed, holes should be cleaned with a jet of compressed air and tamped with a wooden pole. If they are to remain unused for some time after drilling, they should be sealed with paper plugs to keep them clean. Loading a hole consists of placing a primer charge and a number of cartridges in the hole and tamping them into place. A cap is placed in the primer charge

FIGURE 3–66 Series-in-parallel electric blasting circuit.

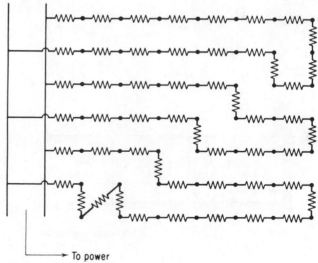

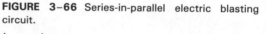

FIGURE 3–65 Diagram of electric blasting cap.

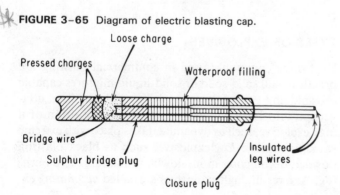

Loose charge

Pressed charges

Waterproof filling

Bridge wire

Sulphur bridge plug

Insulated leg wires

Closure plug

To power

FIGURE 3–67 Horizontal drilling. (*Courtesy Canadian Industries Ltd.*)

FIGURE 3–68 Vertical drilling. (*Courtesy Schramm Inc.*)

by piercing a cartridge with a wooden poker and embedding the cap in the hole, allowing the leg wires to project. The primer charge is lowered into the hole, and additional charges are then placed and tamped. The cartridges are connected by slitting the end of each as it is placed. The pressure caused by tamping causes the slits to open and dynamite to bleed from one cartridge to another. The primer cartridge is not slit. Figure 3–69 illustrates one method of loading a hole.

After the hole has been loaded to the desired depth, it is *stemmed* with an inert material, such as sand or clay. The stemming confines the blast and prevents blowouts through the hole. During the placing, tamping, and stemming operations, the leg wires from the blasting cap must be held against one side of the hole to prevent their being damaged. Figure 3–70 shows a vertical section through a loaded and stemmed hole.

To achieve the best results, an organized and calculated pattern of holes must be set up, loaded with either *instantaneous* or *delay* caps. Instantaneous loading and shooting does not permit control of the direction of rock throw and tends to produce *muck* which is often too large to handle. With the use of delay time caps, the direction of rock throw is controlled, and fragmentation is generally improved. Figure 3–71 shows how the use of delay pattern has controlled the breakage of rock at the excavation face. Figure 3–72 illustrates good rock throw control and good fragmentation achieved with a delay pattern. Figure 3–73 shows two steps in a blasting operation, one at the height of the explosion and the other at its conclusion.

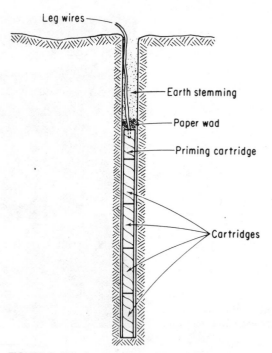

FIGURE 3–70 Section through loaded and stemmed hole.

FIGURE 3–71 Clean rock face made possible with delay pattern blasting. (*Courtesy Canadian Industries Ltd.*)

FIGURE 3–69 Loading a hole with cartridges. (*Courtesy Canadian Industries Ltd.*)

In Fig. 3–74, a typical delay pattern is illustrated. Notice that the longer delays are used in the back rows of the holes and along the sides and that the short delay caps are used in the bore holes close to the working face. The short delay charges move the rock near the face and thus provide space for the material moved by the charges primed with long delay caps.

In some cases, large boulders or rock masses left after the *primary* blast may require *secondary* blasting for complete breakage. Secondary blasting may be done

FIGURE 3-72 Delay pattern blasting. (*Courtesy Canadian Industries Ltd.*)

(a)

FIGURE 3-73 Controlled blasting: (a) explosion at maximum; (b) blast completed.

(b)

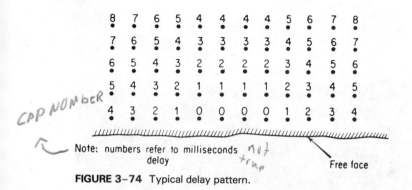

CAP NUMBER

Note: numbers refer to milliseconds *not true* delay

Free face

FIGURE 3-74 Typical delay pattern.

in three ways: (1) by *mudcapping,* (2) by *blockholing,* or (3) by *snakeholing.*

In mudcapping, the required number of cartridges are tied together in a bundle and placed on top of the boulder or rock pile. They are covered with a piece of waterproof paper and a layer of mud about 6 in. thick. The mud contains the blast and directs the shock wave into the rock. Figure 3-75 shows a mudcapping operation before and after the blast.

If extremely hard rock such as granite or taprock does not respond to mudcapping, blockholing must be used. This consists of drilling one or more holes, depending on the size, and loading, stemming, and firing them as in primary blasting.

Snakeholing consists of tunneling under a boulder and placing a charge there in order to lift it and break it up. The excavated earth is replaced in the hole as stemming (see Fig. 3-76).

In primary blasting, the *burden* and the *spacing* of the holes are extremely important. The spacing is the distance between holes in a line, and the burden is the distance of a hole from the free face.

Hole sizes for general blasting range from 1 to 1½ in. and are drilled to a maximum of 20 ft. To illustrate how to plan a drilling pattern and how to determine the amount of explosive required for a given situation, 1⅛-in. drill holes will be considered.

The maximum amount of explosive that can be placed in a hole 1⅛ in. in diameter and 1 ft deep is 0.69 lb. Under free breaking conditions, rock normally requires a load of 1 lb of explosive per cubic yard. At that

FIGURE 3-75 Mudcapping: (a) setting a charge; (b) charge capped; (c) after the explosion. (*Courtesy Canadian Industries Ltd.*)

(a)   (b)

(c)

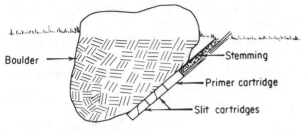

FIGURE 3-76 Snakeholing.

rate, the volume of rock that can be broken up with 0.69 lb of explosive is 0.69 cu yd, or 18.6 cu ft. A slab of rock 1 ft thick that produces 18.6 cu ft will have a square surface dimension of 4.3 ft × 4.3 ft. So if a number of holes 1 ft deep were drilled in a square pattern, the maximum spacing would be 4.3 ft in any direction. The general spacing and burden limit in instantaneous blasting is 4.3 ft, although this space may vary to some degree with the use of delay caps.

A variety of patterns may be developed by the use of delay caps. One basic pattern is illustrated in Fig. 3-77, where a shot is to be fired into a bank with both ends closed, leaving a vertical face.

The three top center holes, loaded with #0 delay caps, open the shot and release the burden on the #1 delay holes below and on each side. The burden of each successive hole is free-faced by the firing of the charge in front of it. This free-facing of the burden of every charge is the basic principle behind delay pattern blasting. The general load factor will be 1 lb of explosive per cubic yard of rock, but the #0 delay charges probably have a load factor of 1.25 lb/cu yd, in order to produce a good, clean start for the pattern. Figure 3-72 shows a bank shot such as that described above being fired.

When there is danger of damage occurring because of rock throw, blasting mats should be used. Two most often used types are woven rope and wire mats. Wire mats must be weighted down to contain rock during a blast, whereas rope mats are heavy enough by themselves. Figure 3-78 shows an area covered with woven rope mats, about to be blasted in a residential section. In Fig. 3-79, rock is being contained by wire mats.

Safety is a prime factor in blasting operations, and following are 15 general safety rules to practice when handling explosives.

1. Smoking must not be tolerated in the blasting area.

2. Only hardwood or nonspark pokers must be used in pierce cartridges.

3. Circuits should always be checked before firing.

4. It must be made certain that the area has been cleared before firing a shot.

5. Shunts should always be used when setting up an electric circuit.

6. When there are electrical storms in the area, all electric blasting should be stopped.

7. Only nonsparking tape should be used for measuring hole depths during loading.

8. The *shucking* of dynamite—removal from the cartridge—should be avoided unless absolutely necessary. Shucked dynamite should not be tamped.

9. Inventory should be taken before and after firing of a shot to guard against mislaid explosives and caps.

10. Only blasting galvanometers should be used for testing the circuit.

11. Loaded hole areas should be clearly marked with appropriate signs.

12. Piles of cartridges should never be left beside a large-diameter hole.

FIGURE 3-77 Typical pattern for bank shot with closed ends.

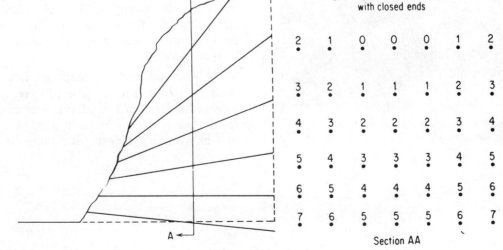

Typical pattern for bank shot with closed ends

Section AA

FIGURE 3-78 Woven rope mats in place. (*Courtesy Canadian Industries Ltd.*)

FIGURE 3-79 Blasting, using wire blasting mats. (*Courtesy Canadian Industries Ltd.*)

13. Holes must be cool prior to loading. This should be checked.

14. The primer cartridge should never be forced into a hole and should never be tamped.

15. Storage or dry-cell batteries should never be used to activate electric blasting circuits.

REVIEW QUESTIONS

1. Explain **(a)** how soil types may influence the kind of excavating equipment to be used, **(b)** the basic difference between a *power shovel* and a *pull shovel*, and **(c)** the operational difference between a *dragline bucket* and a *clamshell bucket.*

2. Answer briefly: **(a)** When is dewatering equipment necessary? **(b)** Name the five basic parts of a dewatering system. **(c)** What is a possible alternative to the use of dewatering equipment?

3. Explain what is meant by **(a)** *special* excavating work, **(b)** soil swell, and **(c)** payline.

4. **(a)** Outline the limitations to the use of shoring as a means of protecting adjacent buildings. **(b)** Outline the procedure for carrying out pretest underpinning.

5. Explain the principle involved in **(a)** the use of bentonite slurry in excavating trenches, and **(b)** the use of tremie for placing concrete in a slurry-filled trench.

6. By means of neat diagrams, illustrate **(a)** the use of internal sloping braces to support a wall of sheet piling, **(b)** belled earth tie backs to support soldier piles, and **(c)** the use of horizontal diagonal braces as support for exavation protection walls.

7. The working face of an open pit consists of soft shale overlaid by hard rock. If the entire face is to be moved in a single blasting operation, describe briefly how the holes should be loaded.

8. A rubber-tired scraper with an efficiency of 50 min/hr has a heaped capacity of 25 cu yd. If it is moving material with a swell factor of 20%, what amount of in-bank material can it move in 1 hr if it has a total cycle time of 6 min?

9. A front-end loader is loading trucks with granular material that has a swell factor of 15%. If the loader has a heaped capacity of 3 cu yd, calculate the total cycle time for the loader if it can move 50 cu yd of in-bank material in 1 hr with an efficiency of 50 min/hr.

Chapter 4

Foundation Layout

Upon completion of the site investigation and design of the foundations, actual site work can begin. In many instances, some preliminary excavation must be done on the site before the actual layout of the building foundations begins. Site access may be required, existing trees may require removal or relocation, and some site leveling may be necessary to provide drainage.

The layout of the foundation structure must be carried out with great accuracy to ensure that its various elements are placed exactly as called for in the building plans. A set of building plans will include a foundation plan, on which is shown the position of piles, footings, column bases, slab stiffeners, etc., with all elevations and necessary details.

Figure 4–2 is the foundation plan for the building shown in Fig. 4–1. This is a typical footing, wall and slab foundation, with the building loads transferred to the earth by the footings.

Figure 4–4 is a pile foundation plan for the office building shown in the site plan, Fig. 4–3. Here the building loads are transferred to solid bearing by perimeter and interior piles.

LOCATING POINTS FROM BATTER BOARD LINES

The first step in laying out the foundation, after completing the excavation, is to rig the batter board lines.

They may be either cord or wire, depending on their length, but they must always be strung very tightly.

A plumb bob dropped from the intersection of two corner batter board lines (see Fig. 4–5) to the excavation floor establishes a corner position. A stake is driven at that point and a nail placed in the top of the stake so that it coincides exactly with the point of the plumb bob. All the outside corners are established in the same way. The perimeter of the building is thus outlined, as shown in Fig. 4–6. The accuracy of the staking should now be checked by measuring the diagonals of the rectangles involved. If there is no error, the diagonals will be exactly equal.

A further check may be made by measuring the angles with the leveling instrument. For example, for the layout shown in Fig. 4–6, proceed as follows:

1. Set up the instrument over stake A, center over the pin, and level it. Backsight on the pin in stake B. Set the horizontal circle to 0.

2. Turn the telescope and sight on stake D. Calculate the size of the angle DAX by trigonometry, using a right triangle and solving for tan angle DAX, from natural trigonometric functions. In this case,

$$\tan DAX = \frac{20}{100} = 0.200; \text{ angle } DAX = 11°19'$$

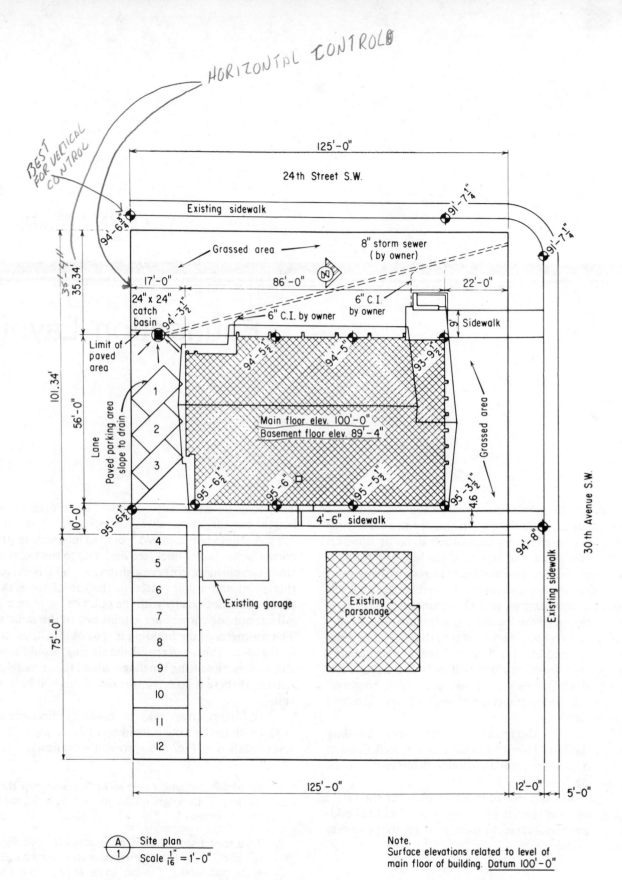

HORIZONTAL CONTROL⊕

BEST FOR VERTICAL CONTROL

125'-0"

24th Street S.W.

Existing sidewalk

9½'-7¼"

9½'-7¼"

94'-6¾"

35'-4"
35.34

Grassed area →

8" storm sewer
(by owner)

17'-0"

86'-0"

22'-0"

24" x 24"
catch
basin

94'-3½"

6" C.I. by owner

6" C.I.
by owner

9'

Sidewalk

Limit of
paved
area

94'-5½"

94'-5"

93'-9½"

101.34'

Lane

56'-0"

Paved parking area
slope to drain

1

2

3

Main floor elev. 100'-0"
Basement floor elev. 89'-4"

Grassed area

95'-6½"

95'-6"

95'-5½"

95'-3½"
4.6"

10'-0"

95'-6½"

4'-6" sidewalk

94'-8"

Existing sidewalk

30th Avenue S.W.

4

5

6

7

8

9

10

11

12

Existing garage

Existing
parsonage

75'-0"

125'-0"

12'-0"

5'-0"

A
─
1

Site plan
Scale 1/16" = 1'-0"

Note.
Surface elevations related to level of
main floor of building. Datum 100'-0"

FIGURE 4-1 Site plan for church.

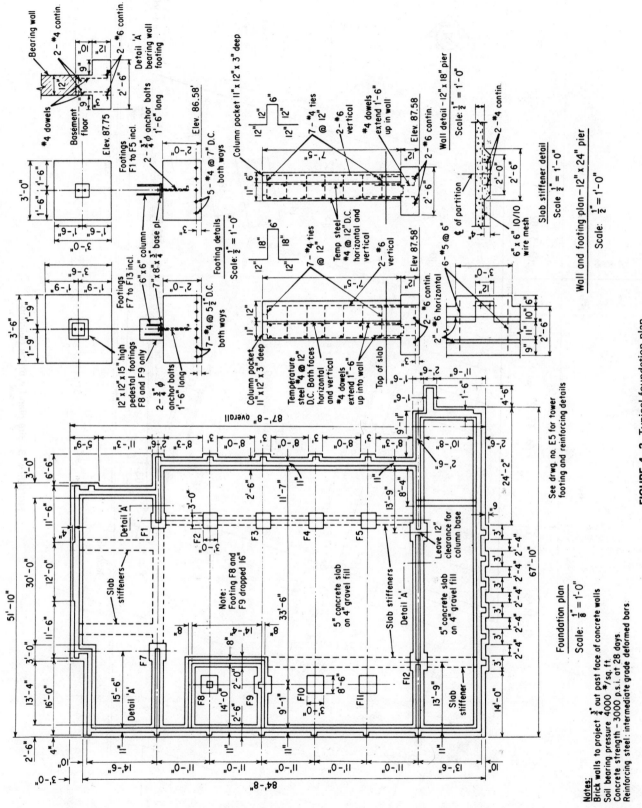

FIGURE 4–2 Typical foundation plan.

73

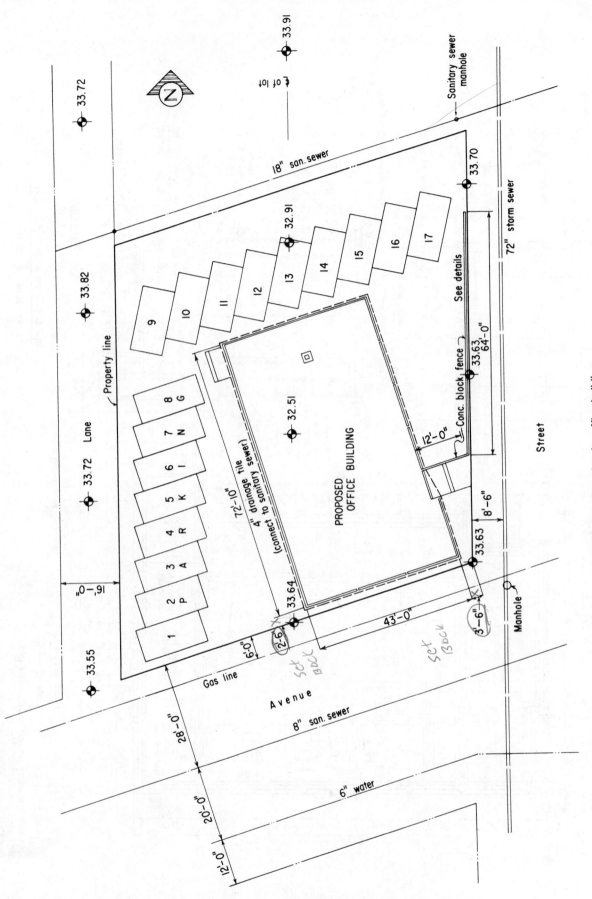

FIGURE 4-3 Site plan for office building.

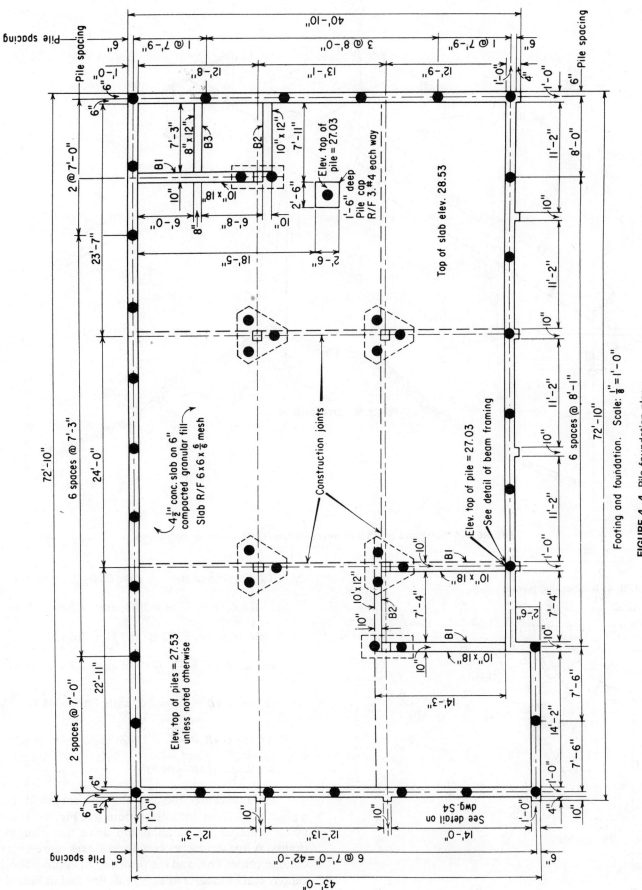

Footing and foundation.   Scale: $\frac{1}{8}'' = 1'-0''$

FIGURE 4-4 Pile foundation plan.

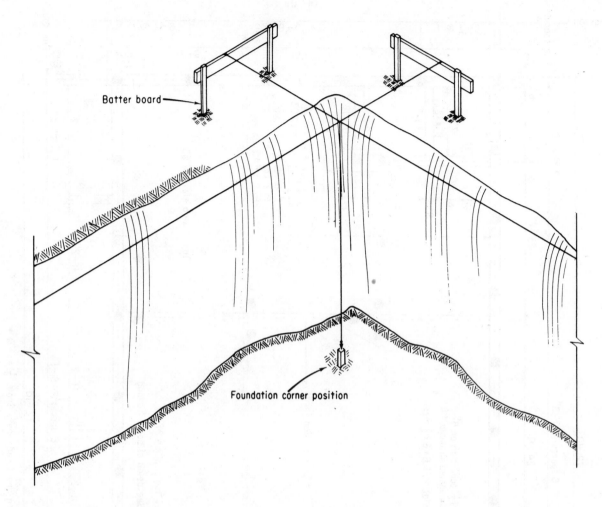

FIGURE 4-5 Establishing a foundation corner position from batter board lines.

FIGURE 4-6 Checking layout.

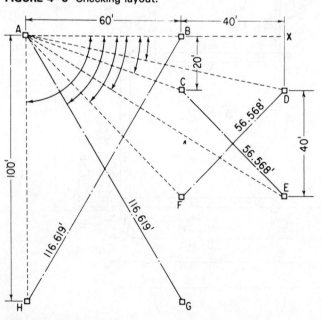

3. Turn again and sight on stake C. By the same process,

$$\tan CAB = \frac{20}{60} = 0.3333; \text{ angle } CAB = 18°26'$$

4. Sight on stakes E, F, G, and H in order:

(a) $\tan EAX = \frac{60}{100} = 0.60; \text{ angle } EAX = 30°58'$

(b) $\tan FAB = \frac{60}{60} = 1.0; \text{ angle } FAB = 45°0'$

(c) $\tan GAB = \frac{100}{60} = 1.666; \text{ angle } GAB = 59°2'$

(d) Angle $HAB = 90°0'$.

Column center lines can now be established. Drop a plumb bob from line intersections (see Fig. 4–7) on opposite sides of the building and drive stakes at those points. A line strung between these stakes represents the column center line, and the positions of column footing centers can be measured along this line and indicated by driving stakes at each point.

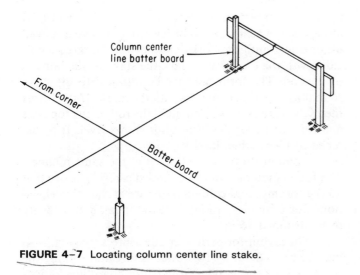

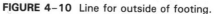

FIGURE 4–7 Locating column center line stake.

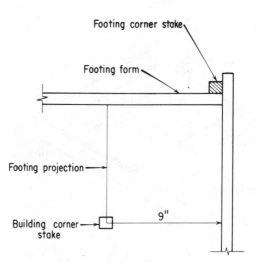

FIGURE 4–9 Footing corner stake position.

LAYOUT FOR FOOTING FORMS

There are a number of methods for establishing the outside line of exterior footing forms. If the footings are not large, the projection of the footing beyond the outside wall line can easily be measured. For example, in Fig. 4–2, the bearing wall footing extends 9 in. beyond the wall. From the corner stake already established, measure off a square with 9-in. sides (see Fig. 4–8), and drive a stake with one face on the line and one the thickness of the footing form material beyond the intersecting line (see Fig. 4–9). This stake should be driven or cut off exactly at the elevation shown for the footings. For example, in Fig. 4–2, the elevation of the bottom of footings $F1$ to $F5$ is shown as 86.58 ft and the footings are 2 ft thick. Therefore, the elevation of the top of the stakes here would be 88.58 ft.

Stakes are similarly placed at each outside corner, and lines strung between them to guide the driving of intermediate stakes. Figure 4–10 illustrates this. Stakes for the inside form are located by measuring from this

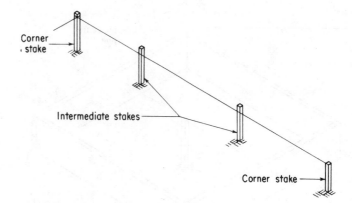

FIGURE 4–10 Line for outside of footing.

line. The inside corner stake must be set away from the line in both directions, as shown in Fig. 4–11.

If footings are wide, their inside and outside extremities may be located on batter boards. The position of inside and outside forms can then be determined from batter board lines (see Fig. 4–12).

FIGURE 4–8 Locating outside footing stake.

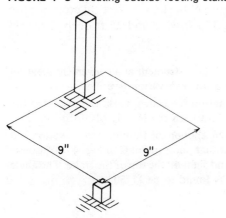

FIGURE 4–11 Inside footing stake position.

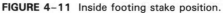

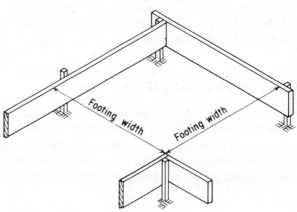

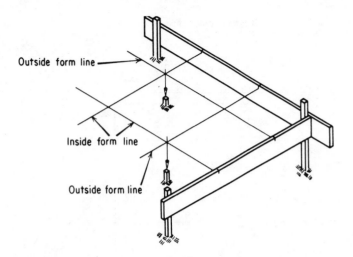

FIGURE 4–12 Double batter board lines.

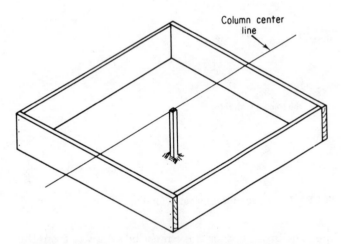

FIGURE 4–13 Positioning column footing form.

If column footings are not large, they can be positioned as in Fig. 4–13. Here the footing form has been made up and set over the center line stake. It is positioned correctly by measuring from the stake and the line, and finally leveled. Large column footings may be located by double batter board lines, in a manner similar to that described above. Double intersecting lines will establish all four corners of a column footing.

LEVELING OF FOOTING FORMS

The first stake driven to a known elevation is now used as a reference point to level the remainder of the footing stakes or the top of footing forms. Set up the instrument in a convenient location within the perimeter of the building. Take a backsight on a rod held on the first stake and

tighten the target in position. Now have the rod held alongside the next stake to be leveled and have it moved up or down until the horizontal line on the target coincides with the cross hair. Mark the stake at the bottom of the rod. The form can then be fastened to the stake so that its top edge coincides with the mark. If the forms themselves are being leveled, hold the rod on the top edge and raise or lower the form till it shows level. It is fastened in place when level.

Column footings are leveled in the same manner. The foundation plan must be studied carefully to see that all the footings and column bases are at the same elevation. For example, in Fig. 4–2, footings *F8* and *F9* are to be dropped 16 in.

The layout for perimeter pile centers (see Fig. 4–4) may be carried out by first establishing and staking the centers of the corner piles from batter board lines. Lines may now be strung between these points and the location of the remainder of the perimeter piles found by measuring along the lines. Pile spacings are shown on the foundation plan. In the particular case shown in Fig. 4–4, 14-in. cast-in-place concrete piles are specified.

The column piles may now be located by measurement. The elevation of the tops of all piles is indicated on the foundation plan and must be checked as the piles are placed.

BENCH MARKS AND DEEP EXCAVATIONS

It is easy to establish an elevation at the bottom of a shallow excavation (no deeper than the rod being used) from a bench mark at ground level. But for deeper excavations, other methods must be used. One method is outlined in the following example. Remember that the *angles of depression* must be read very accurately. Let us suppose that for a particular building the elevation of the top of wall footings is to be + 76.5 ft with relation to a bench mark of + 100.0 ft established at ground level. The excavation is 105 ft wide and 25 ft deep (for a 100-ft-wide building). The elevation of the excavation floor must be established. Figure 4–14 shows a vertical cross section of the excavation, 105 ft wide and 25 ft deep. Proceed as follows:

1. Set up and level the instrument at a convenient location near the edge of the excavation (see Fig. 4–14).
2. Take a backsight on the bench mark. Suppose that the reading is 4.76 ft. Then the H.I. is 104.76 ft.
3. Pick a convenient spot on the floor of the excavation and drive a stake at that point (point C in Fig. 4–14). Lower the telescope and sight on the top of the stake. The angle of depression is found to be 17°30′.

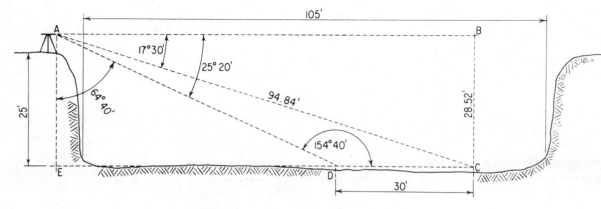

FIGURE 4–14 Elevation in deep excavation.

4. Measure a definite distance (30 ft in this case) in the same line of sight and drive a second stake at D, level with the first. Let the angle of depression to the top of stake D be 25°20′. Both of these readings must be very accurate.

5. Calculate the size of angles *DAC* (7°50′) and *ADC* (154°40′).

6. Apply the sine rule to triangle *ADC* to find the length of side *AC*.

$$\frac{30}{\text{sine } 7°50′} = \frac{AC}{\text{sine } 154°40′}$$
$$30 \times 0.42788 = AC \times 0.13629$$
$$AC = \frac{30 \times 0.42788}{0.13629} = 94.84 \text{ ft}$$

7. With respect to triangle *ABC*, *BC/AC* = sine 17°30′.

Therefore

$$\frac{BC}{94.84} = 0.30071$$
$$BC = 0.30071 \times 94.84 = 28.52 \text{ ft}$$

8. The elevation of the top of stake C, relative to the bench mark, will be 104.76 − 28.52 = 76.24 ft.

Stake C may now be used as a bench mark from which to establish the elevation of footing forms, etc., for the foundation.

REVIEW QUESTIONS

1. Give three reasons it is usually better to establish column center lines rather than outside lines of footing when laying out column footing locations.

2. Explain why it may be preferable to have both inside and outside form lines of wide footings established on batter boards.

3. Why should footing form stakes be cut off to elevation?

4. Suggest an alternative method to the use of batter boards and lines which might be used to transfer points to the footings in an excavation.

5. Suggest an alternative method of establishing the elevation at the bottom of an excavation from a datum point at ground level.

Foundations

NEED FOR DEEP FOUNDATIONS

The foundations of a building are generally regarded as that part of the structure which transmits the load of the building to the earth on which it rests. They must be carried to a depth at which the earth has sufficient bearing strength to carry the load safely.

For lightweight buildings and for heavier buildings on good load-bearing soil, the foundations normally consist of footings and piers, or foundation walls. The footings are spread over a wide enough area to carry safely the load of the piers or walls that must rest on them.

However, tall, heavy buildings present many foundation problems because of their deep basements and heavy column loads. Deep basements bring about excavation problems, particularly where there are other heavy buildings in close proximity, and heavy loads make it uneconomical and impractical to use the conventional piers and footings. Some other type of foundation then becomes necessary, and other alternatives include *piles, caissons, deep wall foundations,* and various types of *spread foundations.*

The type to be used in any particular case will depend on the size and degree of concentration of building loads, the depth from the ground surface to bearing of adequate capacity, the type of subsurface material through which the foundation must pass, the availability of particular equipment, the comparative costs, and the proximity of other heavy buildings (see Fig. 5-1).

TYPES OF PILES

Piles are made from any of several materials, including wood, steel, and concrete. They may also be composite in structure—that is, made up of more than one material. A pile with a wood lower section and concrete upper section is the most common composite type.

There are three main classifications of piles used in the construction industry. The most common is the *bearing pile.* It transfers the heavy loads through the unstable surface soils to the denser, more stable soils below. It should be noted that loads are carried vertically downward to bedrock, or other material of high-bearing strength.

The next is the *friction pile.* This type does not necessarily reach high-bearing materials, but does reach soil resistance to a point where the load is carried by the underlying soil and the soil pressure surrounding the pile. Friction piles thus depend on soil density and side pressure for a great percentage of their load-bearing ability.

The third type of pile is the *sheet pile.* It is not intended to carry vertical loads, but rather to resist horizontal pressure. The principal use for sheet piling is to hold back earth around the perimeter of an excavation.

PARTS OF A TYPICAL PILE

All piles, regardless of the type of material from which they are made, have certain basic parts. Figure 5-2 illus-

FIGURE 5-1 Building site surrounded by heavy buildings. (*Courtesy Spencer, White & Prentis.*)

FIGURE 5-2 Parts of a typical pile

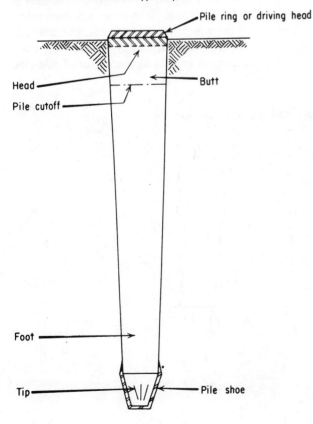

trates a typical pile and its parts. The following definitions apply to any such pile:

- *Head:* The head of a pile is its upper part in final position.
- *Foot:* The foot of a pile is its lower part in final position.
- *Tip:* The tip of a pile is its small end before or after it is placed in position.
- *Butt:* The butt of a pile is its large end before or after it is placed in position.
- *Pile ring:* A pile ring is a wrought-iron or steel hoop that is placed on the head of the pile to prevent cracking, brooming, or splitting.
- *Driving head:* A driving head is a device placed on the head of a pile to receive hammer blows and to protect it from injury while it is being driven. A driving head may be used instead of a pile ring.
- *Pile cutoff:* The portion of the pile that is removed after completion of driving.
- *Pile shoe:* A pile shoe is a metal cone placed on the tip of a pile to protect it from cracking or splitting. It also helps the pile to penetrate such materials as riprap, coarse gravel, shale, or hardpan.

BEARING CAPACITY OF PILE FOUNDATIONS

The design of pile foundations consists of two general steps: (1) the selection and design of the piles and the driving equipment to be used, and (2) the study of the soils to which the loads are transmitted. The allowable load limits for bearing soils range from 1 ton/sq ft for clay up to approximately 30 tons/sq ft for granite bedrock. The average gravel soil will support a compressive load of 5 or 6 tons/sq ft. The supporting strength of a pile is therefore proportional to its size and to the strength of the soil in which it is driven.

The following example will demonstrate the computation of the upward bearing strength of a pile of prescribed dimensions driven into a particular soil.

> *Given:* Soil bearing = 4 tons/sq ft; skin friction = 700 lb/sq ft. A timber pile 35 ft long with an average diameter of 12 in. is to be driven. The pile tip diameter is 8 in.
>
> End area = 22/7 × ⅓ ft × ⅓ ft = 0.35 sq ft
> End bearing = 0.35 × 8000 = 2800 lb
> Pile area = 22/7 × 1 ft × 35 ft = 110 sq ft
> Bearing due to skin friction = 110 × 700 = 77,000 lb
> Total bearing capacity of this pile =
> $$(2800 + 77,000)/2000 = 39.9 \text{ tons}$$

WOOD PILES

Considerable selection is necessary in order to obtain good wooden piles. They should be free from *large* or *loose knots, decay, splits,* and *shakes. Crooks* and *bends* should be not more than one-half of the pile diameter at the middle of the bend. Pile *sweep* should be limited so that (1) for piles less than 70 ft in length, a straight line joining the midpoint of the butt and the midpoint of the tip does not pass through the surface of the pile; (2) for piles 70 to 80 ft in length, a similar straight line does not lie more than 1 in. outside the surface of the pile; and (3) for piles 80 ft in length a similar straight line does not lie more than 2 in. outside the surface of the pile. The *taper* should be uniform from tip to butt.

Pile lengths are in increments of 1 ft. The minimum tip diameter is normally 6 in., and the maximum butt diameter for any length is 20 in.

Various species of trees are used for making piles, the most common being southern pine, red pine, lodgepole pine, Douglas fir, western hemlock, and larch. All these can be pressure-treated, a process designed to protect piles against deterioration. Creosote is the preservative most commonly used for this purpose.

Some of the advantages of using wood piles are as follows:

1. Wood piles have an indefinite life expectancy when placed under water or driven below ground water level.
2. Wood piles are light.
3. In many areas, wood piles are readily available, relatively inexpensive, and easy to transport.

EASY TO CUT OFF AT DESIRED LENGTH.

4. Wood piles produce greater *skin friction* than piles of most other materials.

However, wood piles are subject to attack by insects, marine borers, and fungi unless treated. They have a lower resistance to driving forces than other types of piles and have a tendency to split or splinter while being driven. Wood piles support a smaller load than other types of comparable size, which means using more piles and larger footings.

CONCRETE PILES

There are two principal types of concrete piles, *cast-in-place* and *precast*. The cast-in-place pile is formed in the ground, in the position in which it is to be used. The precast pile is usually cast in a factory, where prestressing techniques can be employed, and after curing is driven or jetted like a wood pile.

Cast-in-place piles are divided into two general groups, the *shell* type and the *shell-less* type. Shell-type piles are made by first driving a steel shell or casing into the ground, filling it with concrete, and leaving the shell in place. The shell acts as a form and prevents mud and water from mixing with the concrete. Such piles may or may not be reinforced, depending on circumstances.

Shell-type piles are useful where the soil is too soft to form a hole for an uncased pile or where the soil is hard to compress and would deform an uncased pile.

Shells may be cylindrical or tapered, with smooth or corrugated outside surfaces. One type of tapered shell produced in sections is known as a *step-taper* pile (see Fig. 5–3).

FIGURE 5-3 Typical step-taper pile.

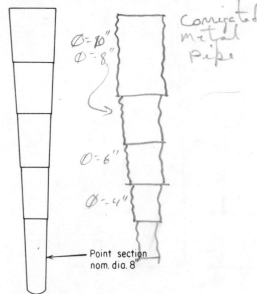

A number of variations of shell-type piles are made. One is a cased pile with a compressed base section, as illustrated in Fig. 5-4. It is used where the side support of the soil is so light that the pile must be used as a column or when the pile is designed to meet lateral forces arising from eccentric loading.

Another variation is the button-bottom pile shown in Fig. 5-5. It is used where an increase in end-bearing area is required. The enlarged bottom eliminates side friction support unless the soil is highly compacted around the pile (see Fig. 5-6).

Swage piles, shown in Fig. 5-7, are used where driving is very hard or where it is desired to leave watertight

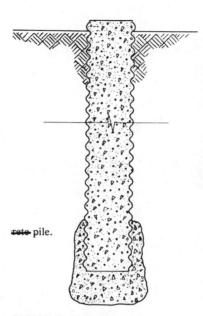

~~rete~~ pile.

FIGURE 5-6 Cased, pedestaled concrete pile.

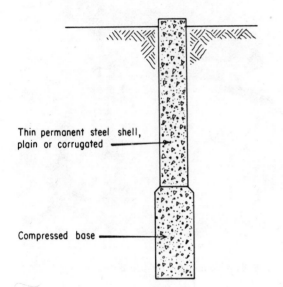

Thin permanent steel shell, plain or corrugated

Compressed base

FIGURE 5-4 Cased pile with compressed base.

FIGURE 5-5 Button-bottom cased concrete pile.

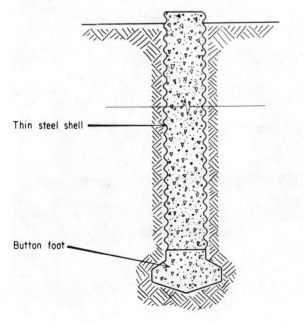

Thin steel shell

Button foot

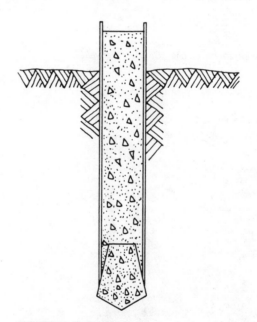

FIGURE 5-7 Swaged pile.

shells for some time before the concrete is placed. This type of pile consists of a steel shell driven over a conical precast concrete end plug.

Shell-less piles are made by driving a steel pipe, fitted with a special end or tapered shoe, into the ground for the full depth of the pile. The pipe is then pulled up, leaving the shoe at the bottom, and the hole is filled with concrete. The type of pile is satisfactory where soil is cohesive enough so that a reasonably smooth inside surface is maintained when the shell is removed. Figure 5-8 illustrates the simplest type of shell-less pile. Sometimes concrete is poured as the shell is being lifted. This

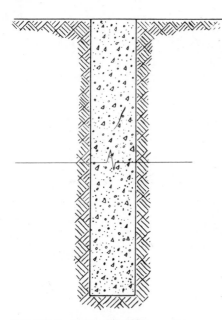

FIGURE 5-8 Simple shell-less pile.

FIGURE 5-9 Drilling for cased pile. (*Courtesy Calweld Inc.*)

eliminates some of the possibility of earth becoming mixed with the concrete.

Variations of the shell-less pile are also made. One involves the use of a bored hole rather than a punched one. In some cases the bottom end of the hold is *belled* out to provide a pedestal for increased end bearing. Figure 5–9 shows a boring rig in place on a job site, while Fig. 5–10 illustrates the belling attachment and Fig. 5–11

FIGURE 5-10 Belling equipment for pile bottoms. (*Courtesy Calweld Inc.*)

FIGURE 5-11 Drill discharging earth. (*Courtesy Calweld Inc.*)

shows an auger discharging earth. Figure 5-12 illustrates a vertical section through a completed belled pile.

Another variation of the shell-less pile involves the use of a very dry concrete mix. The shell is driven by dropping a heavy hammer onto a plug made from this dry mix (see Fig. 5-13). When the shell has reached its full depth, the plug is driven out to form an enlarged base and extra concrete is added as required to make the base as large as necessary. A reinforcing cage is then dropped into place in the shell. As the shell is slowly withdrawn, dry mix is fed into the top and driven out the bottom by the drop hammer, forming a highly compacted pile with

corrugated sides. Figure 5-14 shows the bottom end of a pile such as this after removal from the earth.

Precast concrete piles are normally made in a casting yard under controlled conditions. This allows not only the development of high strength concrete but also flexibility in design, reinforcing, length, etc.

Precast reinforced piles are made in round, square, hexagonal, and octagonal shapes, as illustrated in Fig. 5-15. Very long piles are made by casting hollow reinforced sections, usually 16 ft in length, and joining the sections together by stressed steel cables.

Precast concrete piles usually need assistance when being driven, especially in sand, and one method of providing this is to use a water jet. One type of precast pile has been developed especially for installation by jetting.

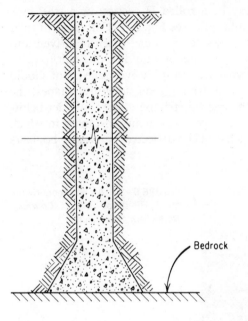

FIGURE 5-12 Section through completed belled pile.

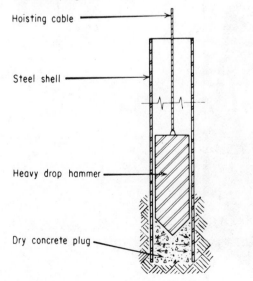

FIGURE 5-13 Driving a shell with drop hammer and concrete plug.

FIGURE 5-14 Bottom end of shell-less pile. (*Courtesy Franki of Canada.*)

FIGURE 5-15 Precast pile shapes.

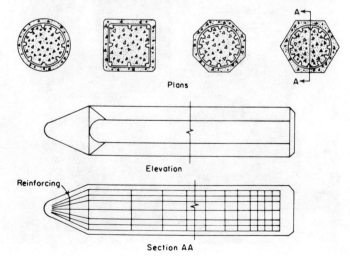

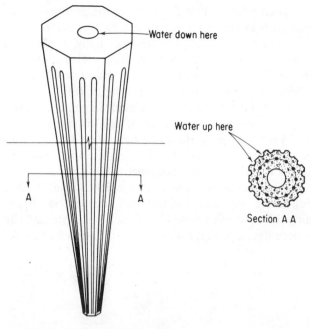

- Water down here
- Water up here
- Section A A

FIGURE 5-16 Precast jetting pile.

It is tapered, octagonal in cross section, and has one or two vertical grooves in each face. The center is hollow, allowing water to be forced down to loosen the soil at the tip. The grooves allow the jetted water to return to the surface (see Fig. 5-16).

Precast piles may be impregnated with asphalt to help prevent damage by seawater, spray, etc. Piles are pressure-treated with hot asphalt at 450 to 500° F, and penetrations of up to 1½ in. are obtained.

STEEL PILES

A steel pile may be a rolled H section or a steel pipe known as a tubular pile (see Fig. 5-17). Because of their small cross section, steel piles can often be driven into dense soils, through which it would be difficult to drive a pile of solid cross section. However, one point should be considered when driving long structural shapes: the possibility of the pile tip striking a large boulder below the surface. There have been many instances in which long steel pilings have been deflected by massive boulders

FIGURE 5-17 Air hammer driving steel HP-shape piles. (*Courtesy Bethlehem Steel Co.*)

below the surface, and as a result of continued driving, the pile tip has worked itself to the surface of the ground again. This, of course, makes it necessary to extract the pile and to determine the limits of the boulder before proceeding. This operation can become rather expensive, and points out the advisability of having a clause in the contract to protect the builder from such contingencies. Steel piles may be driven so that the top end, which extends above the ground, serves as a column. The upper ends of H piles may be encased in concrete to prevent corrosion.

Tubular piles vary in diameter from 8 to 72 in. They may be driven from the top—using a drop or mechanical hammer—or from the bottom, using a drop hammer falling on a concrete plug. The pipe is usually filled with concrete, which means that if the pipe is driven open-ended, the earth must be removed from the inside by a water jet.

COMPOSITE PILES

Two materials may sometimes be combined to make up a single pile. The most common combination is wood and concrete, and the result is known as a composite pile.

This type of pile may be used to advantage under several conditions. One is if the permanent water head is not more than 70 ft below ground level. This is about the length limit for the upper concrete section of the pile. Another is where the use of wood piles alone would necessitate 10 ft or more of dry excavation, or as little as 4 ft of very wet excavation. This can be eliminated by the use of a composite pile. A third situation is when the overall length of the pile is so great that it would be economically impractical to obtain or handle either straight wood or concrete piles. Cast-in-place concrete piles of up to about 70 ft in length can rest on top of wood piles of any obtainable length. Wood piles are difficult to handle and expensive when longer than 80 ft. Precast concrete piles are very expensive in lengths beyond 55 ft.

The wooden pile is driven to ground level, and the head of the pile is fitted with a steel casing. Driving continues by means of a mandrel or core until the required depth is reached. The mandrel is then withdrawn and the shell filled with concrete.

The shell is fitted to the wooden pile head in several different ways. One is to simply set the end of the shell over the slightly tapered end of the pile head. Another method is to make a tenon with a square shoulder at the end of the pile, set a sealing ring over the tenon, and rest the shell on the ring. Another method involves the use of a wedge ring which is forced into the flat top of the pile head. Figure 5–18 illustrates these three methods of connection.

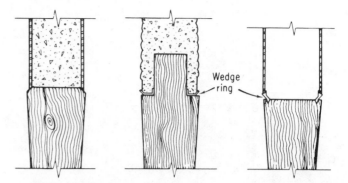

FIGURE 5–18 Composite pile connection.

A number of piles are sometimes driven close together in a group, to be later capped with a common concrete cap. Several points pertaining to this particular operation warrant special consideration.

First, it is generally recognized that when a number of piles are driven close together, the load-bearing value of each is reduced. As a result, a number of building codes have established a minimum center-to-center spacing for grouped piles. These vary from 2 ft 6 in. to 5 ft, with the latter considered less desirable.

There are precautions which must be taken in the use of piles and limitations to their use. For example, when piles are driven close together, those driven last may cause those driven first to heave. This may necessitate some redriving, but, in any case, a check should be made on the elevation of the tops of the piles driven first.

Driving them in the proper order will help to avoid the heaving problem. The piles at the center of the group should be driven first, followed in succession by those closer to the outside. Also, when piles are driven in a group, those driven last are generally more difficult to drive and pull than the earlier ones. This is due to the increased compaction of the soil as it is displaced by the piles.

Driving a large number of piles in a group, in an attempt to accommodate a very heavy load, can cause disturbance to the surrounding area. If they are displacement piles, the heaving caused by this disturbance can be quite destructive to the foundations of buildings in the near vicinity.

When shell-type piles are being used, the increased compaction due to group driving may result in damage (distortion or crushing) to shells already in place. As a result, it may be difficult to place concrete or the passage may even be completely blocked.

One source of trouble with shell-less piles is related to the practice of pulling the shell as the concrete is being placed. Concrete may adhere to the inside of the shell if it is not clean or if the weather is hot and some initial set takes place before the shell is withdrawn. This may

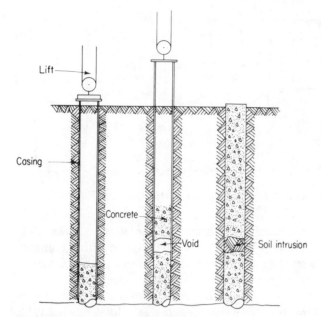

FIGURE 5-19 Void in concrete pile.

FIGURE 5-20 Diesel hammer driving HP-shape bearing pile. (*Courtesy L. B. Foster Co.*)

result in the concrete pile being pulled apart (see Fig. 5-19) and the space filling with soil. Thus, the bearing capabilities of the pile are destroyed.

PILE DRIVERS

Wood, precast concrete, steel H piles, and large tubular piles are driven into the ground by a pile driver striking the pile head. Small tubular piles and shells for cast-in-place concrete piles are more often pulled into the ground by a core. The core is forced into the earth by a tapered drop hammer and pulls the shell or tube with it.

Pile drivers may consist of a drop, mechanical, or vibratory hammer. A drop hammer is the simplest type of machine, consisting of a heavy weight, lifted by a cable and guided by *leads*, which is allowed to drop freely on the pile head.

The hammer of a mechanical driver operates like a piston, actuated by steam, compressed air, (see Fig. 5-17) or the internal combustion of diesel fuel (see Fig. 5-20).

A vibratory hammer is secured to the head of a pile and operates by delivering vibrations to the pile head in up-and-down cycles at the rate of 100 cps (see Fig. 5-21).

FIGURE 5-21 Vibro driver driving steel sheet piling. (*Courtesy L. B. Foster Co.*)

These vibrations set up waves of compression in the pile, which in turn produce minute amounts of expansion and contraction. As the pile expands longitudinally it displaces soil at the pile tip, and the weight of the pile, hammer, and equipment forces the pile into the tiny void. At the same time as the pile expands in length, it contracts slightly in diameter and thus relieves the friction between the earth and the pile surface. This relief of friction helps the pile to move downward.

These movements are very small but occur 100 times every second, and, as a result, penetration may actually be quite rapid. For example, a pile which might require an hour to be set with a steam hammer can be driven in 2 or 3 minutes by a vibratory hammer. This type of equipment performs best in sandy soil but may be used in silt, clay, or soil containing layers of gravel.

Assistance may have to be provided when piles are being driven. This is particularly true when driving precast piles in granular soils. The high friction rate makes driving very difficult. The assistance is usually in the form of water jet which softens the soil at the pile tip. Water is jetted down the hollow center and returns to the surface around the outside of the pile.

Figure 5–13 illustrates the principle involved in *pulling* a tubular pile or pile shell into the ground. A tapered drop hammer is dropped on a core made up of very dry concrete mix, driving it into the ground. At the same time, the hammer's tapered point forces the concrete against the inside wall of the shell, creating tremendous friction. As a result, the friction between the core and the tube pulls the tube down as the core is driven into the ground.

CAISSONS

The increase in column loads in many modern buildings, due to the wider spacing of columns in order to get more uninterrupted floor space and, in some cases, the depth required to reach good bearing strata, has made pile groups impractical in these situations. In their place, many designers are using *caissons*. Strictly speaking, a caisson is a shell or box or casing which, filled with concrete, will form a structure similar to a cast-in-place pile but larger in diameter. Over the years, however, the term has come to mean the complete bearing unit.

Bored Caissons

The development of boring tools for large-diameter holes and the equipment to handle them has made the use of *bored caissons* much more common. As the name implies, a bored caisson is one in which a hole of the proper size is bored to depth and usually a cylindrical casing or caisson is set into the hole. The use of heavy equipment has made it possible to bore 10-ft-diameter holes to a depth of over 150 ft. Figure 5–22 illustrates such

FIGURE 5–22 Six-foot drilling bucket with reamers. (*Courtesy Calweld Inc.*)

FIGURE 5–23 Ten-foot-diameter casing. (*Courtesy Calweld Inc.*)

equipment—in this case a 6-ft drilling bucket equipped with reamers to enlarge the diameter of the hole.

These boring machines are not suitable in areas where there is wet, granular material, silt, or boulders. The bored caisson method works best in cohesive soil in which the drilled hole will remain open until a caisson can be installed in it or until concrete can be placed into the drilled excavation (see Fig. 5–23).

In some cases drilled caissons may be *belled* (see Fig. 5–12) in order to provide greater bearing area at the bottom. Belling equipment (see Fig. 5–10) is used up to a limit of about a 20-ft bell. Hand digging is necessary for larger bells and to clean up machine-dug bells for proper bearing of the concrete on the rock or hardpan.

If there is a problem with water at the top of the rock surface, belling becomes very difficult, and large, straight-sided caissons are used. The bottom edge of such a caisson is fitted with tungsten carbide teeth and becomes a core barrel which can be rotated into the rock to effect a seal and allow men to work at the bottom of the caisson. It is possible to bite into the rock as much as 10 ft to try to seal off water or the inflow of wet silt.

Jacked-in Caissons

In materials such as silt, wet sand, or gravel, which present difficulties in installing bored caissons, a caisson may be forced into the ground by jacks, while at the same time it is rotated back and forth to reduce the side friction. The material inside the caisson is then excavated. In some cases, excavating will be carried out while the caisson is being forced down, thus making the sinking operation easier.

Gow Caisson

One type of caisson is made in tapered sections (see Fig. 5–24), each of which fits inside the one above it. A pit is dug and the first section of steel casing placed in it. It is forced into the ground by driving and excavating inside the cylinder at the same time. A cylinder 2 in. smaller in diameter is then placed inside the first and driven to its full depth. This continues until the full depth of the caisson is reached, at which time the excavation is belled to provide a greater bearing area.

Drilled-in Caissons

Where loads to be carried are very large, the depth to bearing rock excessive, or the type of material encountered unsuitable for bored caissons, *drilled-in* caissons may be used. Such a caisson is installed by driving a steel pipe, 24 to 42 in. in diameter and ½ in. in

FIGURE 5-24 Gow caisson.

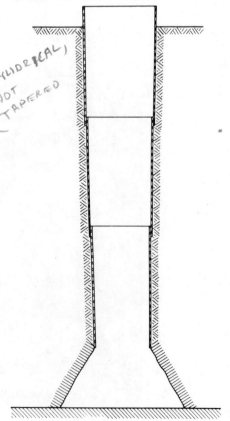

(CYLIDRICAL) NOT TAPERED

thickness, to rock, using a pile hammer. The pipe has a 1¼-in. steel shoe welded to the bottom to reinforce the tip and is excavated by auger, by blowing with compressed air or by a cable drill rig, fitted with suction buckets. Such caissons have been installed to depths of over 200 ft.

When bedrock is reached, a *socket* is churn-drilled into the rock slightly smaller in diameter than the caisson shell and heavy steel HP shape or the equivalent in reinforcing rods is installed to extend from the top of the caisson to the bottom of the socket. The entire assembly is then concreted in place, producing a fixed-end caisson.

Slurry Caisson

The technique of using *slurry walls* (described in detail in Chapter 3) has now been applied to the construction of deep piers or caissons. A rectangle pit is dug to rock, using bentonite slurry to keep the sides of the excavation from collapsing. The pit is then filled with concrete by the tremie method to form a caisson.

COFFERDAMS

Although *caissons* and *cofferdams* fulfill the same general function, a caisson is a permanent structure used for protection while excavating and as a form for concrete, whereas a cofferdam is a temporary box-like structure used to hold back water or earth while work is being done inside it, and is later removed.

A cofferdam consists of sections of sheeting driven side by side to form a watertight unit. The sheeting forms a box large enough to work inside it. The sections are usually driven in water or mud which is to be removed from the inside once the cofferdam is complete. If the bottom is too soft, a layer of concrete may be placed over

it to provide a firm base. Whatever permanent structure is required can then be built inside the cofferdam, which is removed when the work has been completed.

SPREAD FOUNDATIONS

If building loads are not large enough to warrant being carried to bedrock by piles or caissons, but the soilbearing strength is low, a solution may be the use of *spread* foundations. These, as their name implies, are foundations which are extended over large areas in order to distribute the weight and thus reduce the unit load. There are a number of different types, and the one to use in any particular case depends on soil strength, load concentration, depth of the foundation below grade level, etc.

One type of spread foundation is known as a *grillage* footing. It is made up of two or more tiers of steel beam sections, laid at right angles to one another. Each tier has a smaller number of units than the one below it, but the section modulus remains fairly constant.

When a grillage footing rests on compressible soil, a slab of concrete is cast first, 6 to 12 in. deep. The first tier of steel, bolted together, is set on the slab and leveled. The next tier is set on the first at a right angle and fastened down. Succeeding tiers (if any) follow, and a steel slab or billet is finally set on the top tier and fixed in place. A building column will rest on this billet. Finally, the whole grillage is encased in concrete (see Fig. 5-25).

Another type of spread foundation consists of a solid slab of heavily reinforced concrete covering the entire site. It may be from 3 to 8 ft thick, depending on the area to be covered and the loads to be carried. This is known as a *mat* foundation (see Fig. 5-26).

Still another type consists of two series of heavy, continuous ground beams, running at right angles to one another and intersecting under columns. A slab is usual-

FIGURE 5-25 Grillage footing.

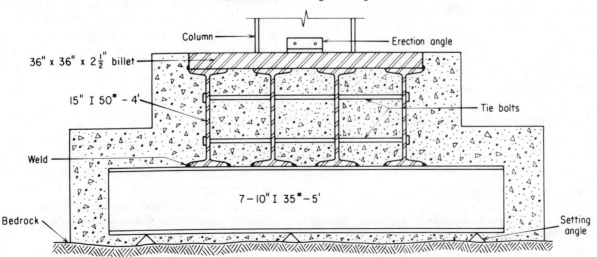

FIGURE 5-26 Preparing a mat foundation. (*Courtesy Bethlehem Steel Co.*)

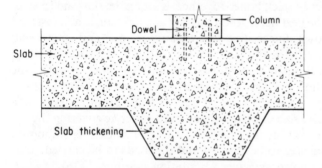

FIGURE 5-27 Mat foundation deepened under column.

ly cast monolithically with the beams, but sometimes the slab is placed at the bottom of the beams, making them and the slab inverted T beams. In other cases, the slab is placed at the top of the beams. In the former case, the space between the slab and the top of the beams is filled with cinders or gravel, while in the latter case, the beams are set into the earth so that the slab is poured on grade.

A fourth type of spread foundation also features a continuous slab, but a much thinner one than the mat foundation described above. The slab thickness is increased under each column, as shown in Fig. 5-27. The whole structure thus resembles a flat slab with drop-panel-type concrete floor (see Chapter 11).

SLURRY WALL FOUNDATIONS

The perimeter slurry walls, described in Chapter 3, may be used as foundation walls as well as a means of protecting an excavation during the excavating stage. If the wall is to be used as part of the foundation, its thickness will have to be designed accordingly, but otherwise the procedure is the same as for an excavation slurry wall.

SOIL COMPACTION FOUNDATIONS

The deep foundation systems described above are usually expensive and time-consuming and the building industry is constantly trying to find ways of eliminating some of the problems involved in providing foundations in poor soil conditions. One of the solutions is to improve the bearing strength of the soil, rather than remove it or drill through it to firmer material.

The bearing strength of soil may be improved by compaction, and various methods have been used to compact insitu soil, or to replace the existing soil with new material and compact it. These methods include rolling with rubber-tired rollers, compacting with vibrating plates, driving short stubby piles, blasting, using sand drains, and, more recently, compacting by deep vibration.

Vibroflotation Foundation

This process of producing a compacted soil foundation, which works best in sandy soils, employs mechanical vibration of soil beneath the surface and simultaneous saturation with water to *move, shake,* and *float* the soil particles into a dense state.

The machine used consists of a long shaft with a vibrating head on the end, mounted on a crane. The head is jetted into the ground to the required depth and slowly withdrawn, compacting the existing soil while additional material is backfilled through the crater created at the surface. Figure 5-28 illustrates a typical machine in action.

It is positioned over the area to be compacted, and the lower water jet is turned on (see Fig. 5-29). Water is pumped in faster than it can drain away into the sub-soil, creating a *quick* condition beneath the vibrator, which allows it to penetrate the soil by its own weight and vibration. The initial action creates a crater at the surface, about 3 ft in diameter, as shown in Fig. 5-30.

On typical sites, the machine can penetrate from 15 to 25 ft in about 2 min. The average depth of compaction is about 15 ft, but depths up to 70 ft have been compacted successfully.

Having reached the required depth, water is switched from the lower to the upper jet and withdrawal begins, with the compaction taking place during this time. The machine compacts by simultaneous vibrations and saturation. The compactor vibrates granular soil with about 10 tons of centrifugal force into a dense mass, while

FIGURE 5-28 Typical vibrating machine in action. (*Courtesy Vibroflotation Foundation Co.*)

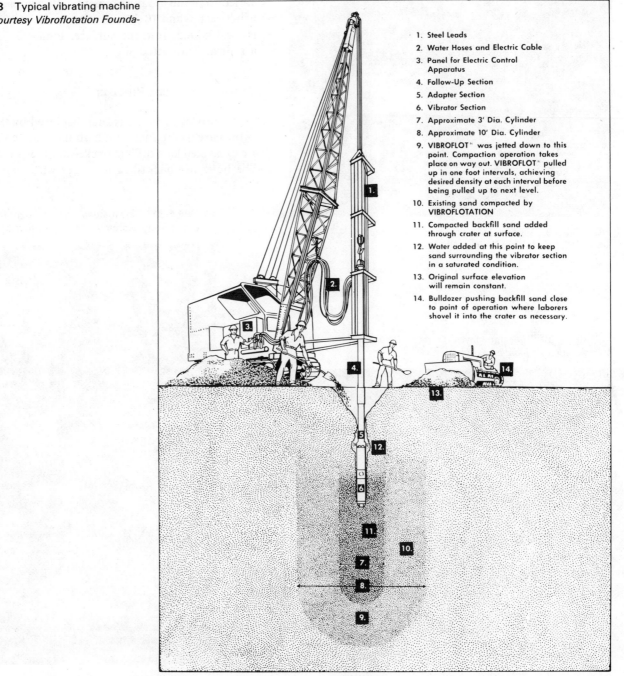

1. Steel Leads
2. Water Hoses and Electric Cable
3. Panel for Electric Control Apparatus
4. Follow-Up Section
5. Adapter Section
6. Vibrator Section
7. Approximate 3' Dia. Cylinder
8. Approximate 10' Dia. Cylinder
9. VIBROFLOT⁺ was jetted down to this point. Compaction operation takes place on way out. VIBROFLOT⁺ pulled up in one foot intervals, achieving desired density at each interval before being pulled up to next level.
10. Existing sand compacted by VIBROFLOTATION
11. Compacted backfill sand added through crater at surface
12. Water added at this point to keep sand surrounding the vibrator section in a saturated condition.
13. Original surface elevation will remain constant.
14. Bulldozer pushing backfill sand close to point of operation where laborers shovel it into the crater as necessary.

FIGURE 5-29 Vibrator ready for action with water jet on. (*Courtesy Vibroflotation Foundation Co.*)

FIGURE 5-30 Vibrator down 3 feet. (*Courtesy Vibroflotation Foundation Co.*)

at the same time the excess water floats the finest particles to the surface and washes them away (see Fig. 5-31). Granular material is added from the surface to replace the material washed away and to compensate for the increased density of the insitu soil. The vibrator is raised slowly at a rate of about 1 ft/min and the hole backfilled at the same time (see Fig. 5-32), so that the entire depth of soil is compacted into a hard core, or column.

A single compaction produces a cylinder of highly densified material with great bearing capacity. Density is uniform above and below the water table and may be varied to fit the job requirement. The compacted cylinder is between 6 and 10 ft in diameter, about 10% of which is sand added from the surface. Figure 5-33 illustrates a typical compacted site.

Vibroreplacement Process

In this process, the soil is first displaced by the vibrator, in the case where there is a high moisture content, holes are excavated by a drilling bucket or other suitable means. The holes are backfilled in stages with coarse material

FIGURE 5-31 Fine material being brought to the surface. (*Courtesy Vibroflotation Foundation Co.*)

FIGURE 5-32 Backfilling during withdrawal. (*Courtesy Vibroflotation Foundation Co.*)

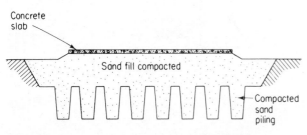

FIGURE 5-33 Vertical section through vibroflotation foundation. (*Courtesy Vibroflotation Foundation Co.*)

(gravel or crushed stone) which is simultaneously compacted by the vibrator. As a result, the soil strength is increased by (1) replacing a portion of the existing soil with a high-strength, compacted column of stone; (2) compacting and adding stone to an outer layer of the existing soil; and (3) some consolidation of further outer volume of soil, due to the horizontal component of the force exerted on the column of stone.

Vibratory Tube Soil Compaction

By this process, a tubular probe is driven into the soil by means of a vibratory driver, similar to that shown in Fig. 5–21, and extracted again by the same means. The vibratory action of the probe causes the soil particles to readjust themselves into a denser state. Soil inside the probe becomes densified as the probe is driven and extracted, and, upon extraction, a highly densified *sand pile* remains. A secondary effect is the compaction which also takes place outside the probe. As a result, properly determined spacing of the probes will allow densification of the soil over the entire area involved (see Fig. 5–34).

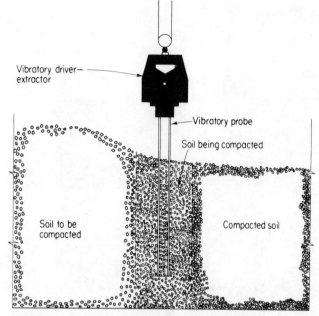

FIGURE 5-34 Vibratory probe soil compaction. (*Courtesy L. B. Foster Co.*)

REVIEW QUESTIONS

1. Name the two major types of piles and explain how they differ in purpose.

2. What is the purpose of **(a)** a pile ring, **(b)** a pile shoe, and **(c)** pile driver leads.

3. Describe briefly how a steel pile shell is driven with drop hammer and concrete plug.

4. List four different types of caissons and explain how each differs from the rest.

5. Explain the purpose of **(a)** a cofferdam, **(b)** jetting a pile, and **(c)** a mat foundation.

6. Explain clearly the difference between the vibroflotation process and the vibroreplacement process.

7. Describe briefly how a *sand pile* is produced. Illustrate by means of a neat diagram.

8. Why is it good practice to check the elevation of the first piles in a group while the subsequent piles are being driven?

9. Outline the advantages of a slurry wall foundation.

Formwork

Concrete structures require the use of *forms* into which the freshly mixed, plastic concrete can be placed. Forms must generally be built so that they may be easily removed after the concrete has gained sufficient strength. This form building may be done in a plant that is producing precast concrete units, but for cast-in-place members, the forms are often built at the site.

FORM MATERIALS

Wood

Wood is the most widely used material for building forms, and no matter what material is used some wood is nearly always required. It may be either lumber or plywood.

Lumber is usually of the softwood variety, because of its availability, relative lightness, and ease of handling. Pine, fir, and western hemlock are the species generally considered most suitable. Partly seasoned lumber is best for formwork because dry lumber swells considerably when wet, while green lumber shrinks and warps in hot weather. When lumber is to be used for concrete surfaces the material should be planed, although rough lumber may be used for frames, bracing, and shoring.

Plywood has replaced boards to a large extent in making forms for concrete surfaces. The large, smooth surface of plywood sheets, its resistance to changes in shape when wet, and its ability to withstand rough usage without splitting have made it a popular form material.

Overlaid plywood is frequently used where very smooth, grainless surfaces are required. This is plywood surfaced with plastic resin which has been fused to the wood under heat and pressure. In addition to producing smoother surfaces, overlaid plywood is generally more resistant to abrasion and moisture penetration.

An additional advantage in the use of plywood for forms is its ability to bend, thus making it possible to produce curved surfaces. The degree of bending depends on the thickness of the sheet and on whether it is curved across the face grain or parallel to it. Table 6–1 gives the minimum bending radii for plywood.

TABLE 6–1   Minimum bending radii for plywood panels

Plywood thickness	Minimum bend radius [a] (in.)	
	Perpendicular to face grain	Parallel to face grain
¼	16	48
⁵⁄₁₆	24	60
⅜	36	72
½	72	96
⅝	96	120
¾	120	144

[a]Average values; no regard for knots, patches, shot grain, etc.
Source: Reprinted by permission of B.C. Plywood Manufacturing Association.

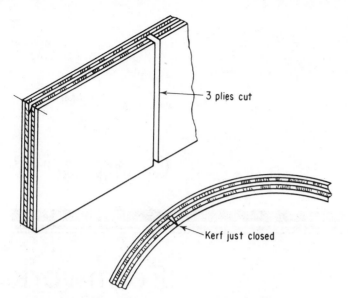

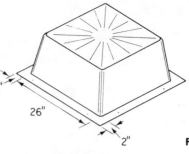

FIGURE 6-3 Dome pan.

FIGURE 6-1 Saw kerfing plywood for bending.

Panels with clear, straight grain will permit greater bending than those shown in the table.

Thicker plywood may be bent by cutting saw kerfs across the inner face at right angles to the curve. The kerf should cut 3 plies of a 5-ply sheet or 5 plies of a 7-ply sheet. They should be spaced so as to just close up when the plywood is bent to the correct radius (see Fig. 6-1).

Another method of curving thick face material is to use two or more thicknesses of ¼-in. plywood, bent one at a time.

A specially treated, tempered, ¼-in. hardboard is also available for form facing material. In addition to being tempered, the surface is coated with plastic, which helps to prevent water from penetrating into the board. Hardboard is essentially a form liner material and should be applied to a supporting backing of lumber. Figure 6-2 illustrates how it may be applied to backing boards.

Steel

Steel is widely used as a material for making forms. Steel angles and bars are used extensively as frames for form panels, which may be faced with plywood or sheet steel. Steel inside and outside corner pieces are used to join panels of this type at corners. Many of the forms used in the precast industry are faced with sheet steel, and standard channels are frequently used as *walers* for large forms. Ribbed slab forms are largely made from sheet steel, and flat slab edge forms are often of steel.

Plastic

Glass-fiber-reinforced plastic is another form facing material in common use. It may be used to fabricate such tings as *dome pans* for *waffle* floors (see Fig. 6-3), or it may be spray-coated over a wooden base form.

Form Liners

Form liners are used mainly to produce a desired texture on exposed concrete surfaces. Plywood, steel, and fiberglass are the usual materials used to provide a smooth finish to the concrete. To produce a certain texture, plastic and rubber liners are commonly used because of their versatility and toughness. In high-rise construction, where one set of formwork must be reused many times, corrugated cardboard covered with polyethylene is used as a form liner to facilitate form removal, thereby reducing the damage to the formwork.

Insulating Board

Various types of insulating board are often used as form liners. The material is fastened to one or both form faces,

FIGURE 6-2 Hardboard form liners.

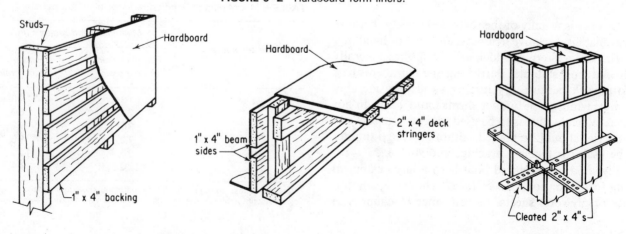

and when the form is removed, the insulation remains, either bonded to the concrete or held in place by clips.

FORM DESIGN

Forms should be so designed that they are practical and economical in both manufacture and use. A number of important factors should be kept in mind when forms are being built:

1. Forms must be strong enough to withstand the pressure of plastic concrete and to maintain their shape during the concrete placing operation.

2. They must be tight enough to prevent wet concrete from leaking through joints and causing unsightly fins and ridges.

3. They must be as simple to build as circumstances will allow.

4. They must be easy to handle on the job.

5. Form sections must be of a size that can be lifted into place without too much difficulty and transported from one job to another if necessary.

6. They must be made to fit and fasten together with reasonable ease.

7. The design must be such that the forms, or sections of them, may be removed without damage to the concrete or to themselves.

8. Forms must be made so that workmen can handle them with safety.

The basic consideration in form design is their strength—the forms' ability to support their own weight and the lateral pressure of concrete. These can be determined with considerable accuracy, but forms are subjected to other loads, such as the live loads imposed during construction, materials stored on the structure, and wind pressure. Generally speaking, the designer must make some reasonable estimates of these factors. The American Concrete Institute Committee 622 recommends a minimum construction live load of 50 psf of horizontal projection to provide for the weight of workmen, equipment runways, and impact.

Two types of strength problems are involved, based on the concrete itself. One deals with the weight of the concrete producing a vertical load and the other with the lateral pressure exerted on forms by fresh concrete in its *liquid* form.

Concrete may vary in weight from 40 to 600 lb/cu ft, depending on the aggregates from which it is made. Concrete made with natural sand and gravel aggregates weighs between 145 and 150 lb/cu ft. The formwork itself will vary from about 3 to 15 psf. With these weights and the addition of a definite imposed live load, it is possible to calculate the size of joists, stringers, and shores required to carry vertical loads.

The lateral pressure exerted by fresh concrete presents a different problem. Freshly placed concrete acts temporarily like a liquid, producing fluid or hydrostatic pressure which acts against vertical forms. This pressure varies, depending on a number of factors. One of them is the weight of the concrete being used. Hydrostatic pressure at any point in a liquid is the result of the weight of fluid above. But because fresh concrete is a composite material, rather than a true liquid, the laws of hydrostatic pressure apply only approximately and for limited periods.

The rate of placement also affects lateral pressure. The greater the height to which concrete is placed while the whole mass remains in the liquid stage, the greater the lateral pressure at the bottom of the form.

The temperatures of concrete and atmosphere affect the pressure because they affect the setting time. When these temperatures are low, greater heights can be placed before the concrete at the bottom begins to stiffen, and greater lateral pressures are therefore built up.

Vibration increases lateral pressures because the concrete is consolidated and acts as a fluid for the full depth of vibration. This may cause increases of up to 20% in pressures over those incurred by spading.

Other factors which influence lateral pressure include the consistency or fluidity of the mix, the maximum aggregate size, and the amount and location of reinforcement.

The American Concrete Institute Committee 622 has developed formulas for calculating the maximum lateral pressure at any elevation in the form. They are based on prescribed conditions of concrete temperature, rate of placement, slump of concrete, weight of concrete, and vibration. They may be used for internally vibrated concrete of normal density, placed at not more than 10 ft/hr, with no more than 4 in. of slump. Vibration depth is limited to 4 ft below the concrete surface. The two formulas for wall form design are as follows:

1. For walls, with rate of pour not exceeding 7 ft/hr,

$$p = 150 + \frac{9000R}{T}$$

Maximum = 2000 psf or 150 h, whichever is less.

2. For walls, with rate of pour from 7 to 10 ft/hr,

$$p = 150 + \frac{43,400}{T} + \frac{2800R}{T}$$

Maximum = 2000 psf or 150 h, whichever is less.

p = maximum lateral pressure, psf

R = rate of placement, ft/hr

T = temperature of concrete in the forms, degrees F

h = maximum height of fresh concrete in the form, ft

TABLE 6-2 Maximum lateral pressure for design of wall forms[a]

Rate of placement, R, (ft/hr)	Maximum lateral pressure, p (psf) for indicated temperature					
	90°F	80°F	70°F	60°F	50°F	40°F
1	250	262	278	300	330	375
2	350	375	407	450	510	600
3	450	488	536	600	690	825
4	550	600	664	750	870	1050
5	650	712	793	900	1050	1275
6	750	825	921	1050	1230	1500
7	850	938	1050	1200	1410	1725
8	881	973	1090	1246	1466	1795
9	912	1008	1130	1293	1522	1865
10	943	1043	1170	1340	1578	1935

[a]Do not use design pressures in excess of 2000 psf or 150 × height (in ft) of fresh concrete in forms, whichever is less.
Source: Reprinted by permission of American Concrete Institute.

Table 6–2 gives the maximum lateral pressure to be used in designing wall forms, based on the foregoing formulas.

Column forms are frequently filled to their full height within the time it takes for concrete to set. In addition, vibration often extends to the full depth of the form. This means that greater lateral pressures are developed than in the case of walls. The A.C.I. Committee 622 has developed a formula for calculating the maxi-

mum pressure to be used in column form design, taking the aforementioned factors into consideration:

$$p = 150 + \frac{9000R}{T}$$

Maximum = 3000 psf or 150 h, whichever is less.

Table 6–3 gives the maximum lateral pressure to be used in designing column forms, based on the above formula.

TABLE 6-3 Maximum lateral pressure for design of column forms[a]

Rate of placement, R, (ft/hr)	Maximum lateral pressure, p (psf) for indicated temperature					
	90°F	80°F	70°F	60°F	50°F	40°F
1	250	262	278	300	330	375
2	350	375	407	450	510	600
3	450	488	536	600	690	825
4	550	600	664	750	870	1050
5	650	712	793	900	1050	1275
6	750	825	921	1050	1230	1500
7	850	938	1050	1200	1410	1725
8	950	1050	1178	1350	1590	1950
9	1050	1163	1307	1500	1770	2175
10	1150	1275	1435	1650	1950	2400
11	1250	1388	1564	1800	2130	2625
12	1350	1500	1693	1950	2310	2850
13	1450	1613	1822	2100	2490	3000
14	1550	1725	1950	2250	2670	
16	1750	1950	2207	2550	3000	
18	1950	2175	2464	2850		
20	2150	2400	2721	3000		
22	2350	2625	2979			
24	2550	2850	3000			
26	2750	3000				
28	2950		(3000 psf maximum)			
30	3000					

[a]Do not use design pressures in excess of 3000 psf or 150 × height of fresh concrete in forms, whichever is less.
Source: Reprinted by permission of American Concrete Institute.

Tables 6-4 to 6-14 are examples of design tables developed by the American Concrete Institute giving allowable spacings of supports for various formwork components under different concrete pressures. The spacing of supports depends on a number of factors, all of which must be considered to ensure that the formwork performs in a satisfactory manner. The type of material, nature of the load, number of reuses, moisture conditions, and deflection limitations all have a bearing on the spacing of supports.

If lumber is used, its grade and species must be known. Different species of lumber have different strengths. Plywood, for example, has a higher strength when the face grain is parallel to the span than when it is perpendicular to the span. The strength of wood is also decreased by absorption of moisture, and this fact must be taken into account in the design of formwork.

Form loading, of course, is the prime concern when designing formwork. It must be established as accurately as possible to ensure that the formwork will support the required loads and yet not be overdesigned to such an extent that it will add unnecessary costs to the job. The duration of loading and the number of reuses anticipated for the forms have a direct bearing on the final allowable loads that can be safely supported.

Deflections in the formwork must also be carefully considered. The final shape of the structure depends on the shape of the forms, and if excessive deflections occur during the placing of the concrete, the final outcome may be embarrassing if not totally unacceptable. In extreme cases, collapse of the formwork during the placing of the concrete may occur because of excessive deflections.

TABLE 6-4 Safe spacing (in.) of supports for board sheathing continuous over four or more supports[a,b]

Maximum deflection is $\frac{1}{360}$ of support spacing, but not more than $\frac{1}{16}$ in.

Pressure or load from concrete, psf	$f = 1000$ psi $H = 150$ psi $E = 1,300,000$ psi				$f = 800$ psi $H = 120$ psi $E = 1,300,000$ psi			
	Nominal thickness of S4S boards, in.				Nominal thickness of S4S boards, in.			
	1	1¼	1½	2	1	1¼	1½	2
75	30	37	44	50	30	37	44	50
100	28	34	41	47	28	34	41	47
125	26	33	39	44	26	33	39	44
150	25	31	37	42	25	31	37	42
175	24	30	35	41	24	30	35	41
200	23	29	34	39	23	29	34	39
300	21	26	31	35	19	25	31	36
400	18	24	29	33	16	22	27	33
500	16	22	27	31	15	20	24	29
600	15	20	25	30	13	18	22	27
700	14	18	23	28	12	17	21	25
800	13	17	22	26	12	15	19	23
900	12	16	20	24	11	15	18	22
1000	12	15	19	23	10	14	17	21
1100	11	15	18	22	10	13	16	20
1200	11	14	18	21	9	13	16	19
1400	10	13	16	20	9	12	15	18
1600	9	12	15	18	8	11	14	16
1800	9	12	14	17	8	10	13	15
2000	8	11	14	16	7	10	12	15
2200	8	10	13	16	7	9	12	14
2400	7	10	12	15	7	9	11	13
2600	7	10	12	14	6	9	11	13
2800	7	9	12	14	6	8	10	12
3000	7	9	11	13	6	8	10	12

[a]Calculations are based on span distances center to center of supports where supports are relatively narrow. Where supports may be wide in relation to the distance between them, such as 2″ × 4″s used flat where the spacing is 8 in., use the tabulated spacing as the clear distance between supports.
[b]Above dashed line span length is controlled by deflection; below dashed line bending governs.
Source: Reproduced by permission of the American Concrete Institute.

STUDS

TABLE 6-5 Safe spacing (in.) of supports for board sheathing on simple span supported at two points[a,b]

Maximum deflection is $\frac{1}{360}$ of support spacing, but not more than $\frac{1}{16}$ in.

Spacing — simple support

Pressure or load from concrete, psf	$f = 1000$ psi $\quad H = 150$ psi $E = 1,300,000$ psi				$f = 800$ psi $\quad H = 120$ psi $E = 1,300,000$ psi			
	Nominal thickness of S4S boards, in.				Nominal thickness of S4S boards, in.			
	1	1¼	1½	2	1	1¼	1½	2
75	25	32	37	43	25	32	37	43
100	24	29	35	40	24	29	35	40
125	22	28	33	38	22	28	33	38
150	21	26	31	36	21	26	31	36
175	20	25	30	35	20	25	30	35
200	19	25	29	33	19	25	29	33
300	17	22	26	30	17	22	26	30
400	15	20	24	28	15	20	24	28
500	14	19	23	27	13	17	22	26
600	13	18	22	25	12	16	20	24
700	12	17	21	24	11	15	18	22
800	12	15	19	23	10	14	17	21
900	11	15	18	22	10	13	16	20
1000	10	14	17	21	9	12	15	19
1100	10	13	16	20	9	12	15	18
1200	9	13	16	19	8	11	14	17
1400	9	12	15	18	8	10	13	16
1600	8	11	14	16	7	10	12	15
1800	8	10	13	15	7	9	12	14
2000	7	10	12	15	7	9	11	13
2200	7	9	12	14	6	8	10	12
2400	7	9	11	13	6	8	10	12
2600	6	9	11	13	6	8	10	11
2800	6	8	10	12	6	7	9	11
3000	6	8	10	12	5	7	9	11

[a]Calculations are based on span distances center to center of supports, where supports are relatively narrow. Where supports are wide in relation to the distance between them, such as a 3-in.-wide support where the spacing is 8 or 9 in., use the tabulated spacing as the clear distance between supports.
[b]Above dashed line span length is controlled by deflection; below dashed line bending governs.
Source: Reproduced by permission of the American Concrete Institute.

STUDS

TABLE 6-6 Safe spacing (in.) of supports for board sheathing continuous over two spans (three supports)[a,b]

Maximum deflection is $\frac{1}{360}$ of support spacing, but not more than $\frac{1}{16}$ in.

DOUBLE

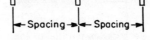

←Spacing→ ←Spacing→

Pressure or load from concrete, psf	$f = 1000$ psi $H = 150$ psi $E = 1,300,000$ psi				$f = 800$ psi $H = 120$ psi $E = 1,300,000$ psi			
	Nominal thickness of S4S boards, in.				Nominal thickness of S4S boards, in.			
	1	1¼	1½	2	1	1¼	1½	2
75	32	39	47	53	32	39	47	53
100	30	37	43	50	29	37	43	50
125	28	35	41	47	26	35	41	47
150	27	33	39	45	24	32	39	45
175	25	32	38	43	22	30	37	43
200	23	31	36	42	21	28	35	41
300	19	25	32	38	17	23	28	34
400	16	22	27	33	15	20	24	29
500	15	20	24	29	13	17	22	26
600	13	18	22	27	12	16	20	24
700	12	17	21	25	11	15	18	22
800	12	15	19	23	10	14	17	21
900	11	15	18	22	10	13	16	20
1000	10	14	17	21	9	12	15	19
1100	10	13	16	20	9	12	15	19
1200	9	13	16	19	8	11	14	17
1400	9	12	15	18	8	10	13	16
1600	8	11	14	16	7	10	12	15
1800	8	10	13	15	7	9	12	14
2000	7	10	12	15	7	9	11	13
2200	7	9	12	14	6	8	10	12
2400	7	9	11	13	6	8	10	12
2600	6	9	11	13	6	8	10	11
2800	6	8	10	12	6	7	9	11
3000	6	8	10	12	5	7	9	11

[a]Calculations are based on span distances center to center of supports, where supports are relatively narrow. Where supports may be wide in relation to the distance between them, such as a 3-in.-wide support where the spacing is 8 or 9 in., use the tabulated spacing as the clear distance between supports.
[b]Above dashed line span length is controlled by deflection; below dashed line bending governs.
Source: Reproduced by permission of the American Concrete Institute.

STUDS, JOIST *(handwritten)*

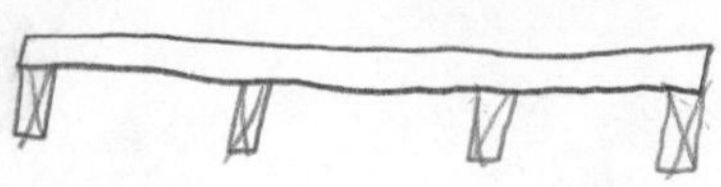

TABLE 6–7 Safe spacing in inches of supports for plywood sheathing, continuous over four or more supports[a]

Maximum deflection 1/360 of span, but not more than 1/16 in.

Pressure or load of concrete, pounds per square foot	Stresses recommended by the American Plywood Association for Class I Plyform, with E modified for deflection calculations. $f = 1930$ psi; rolling shear = 80 psi; $E = 1,500,000$ psi								Stresses recommended by the American Plywood Association for Class I Plyform, with reduction for long term loading, and with E modified for deflection calculations. $f = 1545$ psi; rolling shear = 65 psi; $E = 1,500,000$ psi							
	sanded thickness, face grain parallel to span				sanded thickness, face grain perpendicular to span				sanded thickness, face grain parallel to span				sanded thickness, face grain perpendicular to span			
	½ in.	⅝ in.	¾ in.	1 in.	½ in.	⅝ in.	¾ in.	1 in.	½ in.	⅝ in.	¾ in.	1 in.	½ in.	⅝ in.	¾ in.	1 in.
75	20	23	25	31	12	16	21	27	20	23	25	31	12	16	21	27
100	18	21	24	29	11	15	19	25	18	21	24	29	11	15	19	25
125	16	20	22	27	10	14	17	24	16	20	22	27	10	14	17	24
150	15	18	21	26	10	13	16	23	15	18	21	26	10	13	16	23
175	15	17	20	25	9	12	16	22	15	17	20	25	9	12	16	22
200	14	17	19	24	9	12	15	21	14	17	19	24	9	12	15	21
300	12	15	17	22	8	10	13	18	12	15	17	21	8	10	13	18
400	11	13	15	20	7	9	12	17	11	13	14	18	7	9	12	17
500	10	12	14	18	6	9	11	15	10	11	13	16	6	9	11	15
600	10	12	13	17	6	8	10	14	9	10	12	15	6	8	10	14
700	9	11	12	15	6	8	10	14	8	10	11	14	6	7	9	13
800	9	10	11	14	5	7	9	13	8	9	10	13	5	6	8	12
900	8	9	11	14	5	7	9	13	7	8	10	12	4	6	7	11
1000	8	9	10	13	5	6	8	12	6	8	9	11	4	5	6	10
1100	7	9	10	12	4	6	7	11	6	7	9	10	4	5	6	9
1200	6	8	9	12	4	5	6	10	5	7	8	9	—	4	5	8
1300	6	8	9	11	4	5	6	9	5	6	8	9	—	4	5	7
1400	5	7	9	10	—	4	5	9	4	6	7	8	—	4	4	7
1500	5	7	8	9	—	4	5	8	4	5	7	8	—	—	4	6
1600	5	6	8	9	—	4	5	8	4	5	6	7	—	—	4	6
1700	5	6	7	8	—	4	4	7	4	5	6	7	—	—	4	6
1800	4	6	7	8	—	—	4	7	—	4	6	6	—	—	—	5
1900	4	5	7	7	—	—	4	6	—	4	5	6	—	—	—	5
2000	—	5	6	7	—	—	4	6	—	4	5	6	—	—	—	5
2200	—	5	6	6	—	—	—	6	—	4	5	5	—	—	—	4
2400	—	4	5	6	—	—	—	5	—	—	4	5	—	—	—	4
2600	—	4	5	5	—	—	—	5	—	—	4	4	—	—	—	4
2800	—	4	4	5	—	—	—	4	—	—	4	4	—	—	—	—
3000	—	—	4	5	—	—	—	4	—	—	—	4	—	—	—	—

[a]Above the solid line spans are governed by deflection; below the dashed line spans are governed by shear. Between the lines, bending governs. The spans governed by shear are clear spans; others should be taken as center to center of supports.

Source: Reproduced by permission of the American Concrete Institute.

(handwritten left margin: 2 x 12 147-93 PLF; PDSE 109)

(handwritten bottom: FORMS — STRENGHT, WRECKABILITY, WELL BRACED, ECONOMY, EASE OF ERECTION, SAFTY)

TABLE 6-8 Safe spacing in inches of supports for plywood sheathing with only two points of support[a]

Maximum deflection 1/360 of span, but not more than 1/16 in.

Pressure or load of concrete, pounds per square foot	Stresses recommended by the American Plywood Association for Class I Plyform, with E modified for deflection calculations. $f = 1930$ psi; rolling shear = 80 psi; $E = 1,500,000$ psi								Stresses recommended by the American Plywood Association for Class I Plyform, with reduction for long term loading, and with E modified for deflection calculations. $f = 1545$ psi; rolling shear = 65 psi; $E = 1,500,000$ psi							
	sanded thickness, face grain parallel to span				sanded thickness, face grain perpendicular to span				sanded thickness, face grain parallel to span				sanded thickness, face grain perpendicular to span			
	½ in.	⅝ in.	¾ in.	1 in.	½ in.	⅝ in.	¾ in.	1 in.	½ in.	⅝ in.	¾ in.	1 in.	½ in.	⅝ in.	¾ in.	1 in.
75	16	19	22	26	10	13	17	23	16	19	22	26	10	13	17	23
100	14	17	20	24	9	12	15	21	14	17	20	24	9	12	15	21
125	13	16	18	23	8	11	14	20	13	16	18	23	8	11	14	20
150	12	15	17	22	8	10	13	19	12	15	17	22	8	10	13	19
175	12	14	16	21	7	10	13	18	12	14	16	21	7	10	13	18
200	11	13	16	20	7	9	12	17	11	13	16	20	7	9	12	17
300	10	12	14	18	6	8	11	15	10	12	14	17	6	8	11	15
400	9	11	12	16	6	8	10	13	9	11	12	16	6	8	10	13
500	8	10	11	15	5	7	9	12	8	10	11	15	5	7	9	12
600	8	9	11	14	5	7	8	12	8	9	10	13	5	7	8	12
700	7	9	10	13	5	6	8	11	7	9	10	12	5	6	8	11
800	7	8	10	13	4	6	8	11	7	8	9	12	4	6	8	11
900	7	8	9	12	4	6	7	10	7	8	9	11	4	6	7	10
1000	7	8	9	12	4	6	7	10	6	7	8	10	4	6	7	10
1100	6	8	9	11	4	5	7	10	6	7	8	10	4	5	7	10
1200	6	7	8	11	4	5	7	9	6	7	7	9	4	5	6	9
1300	6	7	8	10	4	5	6	9	5	6	7	9	4	5	6	9
1400	6	7	8	10	4	5	6	9	5	6	7	9	—	4	5	8
1500	6	7	7	9	4	5	6	9	5	6	7	8	—	4	5	8
1600	6	6	7	9	—	5	6	8	5	6	6	8	—	4	5	7
1700	5	6	7	9	—	4	5	8	4	5	6	8	—	—	4	7
1800	5	6	7	9	—	4	5	8	4	5	6	7	—	—	4	6
1900	5	6	7	8	—	4	5	8	4	5	6	7	—	—	4	6
2000	5	6	6	8	—	4	5	7	4	5	6	7	—	—	4	6
2200	5	5	6	8	—	—	4	7	—	4	5	6	—	—	—	5
2400	4	5	6	7	—	—	4	6	—	4	5	6	—	—	—	5
2600	4	5	6	7	—	—	4	6	—	4	5	5	—	—	—	4
2800	4	4	5	6	—	—	—	5	—	—	4	5	—	—	—	4
3000	—	4	5	6	—	—	—	5	—	—	4	4	—	—	—	4

[a]Above the solid line spans are governed by deflection; below the dashed line spans are governed by shear. Bending governs spans between the two lines. Where shear controls, use clear spans; otherwise use center to center of supports.

Source: Reproduced by permission of the American Concrete Institute.

TABLE 6-9 Safe spacing (in.) of supports for joists, studs (or other beam components of formwork), continuous over three or more spans[a]

$\Delta_{max} = l/360$, but not to exceed ¼ in.

$f = 775$ psi $E = 1,100,000$ psi $H = 105$ psi

Uniform load, lb per lineal ft (equals uniform load on forms times spacing between joists or studs, ft)	Nominal size of S4S lumber																			
	2x4	2x6	2x8	2x10	2x12	3x4	3x6	3x8	3x10	4x2	4x4	4x6	4x8	6x2	6x4	6x6	6x8	8x2	8x8	10x2
100	53	84	110	141	171	69	108	140	168	35	81	123	152	43	98	138	174	48	189	52
200	38	59	78	100	121	49	76	101	129	25	58	90	119	31	72	113	147	35	159	40
300	31	48	64	81	99	40	62	82	105	20	47	74	97	25	59	93	126	29	143	33
400	25	40	52	67	81	34	54	71	91	17	41	64	84	22	51	80	109	25	128	28
500	22	34	45	57	70	31	48	64	81	16	36	57	75	20	46	72	98	22	114	25
600	19	30	40	51	62	27	43	57	72	14	33	52	69	18	42	65	89	20	104	23
700	17	27	36	46	56	24	38	51	65	13	31	48	64	16	39	61	83	19	97	21
800	16	25	33	43	52	23	35	46	59	12	28	45	59	15	36	57	77	18	90	20
900	15	24	31	40	49	21	32	43	54	11	26	41	54	15	34	53	73	17	85	19
1000	14	23	30	38	46	19	30	40	51	10	24	38	50	14	32	51	69	16	81	18
1100	14	21	28	36	44	18	28	38	48	10	23	35	47	13	31	48	66	15	77	17
1200	13	21	27	35	42	17	27	36	45	9	21	33	44	13	29	46	63	14	74	16
1300	13	20	26	33	41	16	26	34	43	9	20	32	42	12	28	44	59	14	71	16
1400	12	19	25	32	39	16	25	33	42	8	19	30	40	11	26	41	56	13	68	15
1500	12	19	25	31	38	15	24	31	40	8	18	29	38	11	25	39	53	13	66	15
1600	12	18	24	31	37	15	23	30	39	8	18	28	37	10	24	37	51	12	64	14
1700	11	18	23	30	36	14	22	29	37	7	17	27	35	10	23	36	49	12	61	14
1800	11	17	23	29	36	14	22	29	36	7	16	26	34	9	22	34	47	11	59	13
1900	11	17	22	29	35	13	21	28	35	7	16	25	33	9	21	33	45	11	56	13
2000	11	17	22	28	34	13	21	27	35	7	16	24	32	9	20	32	44	11	54	13
2100	10	16	22	28	34	13	20	27	34	6	15	24	31	8	20	31	42	10	52	12
2200	10	16	21	27	33	13	20	26	33	6	15	23	31	8	19	30	41	10	51	12
2300	10	16	21	27	33	12	19	25	33	6	14	23	30	8	19	29	40	9	49	11
2400	10	16	21	27	32	12	19	25	32	6	14	22	29	8	18	29	39	9	48	11
2500	10	16	21	26	32	12	19	25	31	6	14	22	29	8	18	28	38	9	46	11
2600	10	15	20	26	32	12	18	24	31	6	14	21	28	7	17	27	37	9	45	10
2700	10	15	20	26	31	11	18	24	30	6	13	21	28	7	17	27	36	9	44	10
2800	10	15	20	25	31	11	18	24	30	6	13	21	27	7	17	26	36	8	43	10
2900	9	15	20	25	31	11	18	23	30	5	13	20	27	7	16	26	35	8	42	10
3000	9	15	20	25	30	11	17	23	29	5	12	19	26	7	16	25	34	8	41	9
3200	9	15	19	25	30	11	17	22	29	5	12	19	26	7	15	24	33	8	40	9
3400	9	14	19	24	29	11	17	22	28	5	12	19	25	6	15	23	32	7	38	9
3600	9	14	18	23	28	10	16	21	27	5	12	18	24	6	14	23	31	7	37	8
3800	9	14	18	23	28	10	16	21	27	5	11	18	24	6	14	22	30	7	36	8
4000	8	13	17	22	27	10	16	21	27	5	11	18	23	6	14	22	29	7	35	8
4500	8	12	16	21	26	10	15	20	26	5	11	17	23	6	13	20	28	6	32	7
5000	7	12	16	20	24	9	15	20	25	4	10	16	22	5	12	19	27	6	31	7

[a]Span values above the solid line are governed by deflection. Values within dashed line box are spans governed by shear. Elsewhere bending governs span.

Source: Reproduced by permission of the American Concrete Institute.

(UPPER STRINGER)

TABLE 6-10 Safe spacing (in.) of supports for joists, studs (or other beam components of formwork), single span[a]

$\Delta_{max} = l/360$, but not to exceed $\frac{1}{4}$ in.

$f = 775$ psi $E = 1,100,000$ psi $H = 105$ psi

Uniform load, lb per lineal ft (equals uniform load on forms times spacing between joists or studs, ft)	2x4	2x6	2x8	2x10	2x12	3x4	3x6	3x8	3x10	4x2	4x4	4x6	4x8	6x2	6x4	6x6	6x8	8x2	8x8	10x2
100	48	75	99	126	146	62	97	119	143	30	71	105	130	35	82	118	149	39	161	42
200	34	53	70	89	108	44	68	90	115	22	52	81	107	28	65	99	125	31	135	33
300	28	43	57	73	89	36	56	74	94	18	42	66	87	23	53	83	113	26	122	29
400	24	37	49	63	77	31	48	64	81	16	36	57	75	20	46	72	98	22	114	25
500	21	34	44	56	69	28	43	57	73	14	33	51	68	17	41	64	88	20	102	23
600	19	31	40	51	63	25	39	52	66	13	30	47	62	16	37	59	80	18	93	21
700	18	28	37	48	58	23	37	48	62	12	28	43	57	15	35	54	74	17	86	19
800	17	26	35	45	54	22	34	45	58	11	26	40	53	14	32	51	69	16	81	18
900	16	25	33	42	51	20	32	43	54	10	24	38	50	13	30	48	65	15	76	17
1000	15	24	31	40	48	19	31	40	51	10	23	36	48	12	29	45	62	14	72	16
1100	14	23	30	38	46	19	29	38	49	9	22	35	45	12	28	43	59	14	69	15
1200	14	22	28	36	44	18	28	37	47	9	21	33	44	11	26	41	57	13	66	15
1300	13	21	27	35	43	17	27	35	45	9	20	32	42	11	25	40	54	12	63	14
1400	13	20	26	34	41	16	26	34	43	8	19	31	40	10	24	38	52	12	61	14
1500	12	19	25	33	40	16	25	33	42	8	19	30	39	10	24	37	51	12	59	13
1600	12	19	25	31	38	15	24	32	41	8	18	29	38	10	23	36	49	11	57	13
1700	12	18	24	31	37	15	23	31	39	8	18	28	37	9	22	35	47	11	55	12
1800	11	18	23	30	36	14	23	30	38	7	17	27	36	9	21	34	46	11	54	12
1900	11	17	23	29	35	14	22	29	37	7	17	26	35	9	21	33	45	10	52	12
2000	11	17	22	28	34	14	22	28	36	7	16	26	34	9	20	32	44	10	51	11
2100	10	16	22	27	33	13	21	28	35	7	16	25	33	8	20	31	43	10	50	11
2200	10	16	21	27	33	13	21	27	35	7	15	24	32	8	19	31	42	10	49	11
2300	10	16	21	26	32	13	20	27	34	6	15	24	31	8	19	30	41	9	48	11
2400	10	15	20	26	31	13	20	26	33	6	15	23	31	8	19	29	40	9	47	10
2500	9	15	20	25	31	12	19	25	33	6	15	23	30	8	18	29	39	9	46	10
2600	9	15	19	25	30	12	19	25	32	6	14	22	30	8	18	28	38	9	45	10
2700	9	14	19	24	29	12	19	25	31	6	14	22	29	7	18	28	38	9	44	10
2800	9	14	19	24	29	12	18	24	31	6	14	22	28	7	17	27	37	8	43	9
2900	9	14	18	23	28	11	18	24	30	6	13	21	28	7	17	27	36	8	42	9
3000	9	14	18	23	28	11	18	23	30	6	13	21	28	7	17	26	36	8	42	9
3200	8	13	17	22	27	11	17	23	29	5	13	20	27	7	16	25	35	8	40	9
3400	8	13	17	22	26	11	17	22	28	5	12	20	26	7	16	25	34	7	39	9
3600	8	12	16	21	26	10	16	21	27	5	12	19	25	6	15	24	33	7	38	8
3800	8	12	16	20	25	10	16	21	26	5	12	19	24	6	15	23	32	7	37	8
4000	7	12	16	20	24	10	15	20	26	5	11	18	24	6	14	23	31	7	36	8
4500	7	11	15	19	23	9	14	19	24	5	11	17	22	6	14	21	29	6	34	8
5000	7	11	14	18	22	9	14	18	23	4	10	16	21	5	13	20	28	6	32	7

Nominal size of S4S lumber

[a]Values above solid line are spans governed by deflection. All other values are governed by bending.

Source: Reproduced by permission of the American Concrete Institute.

TABLE 6-11 Safe spacing (in.) of supports for joists, studs (or other beam components of formwork), continuous over two spans[a]

$\Delta_{max} = l/360$, but not to exceed ¼ in.

$E = 1,100,000$ psi $f = 775$ psi $H = 105$ psi

Nominal size of S4S lumber

Uniform load, lb per lineal ft (equals uniform load on forms times spacing between joists or studs, ft)	2x4	2x6	2x8	2x10	2x12	3x4	3x6	3x8	3x10	4x2	4x4	4x6	4x8	6x2	6x4	6x6	6x8	8x2	8x8	10x2
100	48	75	99	126	146	62	97	128	163	31	73	115	151	39	91	144	185	45	200	51
200	34	53	70	89	108	44	68	90	115	22	52	81	107	28	65	102	138	32	162	36
300	28	43	57	73	89	36	56	74	94	18	42	66	87	23	53	83	113	26	132	29
400	24	37	49	63	77	31	48	64	81	16	36	57	75	20	46	72	98	22	114	25
500	21	33	44	56	68	28	43	57	73	14	33	51	68	17	41	64	88	20	102	23
600	19	29	39	50	60	25	39	52	66	13	30	47	62	16	37	59	80	18	93	21
700	17	27	35	45	55	23	37	48	62	12	28	43	57	15	35	54	74	17	86	19
800	16	25	33	42	51	22	34	45	57	11	26	40	53	14	32	51	69	16	81	18
900	15	23	31	39	48	20	31	42	53	10	24	38	50	13	30	48	65	15	76	17
1000	14	22	29	37	45	19	29	39	50	10	23	36	48	12	29	45	62	14	72	16
1100	13	21	28	35	43	18	28	37	47	9	22	34	45	12	28	43	59	14	69	15
1200	13	20	27	34	41	17	26	35	44	9	21	33	43	11	26	41	57	13	66	15
1300	12	19	26	33	40	16	25	33	42	8	20	31	41	11	25	40	54	12	63	14
1400	12	19	25	32	39	15	24	32	41	8	19	29	39	10	24	38	52	12	61	14
1500	12	18	24	31	38	15	23	31	39	8	18	28	37	10	24	37	51	12	59	13
1600	11	18	24	30	37	14	23	30	38	7	17	27	36	10	23	36	49	11	57	13
1700	11	17	23	29	36	14	22	29	37	7	17	26	35	9	22	35	47	11	55	12
1800	11	17	23	29	35	13	21	28	36	7	16	25	33	9	21	34	46	11	54	12
1900	11	17	22	28	34	13	21	27	35	7	16	25	32	9	21	32	44	10	52	12
2000	10	16	22	28	34	13	20	27	34	6	15	24	32	8	20	31	43	10	51	11
2100	10	16	21	27	33	13	20	26	33	6	15	23	31	8	19	30	41	10	50	11
2200	10	16	21	27	33	12	19	26	33	6	14	23	30	8	19	29	40	10	49	11
2300	10	16	21	26	32	12	19	25	32	6	14	22	29	8	18	29	39	9	48	11
2400	10	15	20	26	31	12	19	25	31	5	14	22	29	8	18	28	38	9	46	10
2500	9	15	20	25	31	12	18	24	31	5	14	21	28	7	17	27	37	9	45	10
2600	9	13	19	25	30	11	18	24	30	6	13	21	28	7	17	27	36	9	44	10
2700	9	14	19	24	29	11	18	23	30	6	13	20	27	7	17	26	35	8	43	10
2800	9	14	19	24	29	11	18	23	30	5	13	20	27	7	16	25	35	8	42	10
2900	9	14	18	23	28	11	17	23	29	5	13	20	26	7	16	25	34	8	41	9
3000	9	14	18	23	28	11	17	23	29	5	12	20	26	7	16	25	33	8	40	9
3200	8	13	17	22	27	11	17	22	28	5	12	19	25	6	15	24	32	8	39	9
3400	8	13	17	22	26	10	16	22	28	5	12	19	24	6	15	23	31	7	37	8
3600	8	12	16	21	26	10	16	21	27	5	12	18	24	6	14	22	30	7	36	8
3800	7	12	16	20	25	10	16	21	26	5	11	18	23	6	14	22	30	7	35	8
4000	7	12	16	20	24	9	15	20	26	5	11	17	23	6	14	21	29	7	34	8
4500	7	11	15	19	23	9	14	19	24	5	11	17	22	5	13	20	27	6	32	7
5000	7	11	14	18	22	9	14	18	23	4	10	16	21	5	12	19	26	6	30	7

[a] Values above solid line are spans governed by deflection. Values within dashed-line box are spans governed by shear. All other values are governed by bending.

Source: Reproduced by permission of the American Concrete Institute.

> **TABLE 6-12** Safe spacing (in.) of supports for double wales continuous over three or more spans[a]

Maximum deflection is ⅟₃₆₀ of spacing, but not more than ¼ in.

Equivalent uniform load, lb per lineal ft (equals uniform load, psf, on forms times spacing of wales in ft)	f = 775 psi			E = 1,100,000 psi			H = 105 psi			f = 1000 psi			E = 1,400,000 psi			H = 135 psi		
	Nominal size of S4S lumber used double									Nominal size of S4S lumber used double								
	2x4	2x6	2x8	3x4	3x6	3x8	4x4	4x6	4x8	2x4	2x6	2x8	3x4	3x6	3x8	4x4	4x6	4x8
100	75	119	146	96	135	166	105	147	181	86	126	155	102	143	176	111	156	192
200	53	84	111	69	108	140	81	123	152	61	95	126	78	121	148	93	131	161
300	44	68	90	56	88	116	67	105	137	49	78	102	64	100	132	76	118	146
400	38	59	78	49	77	101	58	91	119	43	67	89	55	87	115	65	103	136
500	34	53	70	44	68	90	52	81	107	38	60	79	49	78	102	59	92	121
600	31	48	64	40	62	82	47	74	97	35	55	72	45	71	94	53	84	111
700	28	44	58	37	58	76	44	68	90	32	51	67	42	66	87	49	78	102
800	25	40	53	34	54	71	41	64	84	30	48	63	39	61	81	46	73	96
900	23	37	48	32	51	67	38	60	80	28	44	58	37	58	76	44	69	90
1000	22	34	45	31	48	64	36	57	75	26	41	54	35	55	72	41	65	86
1100	20	32	42	29	46	61	35	55	72	24	38	50	33	52	69	39	62	82
1200	19	30	40	27	43	57	33	52	69	23	36	47	32	50	66	38	59	78
1300	18	29	38	26	41	53	32	50	66	21	34	45	31	48	64	36	57	75
1400	17	27	36	24	38	51	31	48	64	20	32	42	29	46	61	35	55	72
1500	17	26	35	23	37	48	30	47	62	20	31	41	28	44	58	34	53	70
1600	16	25	33	22	35	46	28	45	59	19	30	39	27	42	55	33	51	69
1700	16	25	32	21	34	44	27	43	56	18	28	37	25	40	53	32	50	66
1800	15	24	31	21	32	43	26	41	54	17	27	36	24	38	51	31	48	64
1900	15	23	30	20	31	41	25	39	52	17	27	35	24	37	49	30	47	62
2000	14	23	30	19	30	40	24	38	50	16	26	34	23	36	47	29	46	60
2200	14	21	28	18	28	38	23	35	47	16	24	32	21	33	44	27	42	56
2400	13	21	27	17	27	36	21	33	44	15	23	31	20	32	42	25	40	53
2600	13	20	26	16	26	34	20	32	42	14	22	30	19	30	40	24	38	50
2800	12	19	25	16	25	33	19	30	40	14	22	28	18	29	38	23	36	47
3000	12	19	25	15	24	31	18	29	38	13	21	27	17	27	36	22	34	45
3200	12	18	24	15	23	30	18	28	37	13	20	27	17	26	35	21	33	43
3400	11	18	23	14	22	29	17	27	35	13	20	26	16	26	34	20	31	41
3600	11	17	23	14	22	29	16	26	34	12	19	25	16	25	33	19	30	40
3800	11	17	22	13	21	28	16	25	33	12	19	25	15	24	32	19	29	38
4000	11	17	22	13	21	27	16	24	32	12	18	24	15	23	31	18	28	37
4200	10	16	22	13	20	27	15	24	31	11	18	24	14	23	30	17	27	36
4400	10	16	21	13	20	26	15	23	31	11	18	23	14	22	29	17	27	35
4600	10	16	21	12	19	25	14	23	30	11	17	23	14	22	29	17	26	34
4800	10	16	21	12	19	25	14	22	29	11	17	23	14	21	28	16	25	34
5000	10	16	21	12	19	25	14	22	29	11	17	22	13	21	27	16	25	33

[a]Within the dashed lines, shear is the governing criterion. Within the solid lines, deflection governs. Within the distance h from supports neglected in computing shear; elsewhere bending strength governs the span. Load within the distance h from supports neglected in computing shear.

Source: Reproduced by permission of the American Concrete Institute.

TABLE 6–13 Safe spacing (in.) of supports for double wales, single span[a]

Maximum deflection is 1/360 of spacing, but not more than 1/4 in.

Equivalent uniform load, lb per lineal ft (equals uniform load, psf on forms times spacing of wales in ft)	f = 775 psi E = 1,100,000 psi H = 105 psi — Nominal size of S4S lumber used double									f = 1000 psi E = 1,400,000 psi H = 135 psi — Nominal size of S4S lumber used double								
	2x4	2x6	2x8	3x4	3x6	3x8	4x4	4x6	4x8	2x4	2x6	2x8	3x4	3x6	3x8	4x4	4x6	4x8
100	67	101	125	80	115	142	89	125	154	73	108	132	86	122	150	95	133	164
200	48	75	99	62	97	119	71	105	130	54	85	111	68	103	127	77	112	138
300	39	61	81	50	79	104	60	94	117	44	70	92	57	90	114	67	101	124
400	34	53	70	44	68	90	52	81	107	38	60	79	50	78	103	59	92	116
500	30	47	63	39	61	81	46	72	96	34	54	71	44	70	92	52	82	109
600	28	43	57	36	56	74	42	66	87	31	49	65	40	64	84	48	75	99
700	26	40	53	33	52	68	39	61	81	29	46	60	37	59	78	44	70	92
800	24	38	49	31	48	64	36	57	76	27	43	56	35	55	73	41	65	86
900	23	35	47	29	46	60	34	54	71	26	40	53	33	52	68	39	61	81
1000	21	34	44	28	43	57	33	51	68	24	38	50	31	49	65	37	58	77
1100	20	32	42	26	41	55	31	49	65	23	36	48	30	47	62	35	55	73
1200	19	31	40	25	39	52	30	47	62	22	35	46	29	45	59	34	53	70
1300	19	29	39	24	38	50	29	45	59	21	33	44	27	43	57	32	51	67
1400	18	28	37	23	37	48	28	43	57	20	32	42	26	42	55	31	49	65
1500	17	27	36	22	35	47	27	42	55	20	31	41	26	40	53	30	47	63
1600	17	26	35	22	34	45	26	40	53	19	30	40	25	39	51	29	46	61
1700	16	26	34	21	33	44	25	39	52	19	29	39	24	38	50	28	45	59
1800	16	25	33	20	32	43	24	38	50	18	28	37	23	37	48	28	43	57
1900	16	24	32	20	31	41	24	37	49	18	28	36	23	36	47	27	42	56
2000	15	24	31	19	31	40	23	36	48	17	27	36	22	35	46	26	41	54
2200	14	23	30	19	29	38	22	35	45	16	26	34	21	33	44	25	39	52
2400	14	22	28	18	28	37	21	33	44	16	25	32	20	32	42	24	38	49
2600	13	21	27	17	27	35	20	32	42	15	24	31	19	30	40	23	36	48
2800	13	20	26	16	26	34	19	31	40	14	23	30	19	29	39	22	35	46
3000	12	19	25	16	25	33	19	30	39	14	22	29	18	28	37	21	34	44
3200	12	19	25	15	24	32	18	29	38	14	21	28	17	27	36	21	32	43
3400	12	18	24	15	23	31	17	28	37	13	21	27	17	27	35	20	32	42
3600	11	18	23	14	23	30	17	27	36	13	20	26	16	26	34	19	31	40
3800	11	17	23	14	22	29	17	26	35	12	20	26	16	25	33	19	30	39
4000	11	17	22	14	22	29	16	26	34	12	19	25	16	25	32	18	29	38
4200	10	16	22	13	21	28	16	25	33	12	19	24	15	24	32	18	28	37
4400	10	16	21	13	21	27	16	24	32	12	18	24	15	23	31	18	28	37
4600	10	16	21	13	20	27	15	24	31	11	18	23	15	23	30	17	27	36
4800	10	15	20	13	20	26	15	23	31	11	17	23	14	22	30	17	27	35
5000	9	15	20	12	19	26	15	23	30	11	17	22	14	22	29	17	26	34

[a]Within the solid line deflection is the governing design criterion; elsewhere bending strength governs the span.

Source: Reproduced by permission of the American Concrete Institute.

110

TABLE 6-14 Safe spacing (in.) of supports for double wales continuous over two spans[a]

Maximum deflection is $\frac{1}{360}$ of spacing, but not more than $\frac{1}{4}$ in.

Equivalent uniform load, lb per lineal ft (equals uniform load, psf, on forms times spacing of wales in ft)	f = 775 psi			E = 1,100,000 psi			H = 105 psi			f = 1000 psi			E = 1,400,000 psi			H = 135 psi		
	Nominal size of S4S lumber used double									Nominal size of S4S lumber used double								
	2x4	2x6	2x8	3x4	3x6	3x8	4x4	4x6	4x8	2x4	2x6	2x8	3x4	3x6	3x8	4x4	4x6	4x8
100	68	106	140	87	137	176	103	156	192	77	121	159	99	152	187	117	166	204
200	48	75	99	62	97	128	73	115	151	54	85	112	70	110	145	83	130	171
300	39	61	81	50	79	104	60	94	123	44	70	92	57	90	118	68	106	140
400	34	53	70	44	68	90	52	81	107	38	60	79	50	78	103	59	92	121
500	30	47	63	39	61	81	46	72	96	34	54	71	44	70	92	52	82	109
600	28	43	57	36	56	74	42	66	87	31	49	65	40	64	84	48	75	99
700	26	40	53	33	52	68	39	61	81	29	46	60	37	59	78	44	70	92
800	24	38	49	31	48	64	36	57	76	27	43	56	35	55	73	41	65	86
900	23	35	47	29	46	60	34	54	71	26	40	53	33	52	68	39	61	81
1000	21	33	44	28	43	57	33	51	68	24	38	50	31	49	65	37	58	77
1100	20	31	41	26	41	55	31	49	65	23	36	49	30	47	62	35	55	73
1200	19	29	39	25	39	52	30	47	62	22	35	46	29	45	59	34	53	70
1300	18	28	37	24	38	50	29	45	59	21	33	43	27	43	57	32	51	67
1400	17	27	35	23	37	48	28	43	57	20	31	41	26	42	55	31	49	65
1500	16	26	34	22	35	47	27	42	55	19	30	39	26	40	53	30	47	63
1600	16	25	33	22	34	45	26	40	53	18	29	38	25	39	51	29	46	61
1700	15	24	32	21	33	43	25	39	52	18	28	37	24	38	50	28	45	59
1800	15	23	31	20	31	42	24	38	50	17	27	35	23	37	48	28	43	57
1900	14	23	30	19	30	40	24	37	49	16	26	34	23	36	47	27	42	56
2000	14	22	29	19	29	39	23	36	48	16	25	33	22	35	46	26	41	54
2200	13	21	28	18	28	37	22	34	45	15	24	32	21	33	43	25	39	52
2400	13	20	27	17	26	35	21	33	43	15	23	30	20	31	41	24	38	49
2600	12	19	26	16	25	33	20	31	41	14	22	29	19	29	39	23	36	48
2800	12	19	25	15	24	32	19	29	39	13	21	28	18	28	37	22	35	46
3000	12	18	24	15	23	31	18	28	37	13	20	27	17	27	35	21	33	44
3200	11	18	24	14	23	30	17	27	36	12	20	26	16	26	34	20	32	42
3400	11	17	23	14	22	29	17	26	35	12	19	25	16	25	33	19	31	40
3600	11	17	23	13	21	28	16	25	33	12	19	25	15	24	32	19	29	39
3800	11	17	22	13	21	27	16	25	32	12	18	24	15	23	31	18	28	38
4000	10	16	22	13	20	27	15	24	32	11	18	24	15	23	30	18	28	36
4200	10	16	21	13	20	26	15	23	31	11	18	23	14	22	29	17	26	35
4400	10	16	21	12	19	26	14	23	30	11	17	23	14	22	28	17	26	34
4600	10	16	21	12	19	25	14	22	29	11	17	23	14	21	28	16	25	34
4800	10	15	20	12	19	25	14	22	29	11	17	22	13	21	27	16	25	33
5000	9	15	20	12	18	24	14	21	28	11	17	22	13	20	27	15	24	32

[a]Within the dashed lines, shear is the governing criterion. Within the solid lines, deflection governs. Load within the distance h from supports neglected in computing shear.

Source: Reproduced by permission of the American Concrete Institute.

Example

A 20-ft-high wall 12 in. thick is to be poured at the rate of 8 ft/hr at a temperature of 60 °F. (a) Calculate the maximum pressure on the formwork. (b) Using ¾-in.-thick plywood with the face grain parallel to the span, determine the maximum allowable spacing of the wales based on the maximum concrete pressure. (c) If the wales are to be double 2″ × 4″s, what maximum spacing of the ties will be allowed if the wales are considered to be continuous over four or more supports? (d) What load will be placed on each tie as a result?

Solution

(a) For walls poured at 7 to 10 ft/hr, maximum pressure is

$$p = 150 + \frac{43,400}{T} + \frac{2800R}{T}$$

$$= 150 + \frac{43,400}{60} + \frac{2800 \times 8}{60}$$

$$= 1247 \text{ psf}$$

Check maximum pressure—the lesser of 2000 psf and $150 \times 20 = 3000$ psf. Therefore, use 1247 psf as maximum pressure.

(b) Based on plywood continuous over four or more supports (Table 6–7), face grain parallel to the span and an $f = 1930$ psi, the maximum spacing of supports or wales is 9 in. o.c.

(c) Load per lineal foot on the wales is

$$1247 \times 9/12 = 935 \text{ plf}$$

From Table 6–12 for double wales continuous over three or more spans, maximum spacing of ties is 24 in. o.c.

(d) The resulting load on each tie is $935 \times {}^{24}\!/_{12} = 1870$ lb.

Example

Plywood ⅝ in. thick is to be used as a form for a concrete floor pour. If the total construction load is anticipated to be 200 psf, determine: (a) The required spacing of the beam supports under the plywood. Assume that the plywood face grain is parallel to the span and the plywood is continuous over four or more supports. (b) If 2″ × 4″ sections are to act as supporting beams for the plywood, what maximum spacing will be allowed if the 2″ × 4″s are simply supported?

Solution

(a) From Table 6–7, assuming an allowable bending stress of 1930 psi, maximum spacing for ⅝-in. plywood supporting 200 psf is 17 in. Use 16-in. spacing for supporting beams.

(b) Assume that material to be used has a maximum allowable bending stress $f = 775$ psi and an allowable horizontal shear stress $H = 105$ psi.

Load per foot on joists = load in psf × joist spacing

Load per foot $= 200 \times {}^{16}\!/_{12} = 267$ plf

From Table 6–10, for simply supported sections (2″ × 4″ in size), maximum spacing by interpolation is 30 in.

The size and spacing of ties should be based on the manufacturer's recommendations or on calculations of the load carried per lineal foot of wale. For example, if the load per lineal foot of wale is found to be 3000 lb and ties are spaced 2 ft o.c., the load per tie will be 6000 lb. Based on an allowable tensile stress of 20,000 lb/sq in., the cross-sectional area of this tie should be $6000/20,000 = 0.30$ sq in. This will require a ⅝-in.-diameter tie with an area of 0.31 sq in.

FOOTING FORMS

There are usually two factors of prime importance to consider in the construction of footings. One is that the concrete must be up to specified strength and the other that the footings be positioned according to plan. The appearance is seldom of importance since footings are usually below grade.

Old or used material may be employed for footing forms, provided it is sound. A certain amount of tolerance is allowed in footing size and thickness, but reinforcing bars and dowels must be placed as specified. Concrete is sometimes cast against the excavation, but care must be taken that this does not give inferior results, caused by the earth absorbing water from the concrete or by pieces of earth falling into it. Trenches may be lined with wax paper or polyethylene film to prevent this. It is sometimes desirable to form the top 4 in. of a footing cast in earth, as shown in Fig. 6–4.

In cases where wall footings are shallow, lateral pressure is small and the forms are simple structures as illustrated in Fig. 6–5. When the soil is firm, the form can be held in place by stakes and braces. If the soil will not hold stakes, the forms may be secured by bracing them against the excavation sides, as shown in Fig. 6–6.

For deeper wall footings and grade beams, (see Fig. 6–7) the formwork must be more elaborate and is constructed in much the same manner as wall forms (see *Wall Forms* on page 115).

FIGURE 6–4 Forming top of trench.

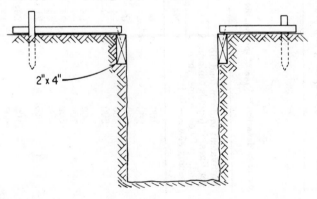

2″ × 4″

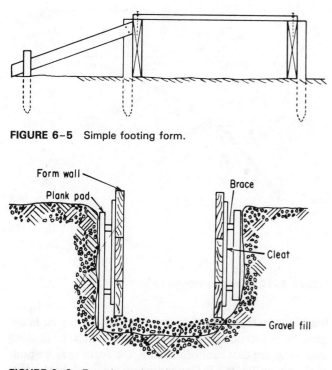

FIGURE 6-5 Simple footing form.

FIGURE 6-6 Form braced against excavation.

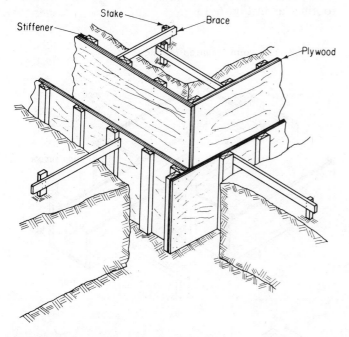

FIGURE 6-7 Heavy grade beam form.

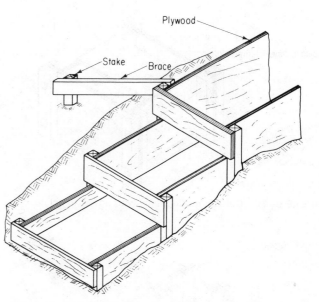

FIGURE 6-8 Stepped wall footing.

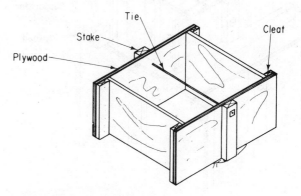

FIGURE 6-9 Simple column footing form.

When a wall footing is supported by sloping ground, the footing may be stepped longitudinally at intervals, the frequency of the steps depending on the slope. Figure 6-8 illustrates a method of forming such steps.

Simple rectangular column footing forms may be made as bottomless boxes, constructed in four pieces—two sides and two end sections. Two sections are made the exact dimensions of the footing and the two opposite ones long enough to take care of the thickness of the material and the vertical cleats used for holding the sections together (see Fig. 6-9). Ties are used to prevent the sides from bulging under pressure. A template for positioning dowels in the footing can be attached to the form box, as shown in Fig. 6-10. Figure 6-10 also illustrates an alternative method of tying the form.

Stepped column footings are shown in Fig. 6-11. The forms are made as a series of boxes stacked one on top of the other. The upper ones can be supported as shown in Fig. 6-11, and if the difference in size between the upper and lower forms is considerable, the bottom one may require a cover. The whole assembly must be weighed or tied down to prevent uplift by freshly placed concrete.

Tapered footing forms are made like a *hopper*. Cleats hold the side and end sections together, and some convenient method must be used to hold the form in

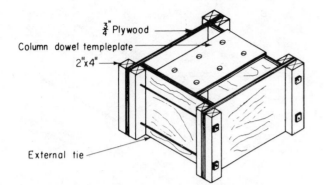

FIGURE 6-10 Column footing form with dowel template.

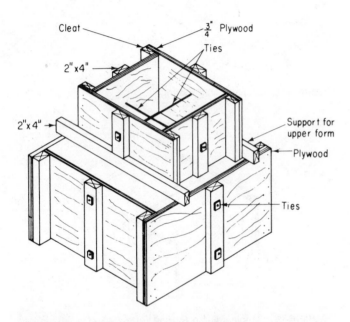

FIGURE 6-11 Stepped column footing form.

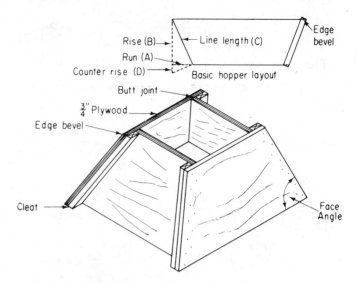

FIGURE 6-12 Tapered footing form.

load is to be transferred to another pad, a strap or beam must be constructed between the two. When the footing and strap are cast monolithically, the form may be built as shown in Fig. 6-13. Depending on the available height, it may not be necessary to set the strap or beam into the footing, as in Fig. 6-13.

FIGURE 6-13 Form for strap footing.

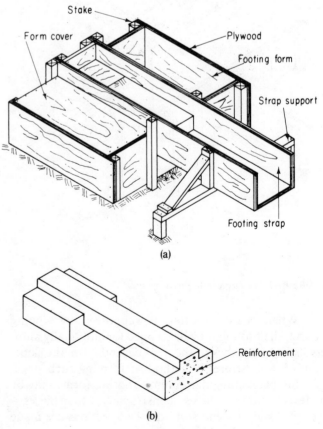

place. Two opposite panels are made to the exact dimensions of the footing face, while the other two must be enough longer to accommodate the thickness of the plywood and the width of the cleats used (see Fig. 6-12). The angles required to cut the panels properly may be obtained as follows, using a framing square to lay them out (see the basic hopper layout):

- *Butt joint:* Use C and D. Mark on D.
- *Face angle:* Use A and C. Mark on C.
- *Edge bevel:* Use B and A. Mark on A.

It is sometimes necessary to transfer part of the load of one footing to another or to support two columns on a single footing. The footing is then known as a *combined*, or *strap*, footing.

When two columns are supported on a single footing, the form is built the same way as a flat or stepped footing, except that it is longer. When part of the footing

WALL FORMS

Wall forms are made up of five basic parts: (1) *sheathing*, to shape and retain the concrete until it sets; (2) *studs*, to form a framework and support the sheathing; (3) *wales*, to keep the form aligned and support the studs; (4) *braces,* to hold the forms erect under lateral pressure; and (5) *ties and spreaders* or *tie-spreader units*, to hold the sides of the forms at the correct spacing (see Fig. 6–14).

Wall forms may be built in place or made up of prefabricated panels, depending on the shape of the structure being formed and on whether the form being built can be reused.

Built-in-Place Forms

Forms are built in place when the design of the structure is such that prefabricated panels cannot be adapted to the shape or when the form is for one use only and the use of prefab panels cannot be economically justified.

The first step in the erection of wall forms of any kind is to locate them properly on the foundation from which the wall will rise. This may be done by anchoring a *sole plate* for either the inside or outside form to the foundation with preset bolts, concrete nails, or power-driven studs. It must be set out from the proposed wall line by the thickness of the sheathing to be used and carefully aligned.

Another method of positioning the form is to anchor a *guide plate* to the foundation in such a position

that the forms will be properly located when they are set behind it. Figure 6–15 illustrates such an arrangement.

If the form is being built on a sole plate and studs are to be used, they are set on the plate, toenailed in place on the required centers, and held vertical by temporary bracing. Plywood sheathing panels are then nailed to the studs, with nails on 12- to 16-in. centers. The first panel should be set and leveled at the highest point of the foundation to establish alignment for the remainder.

Ties are inserted as sheathing progresses, with the tie holes being located so that the tie ends will meet the walers, when they have been installed. Figure 6–16 illustrates some simple types of ties and Fig. 6–17, some typical form tie installations.

When one side of the form has been completed, the other may be built in sections and set in place, with the tie ends being threaded through predrilled holes.

If the guide plate method of locating forms is used, sections of inside and outside form can be built, tied

FIGURE 6–15 Plate anchored to foundation outside of form. (*Courtesy Jahn Concrete Forming.*)

FIGURE 6–14 Parts of typical wall form.

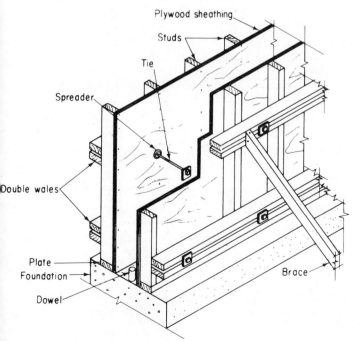

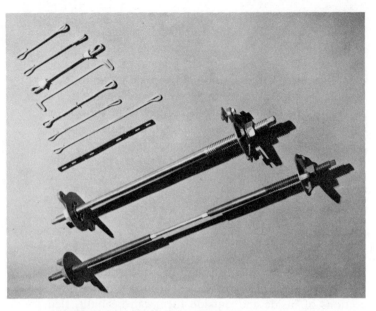

FIGURE 6–16 Simple form ties.

together, and set in position, ready to be bolted or clamped to adjoining sections.

Studs are not always used in form building. In the form shown in Fig. 6–18, for example, plywood sheets are nailed to a sole plate and lined up with single

FIGURE 6–18 Wall form with no studs. (*Courtesy Jahn Concrete Forming.*)

FIGURE 6–17 Some typical form tie installations.

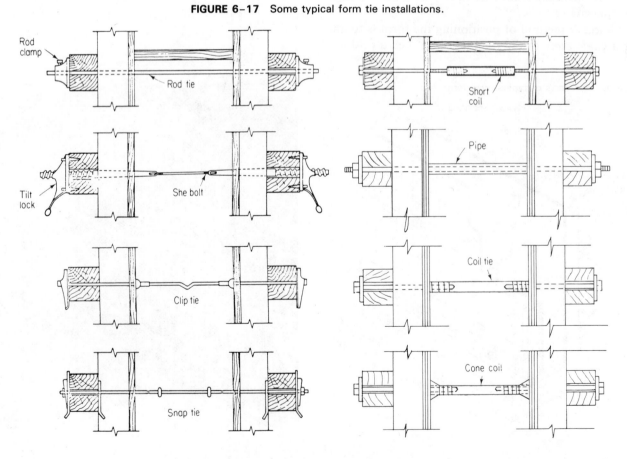

$2'' \times 4''$s. The $2'' \times 4''$s rest on brackets which are carried on the tie ends (see Fig. 6–19).

Walers are usually doubled for greater strength and spaced to allow the passage of tie ends between them. They may be made up of $2'' \times 4''$s or $2'' \times 6''$s with plywood spacers, or, for greater stiffness, a double waler may be made from two *standard channels*, placed back to back and spaced with bars welded to their flanges (see Fig. 6–20). When studs are used in form construction, walers are placed outside of them and held in place by nails, clips, or waler brackets nailed to the studs (see Fig.

6–21). When there are no studs, walers are placed against the plywood sheathing, as shown in Fig. 6–19. In such a case, *strongbacks*—vertical members tied together in pairs with long ties through the form—are set and braced to provide vertical rigidity (see Fig. 6–19).

Braces will consist of single or double—depending on the length—$2'' \times 4''$s or $2'' \times 6''$s, lengths of *pipe*, or small *S or W shapes*. They should extend from the walers or strongbacks to a solid support (see Fig. 6–18). That support may be in the form of a well-driven *stake* (see Fig.s 6–19 and 6–22), an *adjustable brace end* anchored to the floor (see Fig. 6–23), or a *rock anchor* grouted into a rock face (see Fig. 6–24).

Braces may act in compression only, in tension only, or in both—as when forms are braced on one side only. Heavy wire or cable is suitable for bracing which will be in tension only.

Plumbing is the final operation in building a wall form. Many bracing systems have some means of ad-

FIGURE 6–19 Form stiffener brackets in place. *(Courtesy Jahn Concrete Forming.)*

FIGURE 6–20 Double-channel waler.

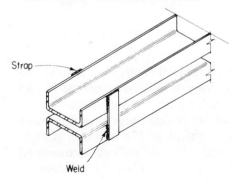

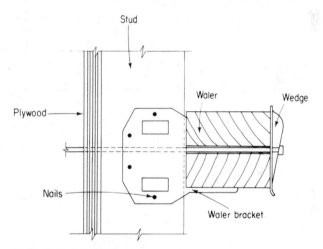

FIGURE 6–21 Waler bracket.

FIGURE 6–22 Outside adjustable brace.

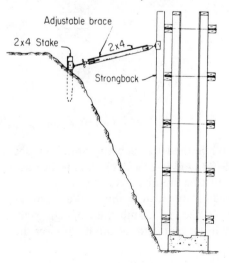

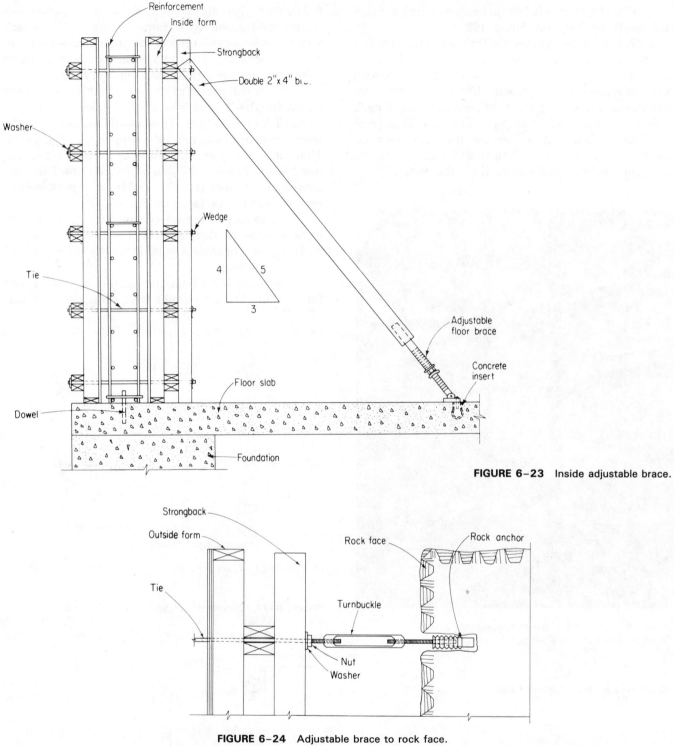

FIGURE 6-23 Inside adjustable brace.

FIGURE 6-24 Adjustable brace to rock face.

justing the braces while the form is being plumbed (see Figs. 6–15, 6–22, 6–23, and 6–24). If braces are not adjustable, the wall must be plumbed as the braces are installed and anchored. If one wall is plumbed as soon as it is built, there is no need to plumb the opposite one. The second form will be plumbed automatically by the ties and spreaders.

When steel reinforcement is placed in a wall form piece by piece, it will have to be done before the second form is erected. In such a case, the second wall can be built in a flat or inclined position and tilted into place after the reinforcing is complete. If reinforcing *mats* or *cages* are prefabricated and then set into a completed form, ties will have to be left out until this is done.

Finally, the ties are tightened. Where through-the-wall tying is impossible or not allowed, external bracing must be provided to securely support both walls.

Prefabricated Forms

A majority of the formwork being erected is done with the use of prefabricated form panels. These can be re-used many times and reduce the time and labor required for erecting forms on the site.

Many types of prefabricated form panels are in use. Contractors sometimes build their own panels from wood framing covered with plywood sheathing. A standard size is $2' \times 8'$, but panels can be sized to suit any particular situation. Figure 6–25 illustrates a typical $2' \times 8'$ panel with metal straps at the corners. Note the grooves in the outside edges to allow ties to pass through. A simple bolt and wedge device for holding adjacent panels tightly together is shown in Fig. 6–26.

Panels made with a metal frame and plywood sheathing are also in common use and are available in a variety of sizes. Special sections are produced to form inside corners, pilasters, etc. Panels are held together by patented panel clamps. Flat bar ties (which lock into place between panels) eliminate the need for spreaders. Forms

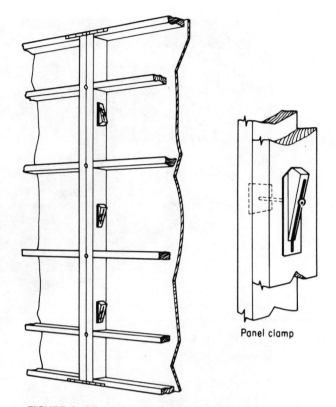

Panel clamp

FIGURE 6–26 Bolt-and-wedge panel clamp.

are aligned by using one or more doubled rows of $2'' \times 4''$s, secured to the forms by a special device which is attached to the bar ties (see Fig. 6–27).

Form panels made completely of steel are also available. $24'' \times 48''$ is a standard size, but various other sizes are also manufactured. Inside and outside corner sections are standard and insert angles allow odd-sized panels to be made up as desired.

Giant Panels and Gang Forms

High walls, in which the concrete will have to be placed in two or more stages or *lifts*, will normally be formed by the use of *giant panels*—panels much larger than the normal $2' \times 8'$ or $4' \times 8'$ wood panels—or by *gang forming*—making up large panels by fastening together a number of stell-framed panels (see Fig. 6–28). A special type of steel-framed panel, shown in Fig. 6–29, is made specifically for gang forming.

These large forms are built or assembled on the ground and raised into place by crane. The basic differences between them and regular form panels are their *size*, the *extra bracing* that is required to withstand handling, and the fact that they require *lifting hardware*. Manufacturers of steel-framed panels also supply lifting brackets which are attached to the top edge of the gang form, but wooden panels have to be fitted with lifting hardware as illustrated in Fig. 6–30.

FIGURE 6–25 Typical prefabricated wooden form panel.

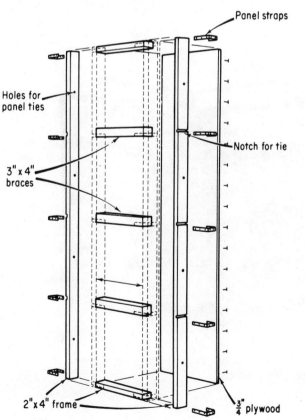

Panel straps

Holes for panel ties

Notch for tie

$3'' \times 4''$ braces

$2'' \times 4''$ frame

$\frac{3}{4}''$ plywood

(a)

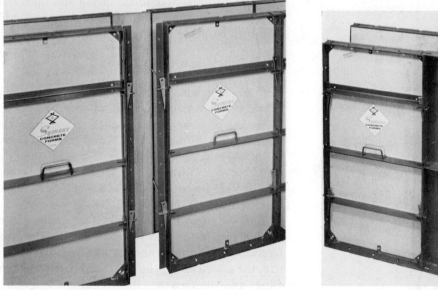

(b)   (c)

FIGURE 6-27 Typical steel frame form panels.

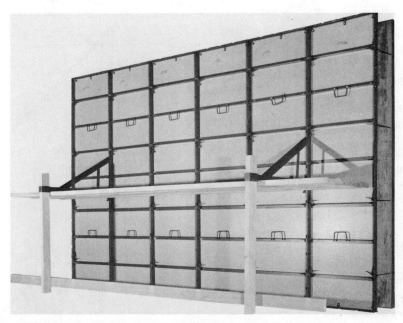

FIGURE 6–28 Wall form section made from gang-ed panels.

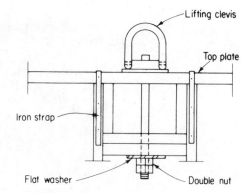

FIGURE 6–30 Panel lifting hardware.

FIGURE 6–29 Panel sections made for gang forming.

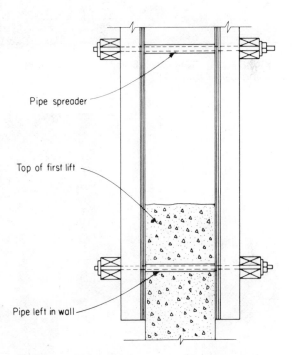

FIGURE 6–31 Form clamped to top of first lift.

These forms are built with removable ties so that when the first lift of concrete has been placed and has gained sufficient strength, the bolt ties are removed and the form lifted a new position at the top of the first lift. It is held in place by clamping with bolts through the top one or two rows of bolt holes left in the first lift (see Fig. 6–31). In the illustration shown in Fig. 6–32, the first lift has been completed and the forms lifted and bolted in place, ready for placing of the next lift.

Special attention must be given to corners when forms are being erected. These are weak points because the continuity of sheathing and wales is broken. Forms must be pulled tightly together at these points to prevent leakage of concrete. One method of doing this is illustrated in Fig. 6–33. Another system involves the use of corner brackets and tie rods, as shown in Fig. 6–34. Vertical rods running through a series of metal eyes fastened to the sheathing panels are also used to secure forms at the corners (see Fig. 6–35). Another way of making a tight corner is to overlap the wales at the corner and to provide a vertical kick strip for each group of wale ends. Wedges behind each wale tighten the corner.

FIGURE 6-32 Giant panels bolted in place. (*Courtesy Jahn Concrete Forming.*)

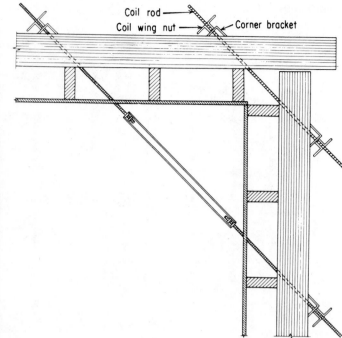

FIGURE 6-34 Corner brackets and tie rods.

FIGURE 6-33 Wedge corner clamps. (*Courtesy Jahn Concrete Forming.*)

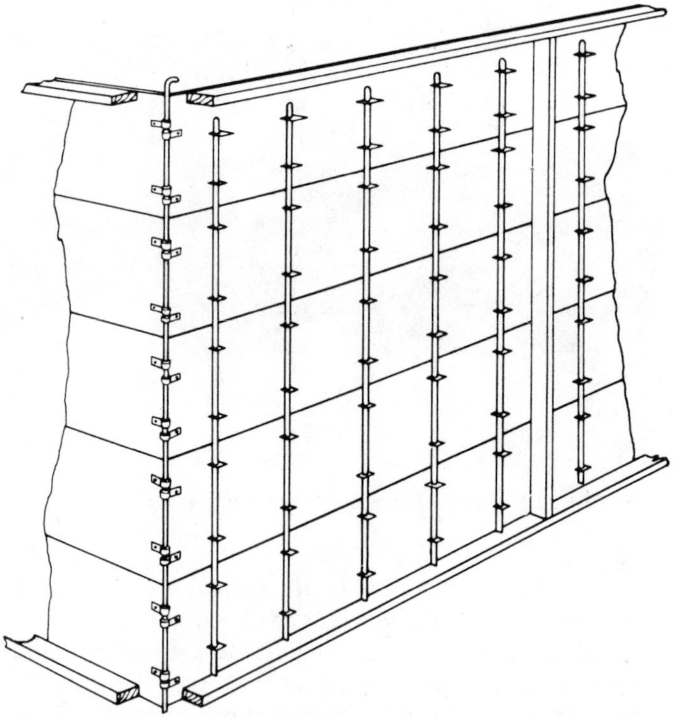

FIGURE 6-35 Corner rods and eyes.

Openings in concrete walls are made by setting *bucks* (see Fig. 6-36) into the form at the required position. Bucks are simply wooden or steel frames which are set between the inner and outer forms and held in place so that the forms can be tightened against them. Bucks should be braced horizontally, vertically, and diagonally so that concrete pressure will not distort their shape. In some cases, the window or doorframe itself is set into the form. Care must be taken that proper anchorage is provided so that the frame will not move in the opening.

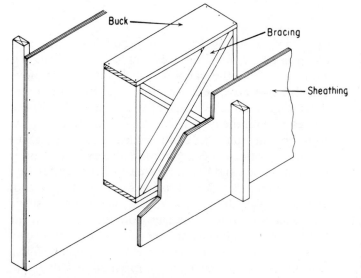

FIGURE 6-36 Simple wood buck.

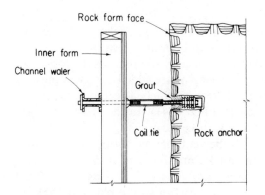

FIGURE 6-37 Wall formed with single form.

FIGURE 6-38 Slip-forming multiple silos. (*Courtesy MacDonald Engineering.*)

SLIP FORMS

Slip forms are a type of form which could be compared to a set of dies, originally designed for curved structures such as bins, silos, and towers (see Figs. 6-38 to 6-40), for which the conventional forming systems were not suited. However, they are finding increasing use for a variety of structures, including rectangular buildings, bridge piers, canal linings, etc.

They consist basically of an inner and an outer form, 3 to 5 ft in height, fabricated from wood or steel to produce the building shape desired, and supported by two strong, vertical *yokes* (see Fig. 6-41). These yokes are tied together across the top to give the form sides the rigidity needed—to apply the pressure required—to produce reasonably smooth surfaces. The bottom ends of the form sheathing are slightly tapered in order to help make the forms self-clearing.

To provide a platform from which men may look after the placing of concrete in the forms, the fabrication of the steel reinforcement, the extension of the jack rods, and the maintenance of the jacks, a working platform is attached to the inner form and rides upward with it. At the same time a finisher's scaffold is suspended from the outer form so that workmen can finish the newly extruded concrete as it emerges.

Instead of remaining stationary as normal forms do, slip forms move continuously upward, drawn by *jacks* climbing on vertical steel *jack rods*. These are anchored at the base of the structure and embedded in the concrete below the forms. The jacks may be hydraulic, electric,

FIGURE 6–39 Slip-forming downtown tower. (*Courtesy of British Lift Slab Ltd.*)

FIGURE 6–40 Slip-forming core of London Stock Exchange Building. (*Courtesy British Lift Slab Ltd.*)

or pneumatic and are capable of producing form speeds of up to 20 in./hr. If the jack rod is to be reused, it is withdrawn from the wall after the forming is complete. This is made possible by sheathing the rod with a thin pipe, which is attached at its top end to the jack base and moves up with the forms. The sheath prevents concrete from bonding to the jack rod and leaves it standing free within the hardened concrete. In some cases the rod is left unsheathed and remains as part of the reinforcing.

Concrete is placed into the forms at the top end, and, as they are drawn upward, they act as dies to shape the concrete so that it emerges from the bottom of the form shaped as planned and set enough to carry its own weight. Thus, a continuous process is carried on, filling and moving the forms upward, often 24 hr a day until the structure is complete. However, the operation may be stopped and resumed later, resulting in the same kind of concrete joint which one would get between lifts in conventional form use.

Vertical reinforcing rods are anchored at the base of the structure and extend upward between the inner and outer form. As the form rises and reaches the top of the first set of rods, new lengths are added as concreting continues. Horizontal steel is attached to the vertical rods as work goes on (see Fig. 6–41). In the same way, jack rods are extended as necessary.

Openings in the wall may be formed by introducing neatly fitting bucks into the form at the proper location. Beam pockets, key slots with dowels, and anchor slots for attaching masonry can also be placed in the same way. If a concrete projection from the wall is required, it must be added after the forming is complete. A pocket is formed in the wall with dowels bent in so as not to interfere with the operation of the forms. After the forming is complete, the dowels can be bent out, the forms for the projection built around them, and the structure cast.

The success of a slip-forming operation depends on good planning, design, and supervision so that the operation may, in fact, be as continuous as possible. Some of the major factors contributing to successful slip form construction are:

1. The proper concrete mix design and careful control of the concrete to maintain the proper slump and set, in spite of changing temperatures.
2. Adequate facilities for supplying concrete to the forms at any height and an adequate concrete supply.
3. A supply of reinforcing steel at hand and experienced workers to do the fabricating as work progresses.
4. Reliable forms, designed to stand the stresses placed on them by the constant head of liquid concrete.
5. Supervisors who thoroughly understand the operation of slip forms.

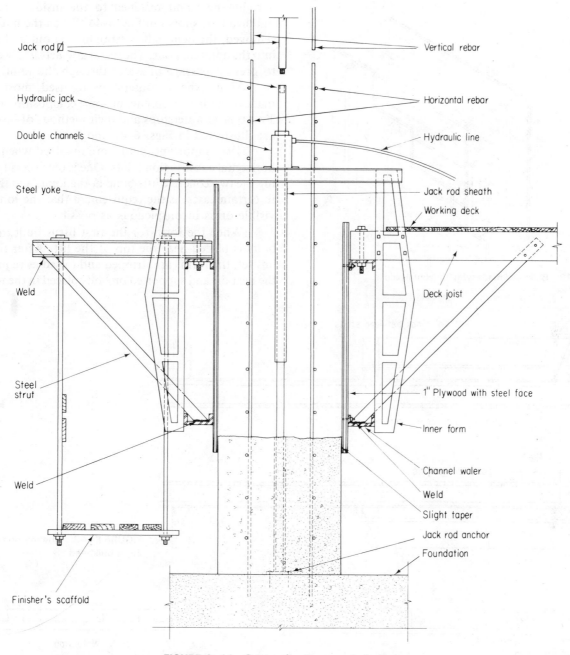

Jack rod Ø

Hydraulic jack

Double channels

Steel yoke

Weld

Steel
strut

Weld

Finisher's scaffold

Vertical rebar

Horizontal rebar

Hydraulic line

Jack rod sheath

Working deck

Deck joist

1" Plywood with steel face

Inner form

Channel waler

Weld

Slight taper

Jack rod anchor

Foundation

FIGURE 6–41   Schematic diagram of slip form.

CONSTRUCTION JOINTS

It is often necessary to place long or high walls in sections. This results in a vertical joint (where two sections of a long wall meet) or in a horizontal joint (when high walls are placed in two or more lifts). Such joints are called *construction* joints and require special attention.

Vertical construction joints are formed by placing a *bulkhead* in the form. It may be one piece of material placed vertically in the form or a number of short pieces placed horizontally. The latter is preferable when horizon-

tal reinforcing bars project beyond the bulkhead, because the pieces can be cut to fit around the bars. In Fig. 6–42, the bulkhead is held in place by vertical strips nailed to the inside faces of the forms. Another method of securing the bulkhead is shown in Fig. 6–43.

It is usually desirable to provide a key between adjoining sections to prevent lateral movement. One method of doing this is to attach a tapered strip to the inside face of the bulkhead, as shown in Fig. 6–43. Another type of key consists of a 6-in. strip of corrugated copper or galvanized iron to bridge the joint. The strip is bent into

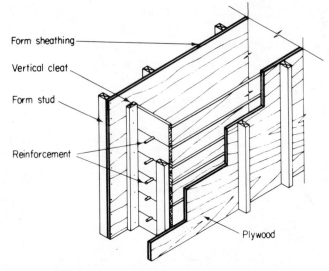

FIGURE 6-42 Bulkhead made with horizontal pieces.

a right angle and fastened to the inside face of the bulkhead, as shown in Fig. 6-44. When the bulkhead is removed, the bent half is straightened out and projects into the adjacent pour. The metal also acts as a *waterstop*, to prevent passage of water through the joint.

Many types of waterstops are used, most of them made of rubber, neoprene, or some composition material. Two typical waterstops and their methods of installation are illustrated in Figs. 6-45 and 6-46.

Two important factors are involved when making horizontal construction joints. One is the necessity of having the two concrete lifts bond at the joint, and the other, in certain cases, is the requirement that the joint be invisible or as inconspicuous as possible.

When the form for the first lift is built, a row of bolts is placed near the top of the lift. After this pour has set, the forms are stripped and raised as required for the next lift and supported on bolts placed in the top holes

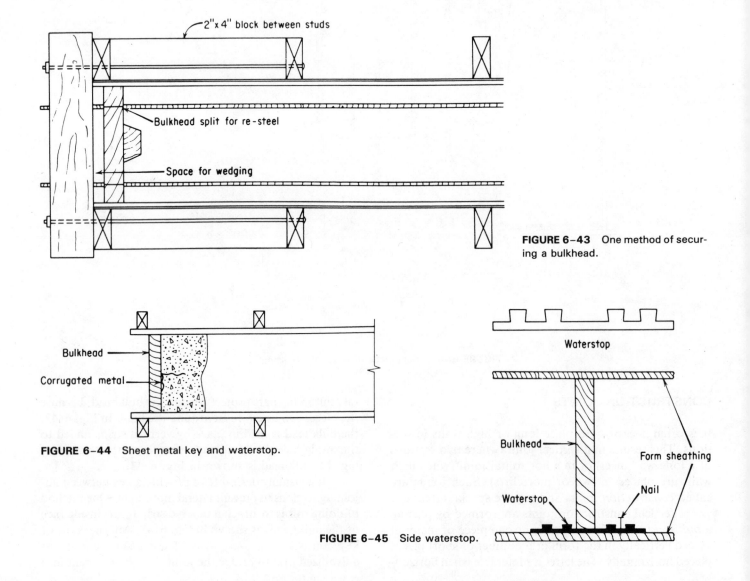

FIGURE 6-43 One method of securing a bulkhead.

FIGURE 6-44 Sheet metal key and waterstop.

FIGURE 6-45 Side waterstop.

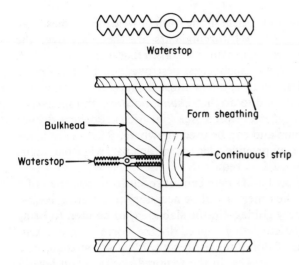

FIGURE 6-46   Center waterstop.

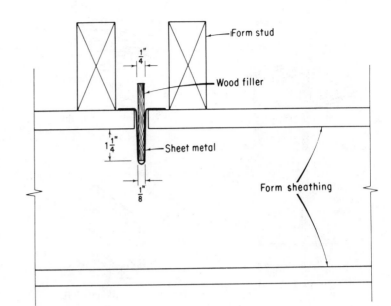

FIGURE 6-48   Control joint form.

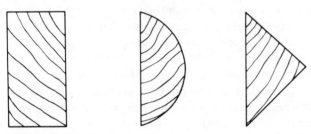

FIGURE 6-47   Rustication strips.

previously cast. A row of ties should be placed about 6 in. above the joint between the lifts to prevent leakage and help make the joint inconspicuous.

Hiding of horizontal construction joints is often accomplished by the use of *rustication strips*. These are narrow strips of various shapes (see Fig. 6-47) nailed to the inside face of the form at the level of the joint. Concrete is placed to a height slightly above that required and then cut back to it.

CONTROL JOINTS

Control joints in concrete are those which are made to control cracking by expansion and contraction. They are formed by fastening a beveled insert of wood, metal, rubber, etc., to the inside face of the form. The insert produces a groove in the concrete which will control surface cracking. After the concrete has set, the insert may be removed and the joint can be caulked. Rubber inserts may be left in place. Figure 6-48 illustrates the forming of control joints. When a waterstop is required at a control joint, a rubber stop like that shown in Fig. 6-46 is often used.

COLUMN, GIRDER, AND BEAM FORMS

See Chapter 9.

FLOOR FORMS

The design of forms for concrete floors depends a great deal on whether the floor is a *slab-on-grade* or a *deck* supported on a steel or concrete structural frame.

Slab-on-Grade Forms

Forms for concrete slabs placed on grade are usually quite simple. Concrete is cast on a compacted earth or gravel base, and forms are required only for the edges. Plywood, planks, or steel edge forms are commonly used for this purpose. Wood forms may be held in place by wooden stakes or by form supports such as the one illustrated in Fig. 6-49(a). In the latter case, the subgrade stake is pushed firmly into the earth and the bolt bracket mounted on the form plank. The plank can then be adjusted for height by turning the bolt up or down. Steel edge forms (see Fig. 6-50) are commonly used on larger jobs and for highway work. Provision is made for the wheels of mechanical strike-off bars and floats to ride on the top flange of steel edge forms.

Polyethylene sheets are usually laid over the compacted base to prevent water in the concrete from being absorbed by the earth and, eventually, to act as a moisture barrier for the slab.

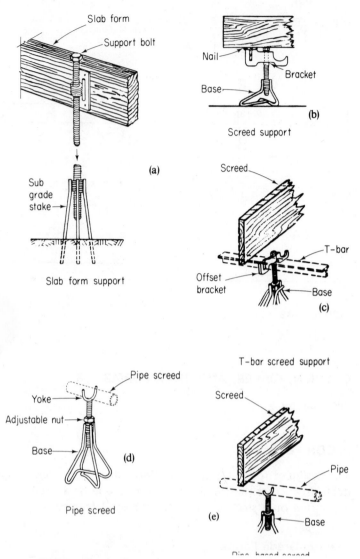

FIGURE 6-49 Typical slab form and screed supports.

Reinforcement in the form of welded wire mesh (see Fig. 6–51) or reinforcing bars are then laid over the polyethylene. Care must be taken that it is not torn and, as a result, its effectiveness destroyed, during this operation. The reinforcement must be supported above the subgrade, and to do this, *chairs, bolsters,* and *spacers* of various types are used (see Fig. 6–52). Bolsters do not space bars and can be used anywhere. Spacers are made to order, to provide a leg under each bar, while bar chairs may be used as required.

If strike-off bars cannot reach from one edge of a slab to the other, it will be necessary to set up *screeds*—temporary guides—in the slab area, to be used to bring the concrete being placed to the correct grade. One method of doing this is to set a $2'' \times 4''$ on edge, held in place by stakes, in the required position and level it to the same height as the edge forms (see Fig. 6–53). Wood screeds may also be supported by *adjustable screed supports*, such as the one shown in Fig. 6–49(b). Straight pipe may be used as a screed or small pipe or T-bars on adjustable supports used to support wooden screeds. When concrete has been placed to the correct level, the screed is removed and the depression filled.

If the slab is to be placed in sections, *construction* joints must be made between them which will transmit shear from one to the other. Forms for construction joints are illustrated in Fig. 6–54.

Forms for *control joints*—joints to control the location of cracks which inevitably occur in a concrete slab as a result of the shrinking of the concrete—may be made in several ways. One common method is to insert a wood or metal strip into the slab, top or bottom, to form a plane of weakness. Another method is to make a cut part way through the slab with a masonry blade, as shown in Fig. 6–55. Still another method is to set two structural

FIGURE 6–50 Steel slab edge form.

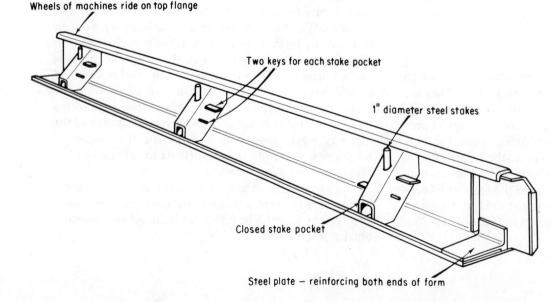

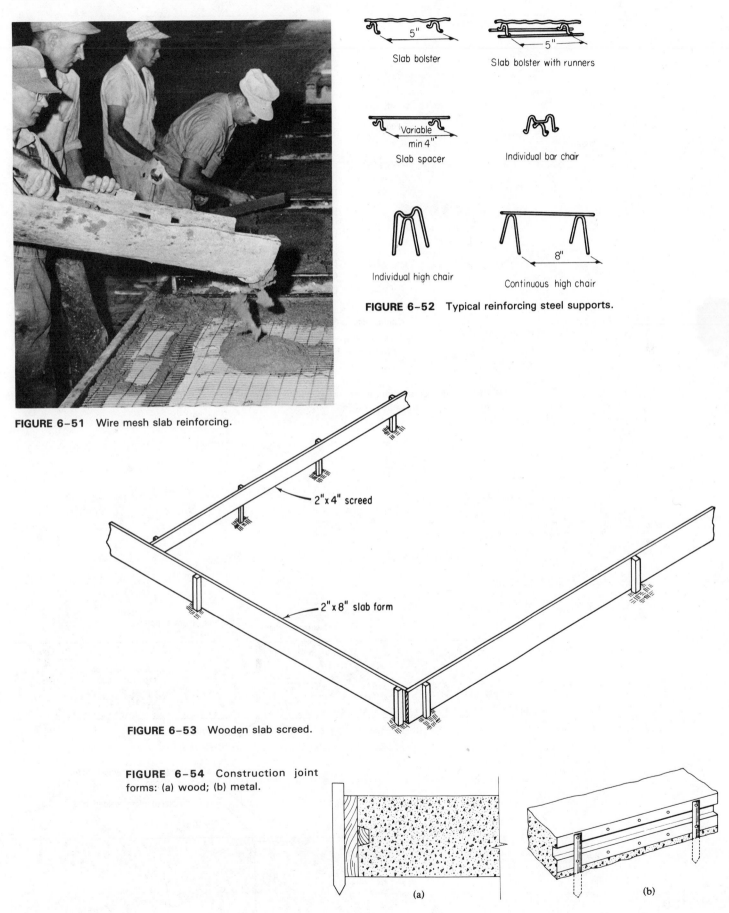

FIGURE 6-51 Wire mesh slab reinforcing.

Slab bolster

5"

Slab bolster with runners

5"

Variable min 4"

Slab spacer

Individual bar chair

Individual high chair

Continuous high chair

8"

FIGURE 6-52 Typical reinforcing steel supports.

2"x 4" screed

2"x 8" slab form

FIGURE 6-53 Wooden slab screed.

FIGURE 6-54 Construction joint forms: (a) wood; (b) metal.

(a)

(b)

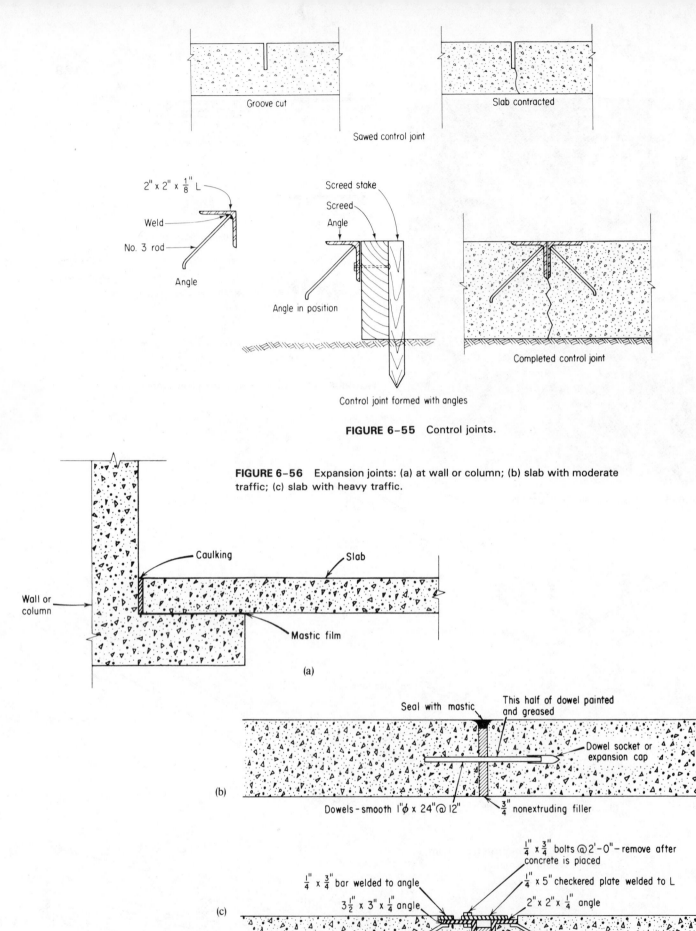

FIGURE 6-55 Control joints.

FIGURE 6-56 Expansion joints: (a) at wall or column; (b) slab with moderate traffic; (c) slab with heavy traffic.

steel angles into the slab surface so that a plane of weakness is established below them.

Expansion joints are necessary in slab construction to provide space for the slab to expand, due to changes in temperature, without exerting damaging pressure on the member adjacent to it. They may be formed around exterior walls, columns, and machine bases by placing a tapered wood strip around the perimeter before concrete is placed, removing it after the concrete has set, and filling the void with some type of caulking material. The same result may be accomplished by using a strip of compressible material—fiberboard, for example—and leaving it in position. Joints similar to those shown in Fig. 6-56(b) or (c) should be used in slabs where the expansion joint will be subjected to heavy traffic.

Deck Forms

Forms for concrete *decks*—floors above grade—supported by a structural frame will be as varied as the systems which support them. In some cases, such as a cast-in-place reinforced concrete frame, columns, beams, and slabs will be cast monolithically, and forms must be designed accordingly. In the case of a precast concrete frame, deck forms may be necessary in some cases, depending on the type of members used. For example, single or double tees provide the floor slab as well as the frame. A structural steel frame will require deck forms for the floors.

Beam-and-Slab Floor Forms

Forms for beam-and-slab floors incorporate the forms for the *spandrel beam*, and regular *girder* and *beam* forms, and the *slab* forms. Support systems for these forms include wood *T-posts*, adjustable *metal posts,* and supports made with *metal scaffolding* (see Fig. 6-57).

Flat Slab Floor Form

With this type of floor system (see Fig. 11-25), in which there are no beams and no drop panels, the first storey columns are usually cast first. Then the slab forms are set up for the floor which they will support (see Fig. 6-58). When that floor is ready, the operation is repeated for successive floors.

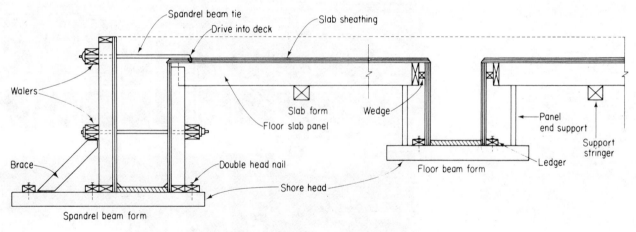

FIGURE 6-57 Beam-and-slab floor beam.

FIGURE 6-58 Flat plate floor form.

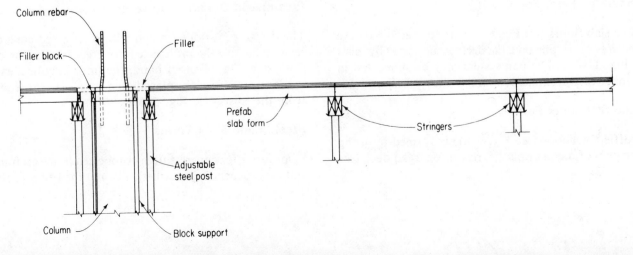

FIGURE 6-59   Ribbed slab floor forms.

FIGURE 6-60   Supports for ribbed slab floor.
(*Courtesy PCM Division, Koehring Company.*)

Ribbed Slab Floor Form

Ribbed slab floors (see Fig. 6–59) are formed by the use of *metal pans*, resting on a flat deck or supported by joists (see Fig. 11–29). The pan system may be supported in a number of ways, one of which is shown in Fig. 6–60.

Waffle Slab Floor Form

A waffle slab floor (see Fig. 6–61) is formed by using *dome pans* of metal or plastic, resting on a flat deck (see Fig. 11–33).

Prestressed Concrete Joist Floor Form

The design of a floor framed with prestressed concrete joists may specify prestressed concrete planks for the floor slab. On the other hand, the floor may be a cast-in-place concrete slab, placed on a slab form suspended from the joists (see Fig. 6–62).

Steel Frame Floor Forms

Floor forms for concrete floors supported by a steel frame will vary, depending on whether or not the fire-proofing

FIGURE 6-61 Waffle slab floor forms.

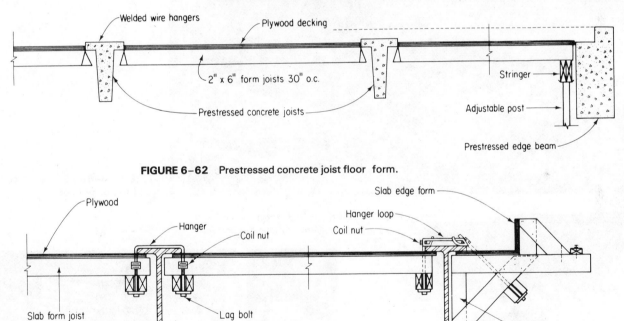

Welded wire hangers — Plywood decking

2" x 6" form joists 30" o.c.

Prestressed concrete joists

Stringer

Adjustable post

Prestressed edge beam

FIGURE 6-62 Prestressed concrete joist floor form.

Plywood

Hanger — Coil nut

Slab edge form

Hanger loop

Coil nut

Slab form joist

Lag bolt

W shape — Spandrel beam

Joist support

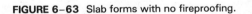

FIGURE 6-63 Slab forms with no fireproofing.

for the steel beams is to be cast monolithically with the slab (see Fig. 6-63). In either case the slab forms will usually be suspended from the steel frame, eliminating the necessity for a shoring system (see Fig. 6-64).

Flying Forms

In many of the present-day high-rise, cast-in-place concrete buildings, construction is identical from bay to bay and from floor to floor. The buildings consist essential-

ly of either a series of columns supporting flat plate floors (see Fig. 6-65) or a number of shear walls doing the same thing (see Fig. 6-68). A floor slab form and its support system, designed for one floor, can be reused for all the others, the main problem being the moving of the forms from one position to another. This problem has been solved by the use of *flying forms*—forms designed as a complete unit, large enough that the entire floor of one bay may be cast on it and capable of being moved as a unit from one position to another.

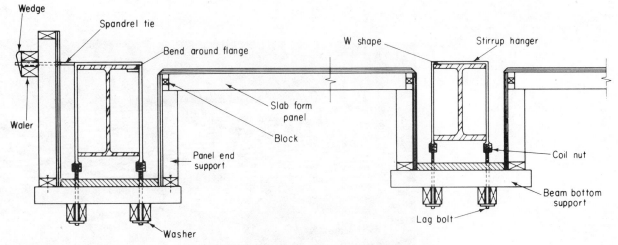

FIGURE 6-64 Slab forms with fireproofing.

FIGURE 6-65 Series of columns and flat plate floors.

FIGURE 6-66 Flying form frame and jacks. (*Courtesy Aluma Building Systems Inc.*)

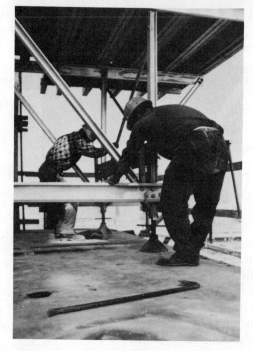

FIGURE 6-67 Forms on rollers, ready for moving. (*Courtesy Aluma Building Systems Inc.*)

(a)

(c)

(b)

(d)

FIGURE 6-68 "Flying" a form: (a) rolling out of bay; (b) suspended, clear of bay; (c) moving up; (d) being guided to new bay; (e) closing in to new bay. (*Courtesy Aluma Building Systems Inc.*)

Forms are made up of a metal frame—steel or aluminum—supporting a wood or metal deck and resting on screw jacks, as shown in Fig. 6-66. The jacks allow the form to be adjusted to exactly the right level for placing the slab, and when it has gained sufficient strength, the form can be lowered away from the slab by turning down the jacks.

The form is then lowered onto rollers (see Fig. 6-67), the jacks folded up into the *flying position*, and the form rolled out of the bay to where it can be hooked to a crane, which will finish moving it out of the bay and swing it into a new position. There it will be used again and moved again, as many times as there are bays in the structure which that form will accommodate. Figure 6-68 shows a sequence of operations in *flying* a form from one location to the next.

(e)

Shoring Systems

Shoring members—members used to support concrete forms and their contents or other structural elements—may be divided into two major categories: *horizontal shoring* and *vertical shoring.* Horizontal shores range from small units such as those which might be used to span between pairs of precast concrete joists to support the plywood deck form to large wood or steel members, with relatively few supports, used to carry much heavier loads with a minimum of interference in the work area below.

Vertical shores are those which support the horizontal ones from a firm base below, such as a concrete slab or a *mudsill,* if there is no solid bearing on which the shore may rest (see Fig. 6–69). They may be of wood or metal and take many different shapes, depending on the particular circumstances.

Vertical wood shores may be *single wood posts,* with wedges at the bottom to adjust the height, *double wood posts,* two-piece *adjustable posts,* or T-head shores. Figure 6–69 illustrates some typical wood shores.

Vertical metal shores may be adjustable pipe shores or shores made up of prefabricated metal scaffolding. Scaffold-type shoring is usually assembled into *towers* by combining a number of units into a single shoring structure (see Fig. 6–70). A typical adjustable pipe shore is illustrated in Fig. 6–71.

Regardless of the type of shoring system to be used, it should be carefully planned in advance by someone who understands the loads and stresses involved. Failure to properly take into account not only the structural loads

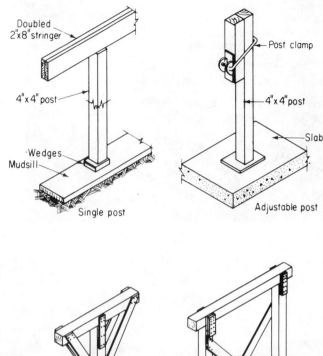

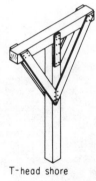
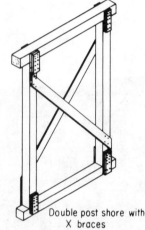

FIGURE 6–69 Typical wood shores.

FIGURE 6–70 Scaffold-type shoring in use. (*Courtesy Sarnia Scaffolds Ltd.*)

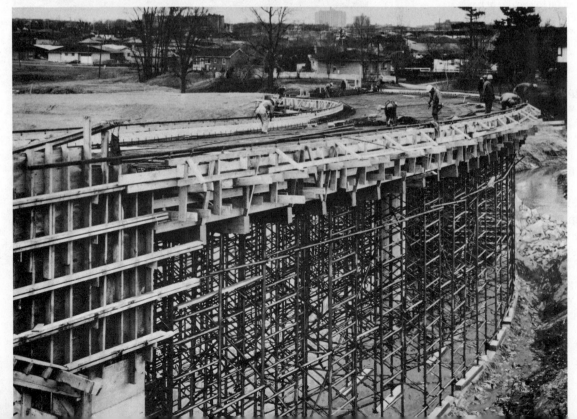

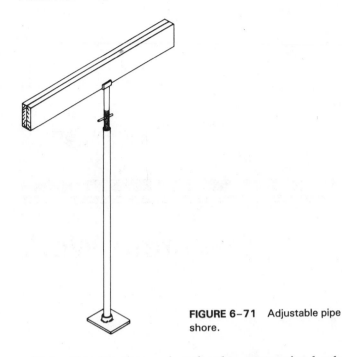

FIGURE 6-71 Adjustable pipe shore.

sisting of liquids which are to be brushed or sprayed on the form. Wooden forms must be treated to minimize absorption of water. Oil is one material used for this purpose. Form sealers which coat the surface of the form with an impervious film are also used for this type of treatment.

Form removal must be carried out without damaging either the forms or the structure being stripped. Levers should not be used against the concrete to pry forms away because green concrete is relatively easy to damage. If levering action is necessary, pressure should be applied against a broad, solid base. Form panels being removed from a wall section should be pulled straight off over the tie rods.

The production of concrete structures of desired size, shape, and strength—yet pleasing to the eye—can be achieved if a number of important points are remembered. Forms must be properly *designed,* carefully *constructed,* and *erected.* They must receive the appropriate *surface treatment. Tying, bracing,* and *shoring* must be adequate for the particular job involved. There must be proper *handling, placing, consolidation,* and *curing* of the concrete, and *form removal* must be carried out in such a way that no damage is incurred by the finished product.

which will be in existence but also the construction loads which may be imposed may result in shoring failure, which is not only very costly but potentially dangerous as well.

FORM ACCESSORIES

A vast array of products is currently available to aid in making forms stronger and erecting them faster. These products include items which have already been mentioned in this chapter, such as *ties, spreaders, wedges, corner brackets, waler clips,* etc. They also include *snap ties* for special conditions, *rock anchors, beam hangers, waterstops* and *keys, masonry ties, column clamps, shores, form rods, concrete inserts, sleeves, slab pans,* and many others. Detailed information on the uses of all these may be obtained from brochures and catalogues published by their manufacturers.

FORM TREATMENT, CARE, AND REMOVAL

In nearly all types of building construction, formwork constitutes a significant part of the cost of the building. To keep this cost at a minimum, forms are often made reusable, either wholly or in part. They must therefore be designed so that removal is simple and can be accomplished without damage to the form sections. Care must be taken in handling and storing these units so they will not be broken or damaged and will be available for reuse.

To facilitate removal, form faces must be treated to prevent concrete from adhering to them. A number of materials are available for this purpose, usually con-

REVIEW QUESTIONS

1. What seven basic features must forms have to fulfill their function?

2. Give three reasons for not including the earth as part of a form.

3. What is meant by *rate of pour?*

4. List three factors used in determining the rate of pour.

5. List the four main factors determining the pressure on forms.

6. Give two basic reasons for treating forms before using them.

7. Give three common methods of providing support against lateral pressure to square or rectangular forms.

8. Differentiate between a *construction joint* and an *expansion joint.*

9. List four factors which determine when forms may be removed.

10. Define the following: **(a)** form clearance, **(b)** form accessories, **(c)** prefabricated forms, **(d)** slip form, **(e)** form buck, and **(f)** flying form.

11. Calculate the maximum pressure developed when a 10-in.-thick wall 10 ft high is to be poured at the rate of 3 ft/hr when the temperature is 70°F. Check your result with Table 6-2.

Concrete Work

The extensive use of concrete is ample proof of its outstanding characteristics as a building material. It is such a familiar material that we take for granted the remarkable process by which cement and water, mixed with a wide range of aggregate materials, are converted into a strong and durable material of almost any desired shape. The development of modern Portland cements which can set quickly, even under water, began over one hundred years ago when the raw materials and processes needed for their manufacture were first recognized. Today, hundreds of scientists, engineers, and technicians are engaged in studying these materials, attempting to understand and to improve them still further.

Many of the physical and chemical reactions which take place during the setting and aging of concrete are so complicated that they are not yet fully understood. This is due in part to the wide range of chemical substances that can exist, partly by design and partly by chance, in any given concrete mix. Additional factors may be introduced in the methods of manufacturer, handling, and curing at the site. All the changes that take place relatively rapidly in the new concrete do not cease at the end of the formal curing period. Some may continue over a long period of time, and some may only begin in the environment to which the concrete is subsequently exposed. Despite all these complications, concrete of predictable properties and performance can be produced— and this does not occur by chance.

It is, fortunately, unnecessary for the designer, specification writer, job engineer, or supplier to keep in touch with the whole field of concrete technology. There are a number of guides to good practice in the form of codes, standards, and specifications from sources such as the Canadian Standards Association, National Building Code, American Society for Testing Materials, and the American Concrete Institute. Nevertheless, it is highly desirable to have some idea of the general nature of the material and its more important properties.

AGGREGATES FOR CONCRETE

Concrete is a product of the combination of portland cement and water paste with some type of aggregate. The paste surrounds the aggregate particles and as it sets— returns to a limestone-like state—binds them solidly together. At the same time, to form a dense concrete, the paste must fill the voids between the particles of aggregate (see Fig. 7-1).

FIGURE 7-1 Voids between aggregate particles.

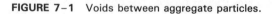
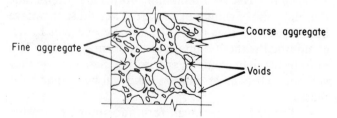

It is desirable to use a maximum amount of aggregate when making concrete of any given strength for a number of reasons. In the first place, cement is up to 10 times more costly than aggregates. Second, cement paste tends to shrink while setting and curing, and consequently the use of more paste than is necessary results in a more expensive concrete and also produces excessive shrinking.

Aggregates fall generally into four classifications. *Sand, gravel, crushed stone* (natural aggregates), and *air-cooled blast furnace slag* are used to produce **normal-weight concrete** (135 to 160 lb/cu ft). *Expanded shale, slate, clay,* and *slag* produce structural **lightweight concrete** (85 to 115 lb/cu ft). Lightweight materials such as *vermiculite, perlite, pumice, scoria,* and *diatomite* are used to produce **insulating concrete** (15 to 90 lb/cu ft), while such heavy materials as *hematite, barite, limonite, magnetite, steel punchings,* and *shot* are used to produce **heavyweight concrete** (180 to 380 lb/cu ft).

The amount of aggregates used in making concrete usually occupies 60 to 80% of the total volume, and, as a result, the kind and quality of the aggregates have a major influence on the properties of concrete. It is therefore important that we know as much as possible about the aggregates being used in order to be able to produce concrete to required specifications.

Natural aggregates are divided into two classes: *fine aggregate* (F.A.) and *coarse aggregate* (C.A.). Fine aggregate includes material which does not exceed ¼ in. in diameter, while coarse aggregate normally ranges from ⅜ to 3 in., though sizes up to 6 in. are sometimes used. Both require study and testing to make sure that they qualify as good material for concrete. It is common practice to carry out tests on both to ensure *soundness, cleanliness* (the presence of excessive amounts of silt and/or fine organic matter), *proper gradation* (distribution of sizes in a given sample) and to determine *specific gravity, moisture content, particle shape,* and *surface texture.*

Soundness

Concrete can only be as strong as the aggregate from which it is made, and aggregate, particularly coarse aggregate, should be inspected to see that it does not contain weak or laminated particles, soft and porous particles, or bits of shale. Aggregate soundness may also be determined by its abrasion resistance, of particular importance when the concrete is to be used for heavy-duty floors. The abrasion resistance test is outlined in ASTM C131, in which a quantity of the material is rotated in a steel drum and then measured to ascertain the amount of material worn away during the test.

Cleanliness

Silt clinging to the surface of course aggregates prevents proper bonding of the cement paste. In fine aggregate, the presence of too much fine material means a significant increase in the surface area to be covered with paste, resulting in either having to use more paste or the thinning out of that in use.

Silt Test

The presence of silt in coarse aggregates can usually be detected visually, but in the case of fine aggregate, a *silt test* is used to determine the amount of very fine material present. To make a silt test, proceed as follows:

Place 2 in. of the fine aggregate to be tested in a standard 1 qt glass jar. Fill the jar ¾ full of water, replace the cover, and shake well. Let the jar stand for several hours until the water has cleared. Upon settling, the fine silt will form a layer on top of the sand. If this layer exceeds ⅛ in. in depth, the aggregate is unfit for making good concrete unless part of this material is removed.

Colorimetric Test

The presence of small amounts of fine organic matter in a cement paste retards its setting action, sometimes so much so that a mix will dry out before setting properly. To test for the presence of harmful amounts of this organic matter in fine aggregate, a *colorimetric test* is used. The test is as follows:

Place 4 oz of the aggregate to be tested into a small jar or bottle. Then fill to 7½ oz with a 3% solution of sodium hydroxide (NaOH). Shake well and allow to stand for 24 hr. The presence of organic matter will cause color changes in the clear solution varying from a very light straw color to a deep chocolate brown. If the color ranges from light to dark straw, the aggregate is suitable. If, on the other hand, the color is darker than this, too much organic matter is present and the aggregate is unsuitable for concrete work.

Gradation

The proper gradation of particle size in the aggregate used to make concrete is essential for a number of reasons. It is apparent that the fewer the number of pieces required to fill a given volume, the smaller their total surface area will be. This can be demonstrated with cubes or spheres, as in Fig. 7–2. Since all these particles must be coated with cement paste, the smaller the surface area, the less paste is required. In other words, the greater the maximum size of the aggregate, within limits, the more economical the mix becomes.

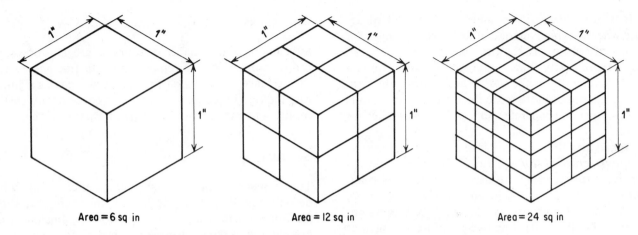

FIGURE 7-2   As particle size decreases per given volume, area increases.

But with a given quantity of aggregate in which all the particles are the same size, it is evident that no matter how these are arranged together, there will always be spaces between them which cannot be filled (see Fig. 7–3). It can also be demonstrated that, regardless of the size of the particle chosen, the amount of voids will remain constant, as shown in Fig. 7–4. If an aggregate of this type was used to make concrete, those spaces would have to be filled with cement paste, necessitating the use of a great deal more than would be required simply to coat the particles and bond them together.

However, if smaller pieces of material are introduced, as illustrated in Fig. 7–4, a considerable portion of those spaces would be filled with aggregate. When still smaller particles are added, more spaces are filled. This process can be continued until all the available particle sizes have been utilized. In other words, the proper gradation of particles, from the largest to the smallest, will effect an economy of paste (see Fig. 7–5).

As previously mentioned, cement paste has a tendency to shrink when it sets. If the volumes of paste are relatively large, as when the particles are all the same size, this shrinkage could result in the breaking of the bond as the paste shrinks away from the aggregates, thus weakening the concrete and also leaving passages through which water could travel. Proper gradation can therefore affect watertightness and strength of concrete. For all these reasons, aggregates are tested to make sure that

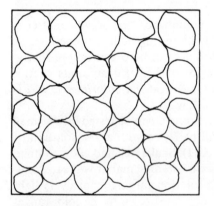

FIGURE 7-3   Unavoidable spaces when particle size is uniform.

FIGURE 7-4   Reducing voids by improving gradation.

1-in. particles $\frac{3}{8}$-in. particles   1-in. and $\frac{3}{8}$-in. particles

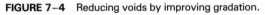

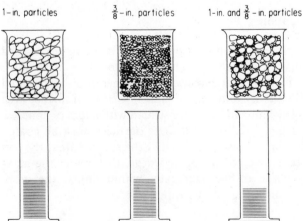

FIGURE 7-5   Well-graded aggregate. (Courtesy Portland Cement Association)

there is a proper gradation of sizes. This test is called a *fineness modulus* test and may be carried out on any aggregate.

Fineness Modulus Test

When making a fineness modulus test on fine aggregate, a set of six standard wire screens is used, including a #4 sieve, having four wires per lineal inch or 16 openings/sq in.; a #8 sieve, with 64 openings/sq in.; a #16; a #30; a #50; a #100; and a pan to retain everything passing through the #100 sieve.

These sieves are stacked as shown in Fig. 7–6 so that aggregate placed in the top can pass through, to be held on any sieve with small enough mesh to retain that particular size. Very fine material passing the #100 sieve will be retained in the pan at the bottom.

Dry out a sample of the fine aggregate to be tested, very carefully weigh out 500 g of it, and place the sample in the top sieve (#4). The sieves may be shaken by hand or by a mechanical shaker. Shaking should continue for at least 1 min. Weigh the material retained on each sieve, as illustrated in Fig. 7–7, and record the weights. Calculate the percentage of the total (500 g) which each size represents and total these percentages as cumulative percentages. This is done by beginning with the percent-

FIGURE 7–7 Weighing sand sample. (*Courtesy Soiltest Inc.*)

FIGURE 7–6 Set of sand screens.

age retained on the #4 sieve and recording it as the cumulative percentage for that grade. The cumulative percentage for the next grade (#8) is found by adding its percentage to the one above. Each grade is treated the same way. Now add these cumulative percentages together and divide the result by 100. This figure represents the fineness modulus number of that aggregate and is an indication of the relative fineness or coarseness of the material. Note that any material passing the #100 sieve is not included in the calculations.

Fineness modulus numbers for fine aggregate should range from 2.20 to 3.20, and any fine aggregate with a fineness modulus number falling within this range will generally be a suitable concrete aggregate, from a gradation standpoint.

The smaller numbers represent fine materials, aggregates with a greater proportion of fine particles, while larger numbers represent aggregates with a preponderance of large particles. Fine aggregate is classed as *fine, medium,* or *coarse*—depending on its fineness modulus. Fine sand ranges from 2.20 to 2.60, medium from 2.61 to 2.90, and coarse from 2.91 to 3.20. In cases where mass concrete is being designed, fine aggregate with a fineness modulus larger than 3.20 may be allowed. Specifications usually define the allowable fineness modulus.

The fineness modulus calculations may be recorded on a chart like that illustrated below. To better understand the procedure, let us go through a sample test. Suppose that a 500 g sample has been put through the standard sieves and the results are as follows:

Retained on the #4 sieve, 12g; on the #8, 55 g; on the #16, 100 g; on the #30, 122 g; on the #50, 118 g; on the #100, 75 g. Totaling these weights, it is found that 18 g passed through the pan below—weigh the contents of the pan to make sure. This information completes the chart as shown below.

Fineness modulus test for sand

Sieve size no.	Weight retained (g)	Percentage retained	Cumulative percentage
4	12	2.4	2.4
8	55	11.0	13.4
16	100	20.0	33.4
30	122	24.4	57.8
50	118	23.6	81.4
100	75	15.0	96.4
Pan	18		284.8

Fineness modulus = 284.8 ÷ 100 = 2.85

The result indicates that this particular aggregate belongs in the medium sand category.

Fineness modulus is not an indication of whether there is the proper amount of each grade of sand in the sample, for a great number of gradings will give the same value for fineness modulus. However, it can be determined from the results of the test whether or not the sample contains the right proportion of each grade of sand.

This is done by plotting the percentages retained on each sieve on a graph such as that shown in Fig. 7–8. On this graph the solid line represents the ideal grading and the dotted lines show the maximum deviations. Consequently, when the test is plotted, the suitability of the sand is indicated. If necessary, recommendations may be made for blending in particular grades which the sample lacks.

FIGURE 7–8 Graphical representation of sieve analysis.

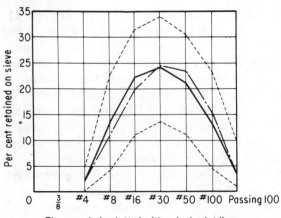

The sample is plotted with a dash–dot line

Moisture Content

Moisture content tests on aggregates, particularly fine aggregates, are carried out for two reasons. One is that the moisture content in a fine aggregate may necessitate adjusting the amount of water added to the mix in order to produce a given strength. The other is that when water is present in certain proportions, a phenomenon known as *bulking* occurs. Volume increases considerably over that at a dry or saturated state because the particles are separated by a film of water. This means that when designing concrete mixes, it is much better to specify the amount of aggregate to be used by weight rather than by volume. It takes only a small compensation to make up for the difference in weight between dry and moist sand, while a large compensation is sometimes required if amounts to be used are specified by volume.

The amount of bulking varies with the fineness of the sand and the amount of moisture present—5 to 6% water, by weight—will produce maximum bulking, and the condition is more pronounced in fine sand than it is in coarse. The sand bulking chart in Fig. 7–9 shows the percentage of increase in volume over dry sand for amounts of water varying from 0 to 25%, for fine, medium, and coarse sands.

The amount of water present in a sample is determined by carrying out a *moisture test*. Proceed as follows:

Weigh out a 500-g sample and spread it on the bottom of a flat metal pan about 12 in. in diameter and 1 in. deep. Set the pan over a heating unit and allow the sand to dry out, stirring continually. A change in the appearance of the sand can be noted as the moisture is driven off. When the material appears to be dry, weigh it again and record the weight. Dry once more and weigh again; it will probably have lost a little more weight this time. Continue this procedure until the weight remains constant. This is the dry weight. The total moisture content is found as follows:

FIGURE 7–9 Sand bulking chart.

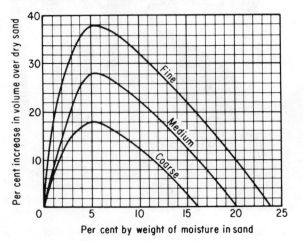

$$\frac{\text{Loss in weight}}{\text{Dry weight}} \times 100 = \text{percentage of moisture}$$

But a portion of the moisture driven off had been present in pores and cracks in the particles—absorbed moisture—and was not available for reaction with cement. For average aggregates, it is assumed to be 1%. The free moisture—that which appears on the surface of the particles—is therefore found by subtracting 1 from the above result. In other words,

$$\frac{\text{Loss in weight}}{\text{Dry weight}} \times 100 - 1 = \text{percentage of free moisture}$$

For example, if a 500-g sample has a constant weight of 478 g after drying, the loss in weight has been $500 - 478 = 22$ g. The percentage of free moisture is then

$$\frac{22}{478} \times 100 - 1 = 3.6\%$$

If this is the same sand which we previously tested for fineness modulus, we can tell by looking at the bulking chart that this medium sand, containing about $3\frac{1}{2}\%$ free moisture, will increase by 24% over its dry volume.

Specific Gravity

The specific gravity (sp. gr.) of an aggregate is the ratio of its weight to the weight of an equal volume of water. It is not a measure of aggregate quality but is used in making calculations involved in mix design, particularly those concerning *absolute volume*. Methods for determining the specific gravity of aggregates are outlined in ASTM C127 and in CSA A23–10A. Most natural aggregates have specific gravities between 2.4 and 2.9, and in concrete calculations the specific gravities of *saturated, surface-dry* aggregates are usually used. These are aggregates in which the pores in the particles are considered to be filled with moisture but with no excess moisture on the surface.

Particle Shape and Surface Texture

The shape of particles of aggregate and the texture of their surface have an effect on the properties of fresh concrete. Rough-surfaced and long, thin, flat particles require more water to produce workable concrete than do rounded or cube-shaped ones. In other words, angular particles need more cement to maintain the same water/cement ratio. Long, slivery particles should be avoided if possible or at least limited to 15% by weight of the total aggregate. It is particularly important to limit such particles in aggregates used in concrete produced for floor toppings.

Water/Cement Ratio

Concrete design is based on the theory that the strength of the concrete is regulated by the amount of water used per unit of cement, provided that the mix is plastic and workable. This is the *water/cement ratio theory,* which says that as the amount of water used per pound of cement is increased, the strength of the concrete will decrease. Table 7–1 gives the probable compressive strengths of concrete at 28 days, based on various water/cement ratios.

Concrete design is also based on the assumption that sound coarse aggregate is absolutely essential to the production of high-quality concrete. This being so, the best concrete is produced by using the largest size of aggregate and the greatest quantity of it, per given volume of concrete, that is compatible with job requirements. However, there is a practical limit to the amount of coarse aggregate which can be used in any given situation, depending on the maximum size of the aggregate and the fineness modulus of the fine aggregate used with it. Table 7–2 gives the volume of coarse aggregate of various sizes that can be used per volume of concrete, for various fineness moduli of fine aggregate.

TABLE 7–1 Compressive strengths for various water/cement ratios[a]

Water/cement ratio (lb/lb of cement)	Probable compressive strength at 28 days (psi)	
	Non-air-entrained concrete	Air-entrained concrete
0.825	2000	1600
0.75	2500	2000
0.68	3000	2400
0.625	3500	2800
0.57	4000	3200
0.525	4500	3600
0.48	5000	4000
0.445	5500	4400
0.41	6000	4800

[a]For concrete subjected to severe exposures such as (1) being wet frequently or continuously, (2) being exposed to freezing and thawing, and, (3) being exposed to sea water or sulfates, water/cement ratios should be restricted as follows: Thin sections and sections with less than 1 in. cover over the steel: (1) and (2) 0.45; (3) 0.40; all other structures: (1) and (2) 0.50; (3) 0.45.

TABLE 7–2 Volumes of C.A., in cu ft/cu yd of concrete

Max. size of C.A. (in.)	Fineness modulus of F.A.			
	2.40	2.60	2.80	3.00
$\frac{3}{8}$	13.5	13.0	12.4	11.9
$\frac{1}{2}$	15.9	15.4	14.8	14.3
$\frac{3}{4}$	17.8	17.3	16.7	16.2
1	19.2	18.6	18.1	17.6
$1\frac{1}{2}$	20.2	19.7	19.2	18.6
2	21.1	20.5	20.0	19.4
3	22.1	21.6	21.1	20.5
6	23.8	23.2	22.7	22.1

TABLE 7-3 Recommended maximum sizes of aggregate for various types of construction

Minimum dimension of section (in.)	Maximum size of aggregate (in.)			
	Reinforced walls, beams, and columns	Unreinforced walls	Heavily reinforced slabs	Lightly or unreinforced slabs
2½-5	½-¾	¾	¾-1	¾-1½
6-11	¾-1½	1½	1½	1½-3
12-29	1½-3	3	1½-3	3
30 or more	1½-3	6	1½-3	3-6

The largest size of coarse aggregate that can be used in any given situation depends on a number of things. Among them are (1) the size of the unit being poured, (2) whether or not it is reinforced, and (3) the spacing of the reinforcement. Table 7-3 gives the recommended maximum sizes of coarse aggregate for various types of construction.

Another consideration in designing concrete is the desired plasticity or flowability of the mix. This will depend on the type of construction, the size of the unit being poured, and whether or not the unit is reinforced. This flowability is known as slump and is measured by testing with a slump cone in a prescribed manner (see p. 149 for the description of a slump test). Table 7-4 gives recommended slumps for various types of construction.

Variations in slump for a given strength of concrete are produced by altering the amount of aggregate used with the paste. However, experience has shown that in order to obtain a given slump, a fairly well fixed amount of water must be used per yard of concrete. This will vary, depending on whether or not air entraining is used and on the maximum size of coarse aggregate used. Table 7-5 shows the approximate amount of water required for each yard of concrete under various conditions.

TABLE 7-4 Recommended slumps for various types of construction

Type of construction	Slump	
	Maximum	Minimum
Reinforced foundation walls and footings	3	1
Plain footings, caissons, and substructure walls	3	1
Slabs, beams, and reinforced walls	4	1
Building columns	4	1
Pavements	2	1
Bridge decks	3	2
Sidewalks, driveways, and slabs on the ground	4	2
Heavy, mass construction	2	1

The amount of water used per bag of cement is influenced not only by the strength of concrete required but also by the climatic conditions to which the concrete will be subjected and by the thickness of the section and the location of concrete—on land or in water. See the note at the bottom of Table 7-1.

TABLE 7-5 Approximate mixing water requirements for various slumps and maximum sizes of C.A.

Slump (in.)	Water (lb/cu yd of concrete of indicated maximum sizes of C.A.)							
	⅜ in.	½ in.	¾ in.	1 in.	1½ in.	2 in.	3 in.	6 in.
	Non-air-entrained concrete							
1 to 2	350	333	308	300	275	258	242	209
3 to 4	383	367	342	325	300	284	267	234
5 to 6	408	384	359	342	317	300	284	250
Approximate amount of entrapped air in %	3	2.5	2	1.5	1	0.5	0.3	0.2
	Air-entrained concrete							
1 to 2	308	300	275	258	242	225	209	184
3 to 4	342	325	300	284	267	250	234	200
5 to 6	359	342	317	300	284	267	250	217
Recommended average total air content in %	7.5	7.5	6	5	4.5	4	3.5	3

PORTLAND CEMENT

The chemical reaction that is produced when cement is mixed with water is known as *hydration*. When the cement and water paste is mixed with aggregates and allowed to cure, the result is a very hard and durable substance.

The raw materials used in the manufacture of portland cement are *lime, silica, alumina,* and *iron.* All of these materials are readily available from natural deposits. From these basic materials four principal compounds make up the composition of cement; *tricalcium silicate, dicalcium silicate, tricalcium aluminate,* and *tetracalcium aluminoferrite.* By varying the proportions of these basic components the properties of the cement can be modified to best suit its particular application.

Five basic types of cement are available.

- *Type I* or normal cement: Used as a general-purpose cement.
- *Type II* or moderate cement: Used where protection against sulfate attack is required. It can also be used in large mass concrete structures, as it produces less heat during the hydration process.
- *Type III* or high early strength: Used where high concrete strengths are required in a very short period of time.
- *Type IV* or low heat of hydration: Used where the heat of hydration must be kept to a minimum. Used in massive concrete structures such as dams.
- *Type V* or sulfate resisting: Used where the concrete will be exposed to severe sulfate attack. Normally used in concrete foundations.

Cement is usually gray in color but can also be white. Coloring can be added to the white cement for special effects. Cement is transported in bulk for large projects or in bags when smaller quantities are required. In the United States, a standard bag of cement weighs 94 lb, while in Canada (metric) a bag of cement has a mass of 40 kg.

CONCRETE DESIGN

Absolute Volume Method

Concrete is made by binding together aggregates with cement paste in such a way as to produce a strong, impervious material containing no voids in its bulk. In other words, a given quantity of concrete is made up of solid material—theoretically there are no air spaces involved. This product is made by mixing four basic ingredients, only one of which (water) contains no voids. The other three are made up of particles which have air spaces between them, and, consequently, in a given volume of any of these, only part is actually usable. That part of the

given volume which is actually cement or stone is known as the *absolute volume*—volume without voids. To get a true picture of the quantities of solid materials to be used to produce concrete, it seems reasonable to calculate these materials in terms of their absolute volume and then convert the absolute volumes to weights. The absolute volume of a loose, dry material is found by the formula

$$\text{Absolute volume} = \frac{\text{weight of loose, dry material}}{\text{sp. gr. material} \times \text{unit wt. water}}$$

For example, the absolute volume of 100 lb of portland cement with a specific gravity of 3.15 will be

$$\text{Absolute volume} = \frac{100}{3.15 \times 62.4} = 0.508 \text{ cu ft}$$

Before calculations can be made, certain data must be known. These items include

1. Specific gravity of F.A. and C.A.
2. Dry-rodded unit weight of C.A.
3. Fineness modulus of F.A.
4. Maximum size of C.A.
5. Cement factor or water/cement ratio for concrete to be designed.
6. Percentage of air entrained or entrapped.
7. Required slump of the concrete.

With this information and with the aid of tables or simple calculations, the quantities (in lbs) of cement, coarse aggregate, water, and air required per cubic yard of concrete can be determined. The absolute volumes (in cu ft) of the four can then be calculated and totaled. When that sum is subtracted from 27 (cu ft in 1 cu yd), the result is the absolute volume of the fine aggregate required. Then, from the absolute volume, the weight of the fine aggregate required may be calculated.

Thus, the quantities of materials required for 1 cu yd of concrete have been estimated and a trial batch can be made based on them. If adjustments are necessary, further batches should be adjusted by keeping the water/cement ratio constant and adjusting the aggregates and air to produce the desired slump and air content.

It is often desirable to make small trial batches, using, for example, 20 lb of cement. In such a case, the amount of each of the materials is reduced accordingly.

Example

Concrete is required for a 10-in. retaining wall which will be subject to frequent wetting by fresh water. 4000-psi concrete is specified. The sp. gr. of the fine aggregate is given as 2.75 and that of the coarse aggregate as 2.65. The dry-rodded weight of the coarse aggregate is 105 lb, with the maximum size 1½ in. Air content is to be 5($\pm$ 1)% and the slump 3 to 4 in.

Data required:

1. From Table 7-1, water/cement ratio   = 0.48
2. Max. size C.A.   = 1½ in.
3. Sp. gr. C.A.   = 2.65
4. Dry-rodded wt. C.A.   = 105 lb
5. Sp. gr. F.A.   = 2.75
6. Slump   = 3-4 in.
7. Air content   = 5($\pm$ 1)%

Solution:

1. Weights:

   From Table 7-5, water required/cu yd = 267 lb
   Cement factor = 267/0.48 = 556 lb/cu yd of concrete
   From Table 7-2, vol. of C.A. required/cu yd = 19.3 cu ft
   Wt. of C.A. required/cu yd = 19.3 $\times$ 105 = 2026 lb

2. Absolute volumes:

   $$\text{Cement} = \frac{556}{3.15 \times 62.4} = 2.83 \text{ cu ft}$$

   $$\text{Water} = \frac{267}{62.4} = 4.28 \text{ cu ft}$$

   $$\text{Coarse aggregate} = \frac{2026}{2.65 \times 62.4} = 12.25 \text{ cu ft}$$

   $$\text{Air} = 0.05 \times 27 = \underline{1.35 \text{ cu ft}}$$
   $$\text{Total} \quad 20.71 \text{ cu ft}$$

   $$\text{Fine aggregate} = 27 - 20.71 = 6.29 \text{ cu ft}$$

3. Wt. of F.A. = 6.29 $\times$ 2.75 $\times$ 62.4 = 1079 lb

4. Trial batch, using 20 lb of cement:

   Cement, 20 lb.; C.A., 20/556 $\times$ 2026 = 73 lb; F.A., 20/556 $\times$ 1079 = 39 lb; water = 0.48 $\times$ 20 = 9.6 lb

Trial Mix Method

An alternative method of designing a concrete mix is to use the suggested trial mixes shown in Tables 7-6 and 7-7. It is quite likely that, having made a trial batch from amounts given in one of the tables, some adjustment will have to be made to produce the desired slump. When making a trial batch, enough water should be added to produce the desired slump, whether or not that is the amount of water given in the table. The *slump, air content,* and *unit weight* of the fresh concrete should then be measured.

Example

Concrete is to be designed with a water/cement ratio of 0.50, a slump of 2 to 3 in., and an air content of 5($\pm$ 1)%. Fine aggregate has a fineness modulus of 2.75 and a moisture content of 5%. Maximum size of coarse aggregate is 1 in., and its moisture content is 1%.

Solution:

For the above specifications, quantities per cubic yard may be calculated from Table 7-6 as follows:

Water (reduced by 3% for 1 in. slump decrease) =   276 lb
Cement (maintained at same W/C ratio)   =   554 lb
Fine aggregate (interpolate for 2.75 F.M.)   = 1075 lb
Coarse aggregate (interpolate for 2.75 F.M.)   = $\underline{1885 \text{ lb}}$
                                         Total = 3790 lb
Free moisture in F.A. = 0.05 $\times$ 1075   =   54 lb
Free moisture in C.A. = 0.01 $\times$ 1885   =   19 lb

The correct weights to allow for moisture in the aggregates are as follows:

Water = 276 - 54 - 19 =   203 lb
Cement   =   554 lb
F.A.   = 1075 + 54 = 1129 lb
C.A.   = 1885 + 19 = $\underline{1904 \text{ lb}}$
           Total = 3790 lb

From a trial mix made with these quantities, the following measurements were taken:

Slump = 1 in.; air content = 5.5%; unit wt. = 145 lb
20 lb of added water were required to produce a 3 in. slump.
Total wt. of water used   = 276 + 20 = 296 lb
Total wt. of concrete produced = 3790 + 20 = 3810 lb

$$\text{Volume of concrete produced} = \frac{3810}{145} = 26.27 \text{ cu ft}$$

To produce 27 cu ft per trial batch, adjust the quantities as follows:

$$\text{Water} = \frac{27}{26.27} \times 296 = 304 \text{ lb}$$

$$\text{Cement} = \frac{304}{0.50} = 608 \text{ lb}$$

Total wt. of materials/cu yd = 145 $\times$ 27   = 3915 lb
Percentage of F.A. (interpolated from Table 7-6) = 36%

Wt. of F.A./cu yd = 0.36 $\times$ 3003 = 1080 lb
Wt. of C.A./cu yd = 3003 - 1080 = 1923 lb

To obtain the best aggregate proportions for workability and economy of materials, further trial batches should be made, altering the ratio of fine and coarse aggregates.

Tests on Concrete

The presence of air in a concrete mix, either incidental or purposeful, has been noted in Table 7-5. Air is added intentionally in the form of millions of tiny bubbles which

TABLE 7–6 Suggested trial mixes for air-entrained concrete of medium consistency (3- to 4-in. slump)[a]

Water/ cement ratio (lb/lb)	Max. size of aggregate (in.)	Air content (%)	Water (lb/cu yd of concrete)	Cement (lb/cu yd of concrete)	With fine sand— fineness modulus 2.50			With coarse sand— fineness modulus 2.90		
					F.A.(% of total aggregate)	F.A. (lb/ cu yd of concrete)	C.A. (lb/ cu yd of concrete)	F.A.(% of total aggregate)	F.A. (lb/ cu yd of concrete)	C.A. (lb/ cu yd of concrete)
0.40	⅜	7.5	340	850	50	1250	1260	54	1360	1150
	½	7.5	325	815	41	1060	1520	46	1180	1400
	¾	6	300	750	35	970	1800	39	1090	1680
	1	6	285	715	32	900	1940	36	1010	1830
	1½	5	265	665	29	870	2110	33	990	1990
0.45	⅜	7.5	430	755	51	1330	1260	56	1440	1150
	½	7.5	325	720	43	1140	1520	47	1260	1400
	¾	6	300	665	37	1040	1800	41	1160	1680
	1	6	285	635	33	970	1940	37	1080	1830
	1½	5	265	590	31	930	2110	35	1050	1990
0.50	⅜	7.5	340	680	53	1400	1260	57	1510	1150
	½	7.5	325	650	44	1200	1520	49	1320	1400
	¾	6	300	600	38	1100	1800	42	1220	1680
	1	6	285	570	34	1020	1940	38	1130	1830
	1½	5	265	530	32	980	2110	36	1100	1990
0.55	⅜	7.5	340	620	54	1450	1260	58	1560	1150
	½	7.5	325	590	45	1250	1520	49	1370	1400
	¾	6	300	545	39	1140	1800	43	1260	1680
	1	6	285	520	35	1060	1940	39	1170	1830
	1½	5	265	480	33	1030	2110	37	1150	1990
0.60	⅜	7.5	340	565	54	1490	1260	58	1600	1150
	½	7.5	325	540	46	1290	1520	50	1410	1400
	¾	6	300	500	40	1180	1800	44	1300	1680
	1	6	285	475	36	1100	1940	40	1210	1830
	1½	5	265	440	33	1060	2110	37	1180	1990
0.65	⅜	7.5	340	525	55	1530	1260	59	1640	1150
	½	7.5	325	500	47	1330	1520	51	1450	1400
	¾	6	300	460	40	1210	1800	44	1330	1680
	1	6	285	440	37	1130	1940	40	1240	1830
	1½	5	265	410	34	1090	2110	38	1210	1990
0.70	⅜	7.5	340	485	55	1560	1260	59	1670	1150
	½	7.5	325	465	47	1360	1520	51	1480	1400
	¾	6	300	430	41	1240	1800	45	1360	1680
	1	6	285	405	37	1160	1940	41	1270	1830
	1½	5	265	380	34	1110	2110	38	1230	1990

[a]Increase or decrease water per cubic yard by 3% for each increase or decrease of 1 in. in slump; then calculate quantities by absolute volume method. For manufactured fine aggregate, increase percentage of fine aggregate by 3 and water by 17 lb/cu yd of concrete. For less workable concrete, as in pavements, decrease percentage of fine aggregate by 3 and water by 8 lb/cu yd of concrete.

Source: Courtesy Portland Cement Association.

serve a number of useful purposes. Air-entrained concrete has greater flowability than plain concrete of similar composition. In other words, the water content can be reduced, thus increasing strength while retaining the same amount of slump. Air-entrained concrete is also more resistant to freezing and thawing cycles than plain concrete. Air is entrained into the concrete by the addition of *air-entraining agents* into the mix. These agents are one of a series of additives commonly used in concrete design today.

One type of additive is called an *accelerator,* a chemical which speeds up the initial set of concrete. The most common material for this purpose is calcium chloride, used at the rate of not more than 2 lb/bag of

TABLE 7-7 Suggested trial mixes for non-air-entrained concrete of medium consistency (3- to 4-in. slump)[a]

Water/ cement ratio (lb/lb)	Max. size of aggre- gate (in.)	En- trapped air con- tent(%)	Water (lb/cu yd of con- crete)	Cement (lb/cu yd of con- crete)	With fine sand— fineness modulus 2.50			With coarse sand— fineness modulus 2.90		
					F.A.(% of total aggre- gate)	F.A. (lb/ cu yd of con- crete)	C.A. (lb/ cu yd of con- crete)	F.A.(% of total aggre- gate)	F.A. (lb/ cu yd of con- crete)	C.A. (lb/ cu yd of con- crete)
0.40	⅜	3	385	965	50	1240	1260	54	1350	1150
	½	2.5	365	915	42	1100	1520	47	1220	1400
	¾	2	340	850	35	960	1800	39	1080	1680
	1	1.5	325	815	32	910	1940	36	1020	1830
	1½	1	300	750	29	880	2110	33	1000	1990
0.45	⅜	3	385	855	51	1330	1260	56	1440	1150
	½	2.5	365	810	44	1180	1520	48	1300	1400
	¾	2	340	755	37	1040	1800	41	1160	1680
	1	1.5	325	720	34	990	1940	38	1100	1830
	1½	1	300	665	31	960	2110	35	1080	1990
0.50	⅜	3	385	770	53	1400	1260	57	1510	1150
	½	2.5	365	730	45	1250	1520	49	1370	1400
	¾	2	340	680	38	1100	1800	42	1220	1680
	1	1.5	325	650	35	1050	1940	39	1160	1830
	1½	1	300	600	32	1010	2110	36	1130	1990
0.55	⅜	3	385	700	54	1460	1260	58	1570	1150
	½	2.5	365	665	46	1310	1520	51	1430	1400
	¾	2	340	620	39	1150	1800	43	1270	1680
	1	1.5	325	590	36	1100	1940	40	1210	1830
	1½	1	300	545	33	1060	2110	37	1180	1990
0.60	⅜	3	385	640	55	1510	1260	58	1620	1150
	½	2.5	365	610	47	1350	1520	51	1470	1400
	¾	2	340	565	40	1200	1800	44	1320	1680
	1	1.5	325	540	37	1140	1940	41	1250	1830
	1½	1	300	500	34	1090	2110	38	1210	1990
0.65	⅜	3	385	590	55	1550	1260	59	1660	1150
	½	2.5	365	560	48	1390	1520	52	1510	1400
	¾	2	340	525	41	1230	1800	45	1350	1680
	1	1.5	325	500	38	1180	1940	41	1290	1830
	1½	1	300	460	35	1130	2110	39	1250	1990
0.70	⅜	3	385	550	56	1590	1260	60	1700	1150
	½	2.5	365	520	48	1430	1520	53	1550	1400
	¾	2	340	485	41	1270	1800	45	1390	1680
	1	1.5	325	465	38	1210	1940	42	1320	1830
	1½	1	300	430	35	1150	2110	39	1270	1990

[a]See Table 7-6 footnote.
Source: Courtesy Portland Cement Association.

cement. Another additive is used as a *dispersal agent*—a material which acts to separate individual particles of cement, thus allowing each one to come in full contact with water and hydrate completely. The cement can thus develop its full potential of strength.

Other additives act as *retarders* and slow up the initial set of cement. This action is important when placing deep girders in hot weather, for example. The first lift tends to set up before the next one can be placed, thus making it difficult to integrate the two and produce a monolithic member. A retarder used in the concrete placed at the bottom of the girder will delay its initial set until the remainder of the concrete has been placed. As a result, the set will be uniform throughout.

Additives are also available for use as hardeners and waterproofers. It is a good idea to contact the manufacturer or his agent and obtain full details on the proper use of these particular products prior to initial usage.

Regardless of tests made on materials which go into the making of concrete and of the care taken in designing a mix, tests are carried out on the concrete itself to make sure that it conforms to specifications. These tests normally consist of a *slump test,* a *compression test,* and sometimes a *flexure test.*

A *slump test* is made to ensure that the concrete has the flowability required for placing. A slump or slump range will usually be stipulated in specifications. The test is carried out as soon as a batch of concrete is mixed, and a standard slump cone and tamping rod are required to carry it out. The cone is made of sheet metal, 4 in. in diameter at the top, 8 in. in diameter at the bottom, and 12 in. high. The rod is a ⅝-in. bullet-nosed rod about 24 in. long. The cone is filled in three equal layers, each tamped 25 times with the rod. After the third layer is in place and has been tamped, the concrete is struck off level, the cone is lifted carefully and set down beside the slumped concrete, and the rod is laid across the top. The distance from the underside of the rod to the average height of the concrete is measured and registered as the amount of slump in inches.

A *compression test* is made to determine whether or not the concrete will withstand the loads to be placed on it. It shows whether the quality of the paste is such

that it will not crumble under the load but rather bind the aggregates together until they break. It also helps to demonstrate the quality of the C.A.—its ability to withstand the specified load.

In order that concrete may demonstrate its quality, it must be allowed time to gain strength—to *cure.* Concrete is assumed to have gained its full working strength within 28 days under ideal curing conditions, and that period is taken as a standard for testing purposes. But experience has shown an age-strength relationship for concrete cured under ideal conditions. Tests taken at other times, usually 1, 3, or 7 days, will accordingly show meaningful results. Figure 7–10 illustrates the strengths which can be anticipated from concrete made with normal portland cement, using various W/C ratios and cured under ideal conditions for test periods of 1, 3, 7, and 28 days.

A standard size of test specimen is used to make the test. It is molded in a cylindrical container, 6 in. in diameter and 12 in. high, (see Fig. 7–11). One commonly used type of container is made of stiff waxed cardboard with a metal bottom and cardboard top.

The cylinder is filled in three equal layers, each layer having been tamped 25 times with the standard tamping rod. The top is struck off level, the cap is placed on, and

FIGURE 7–10 Typical strength-age relationships in concrete, based on compression tests. (*Courtesy Portland Cement Association.*)

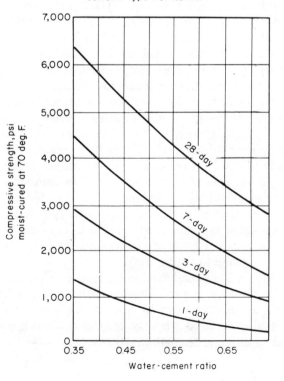
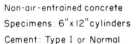

FIGURE 7–12 Test cylinders filled and covered.
(*Courtesy Soiltest Inc.*)

the specimen is weighed (see Fig. 7–11). It is then set aside for 24 hr. Several specimens (at least four) should be taken of each mix to be tested (see Fig. 7–12). Two test cylinders can set under actual conditions and two can be taken to a laboratory for curing under ideal conditions.

After 24 hr, those cylinders to be cured in the lab will be removed from the containers and placed in warm, moist conditions—preferably a curing cabinet—to cure for the designated period. Those cylinders left to cure under job conditions may be left in their containers but should preferably be removed when the forms are taken off or, in the case of slabs, after 24 hr—to simulate job conditions as closely as possible.

At the end of the curing period the cylinders are weighed again and capped with a thin layer of capping compound usually made of a sulfur-base material. Caps should be plane to within 0.002 in.

After a 2 hr period to allow the caps to harden, each cylinder is placed in a compression testing machine and broken, as illustrated in Fig. 7–13. A dial registers the total load required to break the cylinder and this is converted to ultimate strength in psi (see Fig. 7–14).

Flexure tests are sometimes made to determine the bending strength of a particular concrete. Forms used must provide a beam with a length at least 2 in. greater than 3 times its depth as tested. This minimum cross-sectional dimension must be at least 3 times the maximum nominal size of the coarse aggregate used, but in no case less than 6 by 6 in. After proper curing, the beam is tested for flexural strength by third-point or center-point loading.

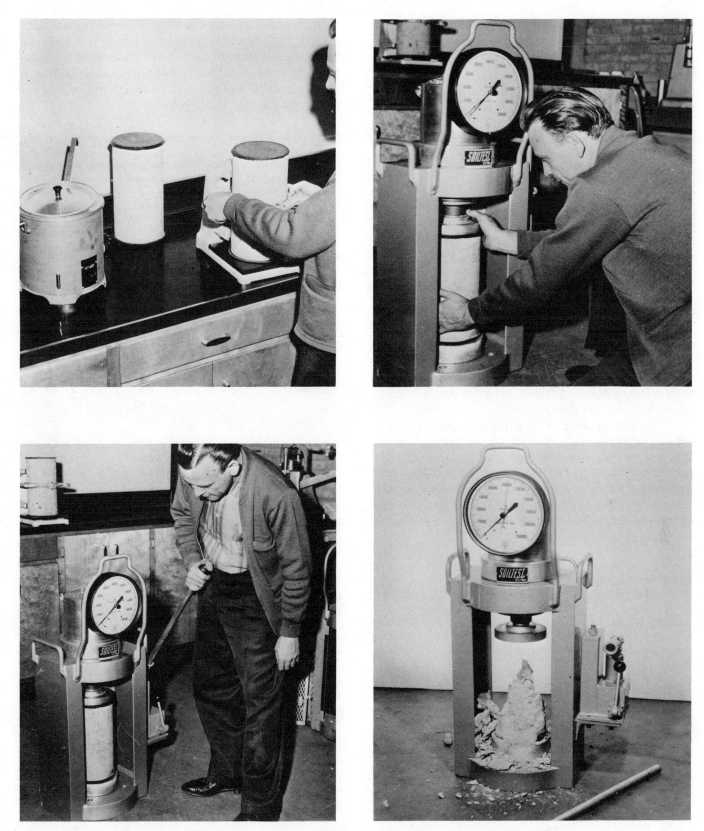

FIGURE 7–13 Testing cylinders in compression tester. (*Courtesy Soiltest Inc.*)

FIGURE 7-14 Checking compression test results.

USE OF FAST-SETTING CEMENTS

It is sometimes necessary to use concrete in situations when it is impossible to wait the required period for concrete made with normal portland cement to set and strengthen. In these situations it is possible to use a high-early-strength portland cement. The difference in early strength is illustrated in Fig. 7-15.

It is also possible to use aluminous cement, which gains its strength in a very short time compared to portland cement. Concrete made with portland cement will gain about 10% of its 28-day strength in 1 day at a temperature of about 65°F. On the other hand, concrete made with aluminous cement attains the 28-day strength of that made from portland cement in approximately 24

hr, at the same temperature. At 36°F, concrete made with normal portland cement will have gained very little—if any—of its 28-day strength, while under similar conditions, concrete made with aluminous cement will have gained about 50% of the 28-day strength of portland cement concrete (see Table 7-8).

TABLE 7-8 Effect of temperature on the strength of concrete

Crushing strength of 1 : 2 : 4 concrete expressed as a percentage of 28-day strength of normal portland cement concrete.

Type of cement	Age (days)	Low temperature (36°F)	Normal temperature (64°F)
Normal	1	0	10
portland	3	10	37
cement	7	25	64
	14	55	80
	28	92	100
	56	108	110
Rapid	1	0	20
hardening	3	12	59
portland	7	37	93
cement	14	70	107
	28	118	126
	56	135	135
High	1	54	146
alumina	3	164	188
cement	7	188	197
	14	197	212
	28	207	218
	56	222	229

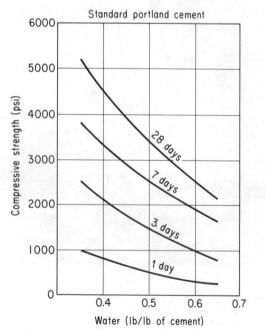

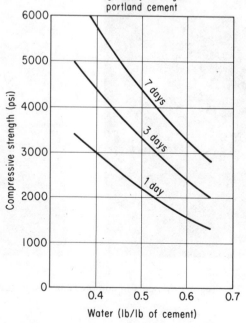

FIGURE 7-15 Early strength of normal portland and high-early-strength cements.

This fast-setting feature makes aluminous cement particularly valuable when concrete must be placed in cold weather. A considerable amount of heat is generated during the hydration process, and that heat is usually sufficient to ensure the continuation of the hardening process if the concrete is protected from freezing for the first 6 hr.

PREPARATION FOR PLACING CONCRETE

Before placing concrete, the subgrade must be properly prepared, and forms and reinforcing must be erected according to specifications. Subgrades must be trimmed to specific elevation, thoroughly compacted (see Fig. 7-16) and should be moist when the concrete is placed. A moist subgrade is particularly important to prevent overly rapid extraction of water from concrete when flat work—such as floors and pavements—is being placed in hot weather. When concrete is being placed on a rock foundation, all loose material should be removed before it is placed. If rock has to be cut, the rock faces should be vertical or horizontal, not sloping.

Forms must be made of material that will impart the desired texture to the concrete and should be clean, tight, and well braced. They should be oiled, or treated with some type of form seal that will prevent them from absorbing water from the concrete and will facilitate form removal. Care must be taken to see that sawdust, nails, or other debris are removed from the bottom of the forms. Column and tall wall forms should have "windows" at the bottom to facilitate debris removal. It is also important that reinforcing steel be clean and free from scales or rust.

When fresh concrete is to be placed on hardened concrete, it is important to secure a good bond and a watertight joint. The hardened concrete should be fairly level, rough, clean, and moist. Some of the aggregate particles should be exposed by cutting away part of the existing surface by sand blasting, by cleaning with hydrochloric acid, or by using a wire brush. Any laitance or soft layer of mortar must be removed from the surface.

When concrete is to be placed on a hardened concrete surface or on rock, a layer of mortar should be placed on the hard surface first. This provides a cushion

FIGURE 7-16 Compacting subgrade.

on which the new concrete can be placed and stops aggregate from bouncing on the hard surface and forming stone pockets. The mortar should be about 2 in. deep, with the same water content as the concrete being placed and a slump of 5 to 6 in.

Preparations must also include the building of adequate runways for wheelbarrows or buggies. Runways should be reasonably smooth and stiff enough to prevent the wheelbarrows or buggies from bouncing or jarring as they travel. Some means of transferring concrete from the wheelbarrow, buggy, or bucket into the forms must also be provided. This may consist of a baffle board, hopper, downspout, or chute. These will be discussed in another section.

MIXING CONCRETE

Concrete should be mixed until it is uniform in appearance and consistency. The time required for mixing depends on the volume and stiffness of the mix, to some extent on the size of the C.A. and the fineness of the sand, and on the type of mixer being employed.

Drum Mixer

A drum mixer, one size of which is shown in Fig. 7–17, consists of a rotating drum with stationary blades fixed on the inside. Mixing is accomplished by the action of the blades passing through the fluid concrete. For this type of mixer, specifications usually require a minimum

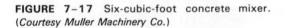

FIGURE 7–17 Six-cubic-foot concrete mixer. (*Courtesy Muller Machinery Co.*)

of 1 min for mixers of up to 1-cu yd capacity, with an increase of at least 15 sec for every ½ cu yd (or fraction thereof) of additional capacity. If mixes are stiff, additional time will be required, and the same is true for mixes containing fine sand or small C.A. The mixing time begins when all materials are in the mixer drum.

Batch mixers are available in sizes varying from 1½-cu ft to 4-cu yd capacity. For general construction work, standard mixers are rated at 3½, 6, 11, 16, and 28 cu ft of mixed concrete. For larger jobs and central mixing plants, mixers of 2- to 8-cu yd capacity are used.

Mixers may be tilting or nontilting with either high or low loading skips. High loading skips are suitable when the skip is being charged from a bin batcher, but the low loading type is preferable if the skip has to be charged by shovel or wheelbarrow. Nontilting mixers may be equipped with a swinging discharge chute. Many mixers are provided with timing devices so that they can be set for a given mixing period and cannot be opened until the designated mixing time is up.

Mixers should not be loaded beyond their rated capacity and should not be operated at speeds other than those for which they were designed. Overloading or running either too fast or too slowly prevents the proper mixing action from taking place. When increased output is required, a larger mixer or additional ones should be used.

Under usual conditions, no more than about 10% of the mixing water should be placed in the drum before the dry materials are added. Water should then be added uniformly with the dry materials, leaving about 10% to be added after the dry materials are in. If heated water is used during cold weather, this order may have to be changed. In this case, addition of cement should be delayed until most of the aggregate and water are mixed in the drum. This is done to prevent a flash set of the paste, a set brought about by too much heat when the water and cement combine.

Countercurrent Mixer

A *countercurrent mixer* (see Fig. 7–18) is so called because of the mixing system employed. It consists of a horizontally revolving pan, into which is suspended one or more vertical, three-paddle mixing tools, offset from the center of the pan. The pan rotates in a clockwise direction, while the mixing tools turn counterclockwise, thus producing a high degree of agitation in the materials being mixed. Additional agitators are optional equipment which intensify the movement of the particles in the mix and bring about rapid homogenization of otherwise difficult mixes. Such a mixer is particularly efficient at mixing additives, colors, etc., to a concrete mix (see Fig. 7–19).

FIGURE 7-18 Countercurrent concrete mixer. (*Courtesy Almar Specialty Machines Inc.*)

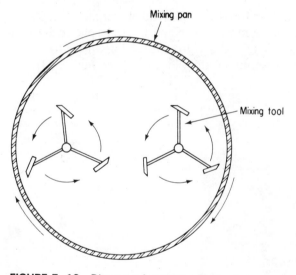

FIGURE 7-19 Diagram of countercurrent mixer.

Temperature and Slump Control

It is important that the maximum and minimum temperatures of the concrete in the mixer be controlled in all seasons. Significant temperature changes may bring about changes in water requirement, slump, or air content. (See the sections on hot and cold weather concreting, pages 167 and 168.)

Compensation for the loss of slump due to a change in concrete temperature or to delays in delivery to the forms is sometimes brought about by the addition of some water to the mix (retempering). This practice should be carefully controlled and should be allowed only to the degree that the specified water/cement ratio is not exceeded.

It is also important that the mixer be of a type that is capable of mixing and discharging concrete at the slump designated for the job at hand. In other words, the mixer must be suitable for the job—not the concrete made suitable for the mixer.

MACHINERY FOR TRANSPORTING CONCRETE

Concrete may be mixed on the job site or brought to it from a mixing plant. In either case, a variety of methods may be used to transfer the concrete from mixer or truck to the forms, and one of the prime considerations is the prevention of segregation—the separation of coarse aggregate from the mortar or of water from the other ingredients. Unless proper care is taken, the uniformity of concrete achieved by good mixing may be destroyed by failure to discharge properly from mixer or truck or in the transfer to the forms.

Concrete is transferred to the forms by *wheelbarrows, push buggies, power buggies, chutes, crane-hoisted buckets, belt conveyors,* and *pumps.*

Wheelbarrows

On small jobs, concrete may be transported by wheelbarrow (see Fig. 7-20) and the important thing is to provide smooth runways for their travel. Bouncing results in segregation in the wheelbarrow, thus causing concrete to be placed in a separated state.

FIGURE 7-20 Placing concrete by wheelbarrow. (*Courtesy Gar/Bro Mfg. Co.*)

Buggies

Pushcarts (see Fig. 7–32) and power buggies are made in a variety of sizes, from about 3½ to 11 cu ft of capacity, and are equipped with pneumatic tires for smoother operation (see Fig. 7–21). They must also be provided with smooth and rigid runways to minimize segregation.

When pushcarts or power buggies are used to transfer concrete in high-rise construction, they will probably be carried to the floor under construction by means of a materials hoist like that shown in Fig. 7–22. When concrete is being transferred in such a manner, it is important to ensure that the lift is operated smoothly so that the concrete in the buggies will not be segregated in transit.

Chutes

Chutes may be used to carry concrete directly from mixer to forms or from a hopper conveniently situated to allow chuting. Chutes should be metal or metal-lined, with rounded bottoms, and of sufficient size to guard against overflow. They should be designed so that concrete will travel fast enough to keep the chute clean but not fast enough to cause segregation. It is generally recommended that the slope of chutes be between 3 to 1 and 2 to 1, although steeper chutes may be used to carry stiff mixes.

Buckets

Buckets lifted and moved about by crane or cable are commonly used where concrete has to be placed at a considerable height above ground level or where forms are in otherwise inaccessible locations (see Fig. 7–23).

FIGURE 7–21 Concrete power buggy. (*Courtesy Whiteman Mfg. Co.*)

FIGURE 7–22 Materials hoist for high-rise building. (*Courtesy Lafarge Cement & Zajac Construction Ltd.*)

(a)

FIGURE 7-23 Tower cranes in operation: (a) crane mounted on building; (b) crane mounted outside building. (*Courtesy Heede International Inc.*)

(b)

(a)

(b)

(c)

FIGURE 7–24 Concrete buckets: (a) bucket with short drop chute (*Courtesy Gar/Bro Mfg. Co.*); (b) bucket with metal chute (*Courtesy Camlever*); (c) bucket with regulated gate.

Buckets vary in size from about ½ to 8 cu yd of capacity and may be circular or rectangular in cross section (see Fig. 7–24). The load is released by opening a gate which forms the bottom of the bucket. Gates which can regulate the flow and close when only part of the load has been discharged are preferred where sections are small and it is not desirable to place a full load in one location [see Fig. 7–24(c)].

Care should be taken to prevent jarring or shaking of buckets while they are in transit. This will cause segregation, particularly if the concrete has a relatively high rate of slump.

Belt Conveyors

Belt conveyors used to transfer concrete may be classified into three types: (1) *portable*, (b) *series*, and (c) *side discharge* conveyors. Series (feeder) conveyors operate at fairly high belt speeds—500 ft/min or better—while the other two conveyors operate best at considerably slower speeds.

Concrete should be fed onto a conveyor belt from a hopper (see Fig. 7–25) in order to get an even distribution of material along the belt, and they should be well enough supported that they will not vibrate and cause

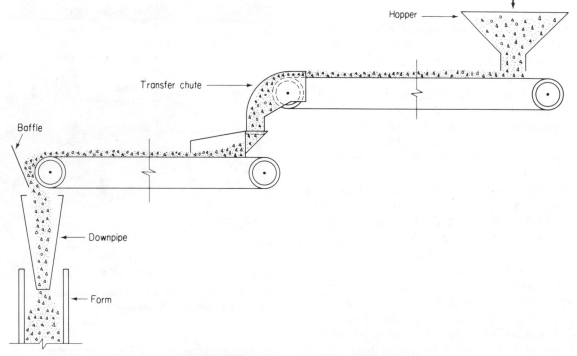

FIGURE 7-25 Series conveyor with hopper.

segregation of the concrete. The slope which can be used will vary with the concrete mix and with the type of belt used. Those with straight ribs on their surface will work best on steep slopes. Conveyors should be covered to prevent climatic conditions (sun, rain, or wind) from affecting the concrete during its transfer.

Pumps

Concrete pumps are used to transport concrete under pressure through some type of piping system. Pumps are of three types: *piston pump, pneumatic pump,* and *squeeze pressure pump.* A piston pump consists of an inlet and valve, a piston and a cylinder connected to a hopper on the intake end and to a hose or pipe on the discharge end. The cylinder receives concrete from the hopper, and the piston forces it out into the hose and, by continuous action, eventually to the form.

A pneumatic pump consists of a pressure vessel and equipment for supplying compressed air. Concrete is taken into the pressure vessel, the intake valve closed, and compressed air supplied into the top end of the vessel. The pressure forces the concrete out through a pipe at the bottom and into the delivery system (see Fig. 7-26).

A squeeze pressure pump consists of a steel drum maintained under high vacuum, inside which hydraulically powered rollers operate. A flexible hose runs from a hopper, enters the bottom of the drum, and runs around the inside surface and out the top. The vacuum main-

tains a supply of concrete from the hopper in the hose, while the rollers, rotating on the hose inside the drum, force the concrete out at the top. A delivery system carries the concrete to the job.

FIGURE 7-26 Hydraulic concrete pump. (*Courtesy Koehring PCM Division.*)

The delivery line may be either rigid pipe or flexible hose (see Fig. 7–27 and 7–28). Depending on the equipment, pumps will deliver from 10 to 90 cu yd/hr, through lines which can range up to 2000 ft horizontally and up to 300 ft vertically. One foot of vertical lift is considered to be equal to 8 ft of horizontal run, a 90° bend in rigid pipe equal to 40 ft of horizontal run, and a 45° bend equal to 20 ft.

Pumping may be used in nearly all types of concrete construction but is especially useful where space for construction equipment is at a premium. The pump can be set in an out-of-the-way location where concrete can readily be delivered to it (see Fig. 7–27), while hoists and cranes are used to deliver other materials to the job. Pumping will generally be restricted to concrete which has a maximum coarse aggregate size of 1½ in. and ultimate strength of 2500 psi or better.

It should be remembered that there is a limit to the amount of pressure that can be applied to the concrete. This will vary, depending on *slump, water/cement ratio, gradation, size* and *type* of aggregates, but all concretes have their *bleed-out point,* the pressure point beyond which the cement and water will be forced (bled) out of the concrete at the pump by applied pressure into the concrete ahead of it. The concrete that has bled out becomes dry and becomes what is called a *slug* in the line, which

FIGURE 7–27 Rigid pipe from concrete pump. (*Courtesy Koehring PCM Division.*)

makes pumping very difficult. Concretes which have poor gradation of aggregates will bleed at low pressures, and such concrete should be avoided if pumping is to be the method of transfer to be used.

FIGURE 7–28 Concrete delivered by flexible hose from concrete pump. (*Courtesy Koehring PCM Division.*)

CONCRETE TRANSPORTATION

When concrete is brought to the job from a mixing area some distance away, any of a number of methods may be used. They include *dump trucks, transit mix trucks, agitator trucks,* and *rail cars.*

Dump trucks should have a body of special shape, with rounded and sloping front and rounded bottom. The rear end should taper to a discharge gate to facilitate delivery of the load.

A transit mix truck is essentially a heavy-duty truck chassis and motor on which is mounted a large, drum-type concrete mixer. It is equipped with a water tank and auxiliary engine which operates the mixer (see Fig. 7–29). The dry ingredients for a concrete mix are charged into the truck from a batching plant (see Fig. 7–30). If the distance to be traveled from batching plant to job site can be covered within initial setting time of the cement paste, water may also be added at the batcher. The concrete is then mixed and agitated en route. But when long distances are involved, the driver must add the water at the appropriate time along the way.

Transit mix trucks are available in sizes ranging from 1 to 12 cu yd of capacity. Each batch of concrete should be mixed not less than 50 or more than 100 revolutions of the drum or blades at the prescribed rate of rota-

FIGURE 7–29 Twelve-yard transit mix truck. (*Courtesy Miron Co. Ltd.*)

FIGURE 7–30 Concrete batching plant.

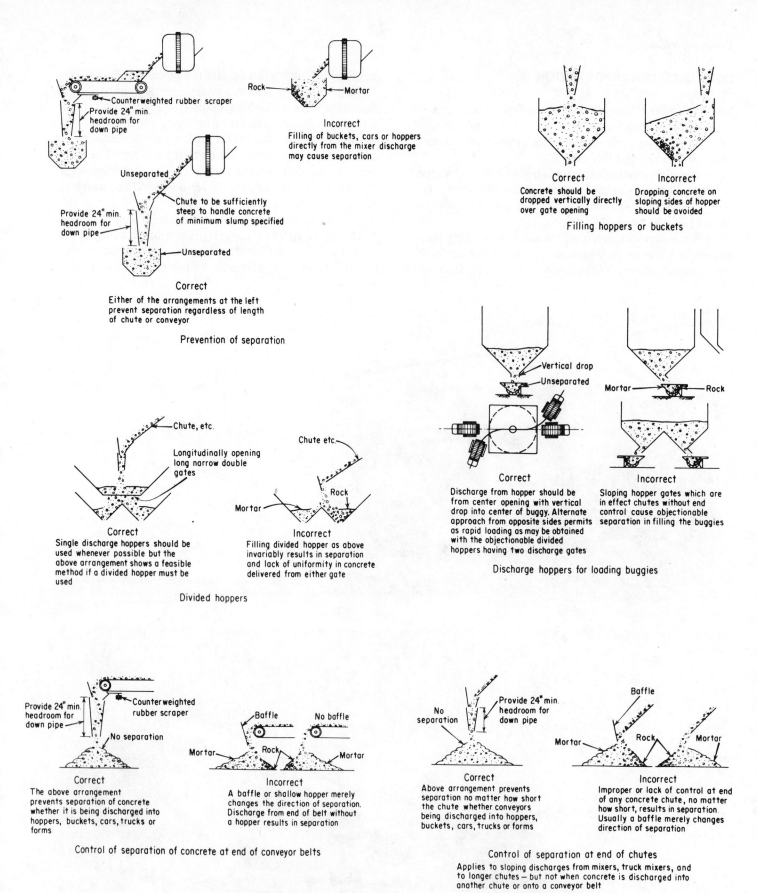

Incorrect
Filling of buckets, cars or hoppers directly from the mixer discharge may cause separation

Correct
Concrete should be dropped vertically directly over gate opening

Incorrect
Dropping concrete on sloping sides of hopper should be avoided

Filling hoppers or buckets

Correct
Either of the arrangements at the left prevent separation regardless of length of chute or conveyor

Prevention of separation

Correct
Single discharge hoppers should be used whenever possible but the above arrangement shows a feasible method if a divided hopper must be used

Incorrect
Filling divided hopper as above invariably results in separation and lack of uniformity in concrete delivered from either gate

Divided hoppers

Correct
Discharge from hopper should be from center opening with vertical drop into center of buggy. Alternate approach from opposite sides permits as rapid loading as may be obtained with the objectionable divided hoppers having two discharge gates

Incorrect
Sloping hopper gates which are in effect chutes without end control cause objectionable separation in filling the buggies

Discharge hoppers for loading buggies

Correct
The above arrangement prevents separation of concrete whether it is being discharged into hoppers, buckets, cars, trucks or forms

Incorrect
A baffle or shallow hopper merely changes the direction of separation. Discharge from end of belt without a hopper results in separation

Control of separation of concrete at end of conveyor belts

Correct
Above arrangement prevents separation no matter how short the chute whether conveyors being discharged into hoppers, buckets, cars, trucks or forms

Incorrect
Improper or lack of control at end of any concrete chute, no matter how short, results in separation. Usually a baffle merely changes direction of separation

Control of separation at end of chutes
Applies to sloping discharges from mixers, truck mixers, and to longer chutes — but not when concrete is discharged into another chute or onto a conveyor belt

FIGURE 7–31 Correct and incorrect methods of handling concrete. (*Courtesy Portland Cement Association.*)

162

tion. Any additional mixing should be done at the designated agitating speed. Concrete should be delivered and discharged from the truck mixer within 1½ hr after the introduction of water to cement and aggregate.

An agitator truck is similar to a transit mix truck, except that it carries no water tank. This means that the complete mix is made up at the batching plant and charged into the truck drum. The truck simply keeps the concrete agitated until it is delivered. As a result, the distance that may be traveled is limited to that which can be covered within the initial setting time of the paste. In extremely hot weather, it may be necessary to use ice to keep the temperature down. This is done to prevent the initial set of concrete from taking place before it can be delivered.

Rail cars especially designed for transporting concrete are used only on large projects. Some are tilted to discharge through side or end gates, while others discharge through bottom gates. Concrete is normally dumped into a large hopper from which short chutes or downspouts direct it to the forms. It is essential to closely supervise this operation in order to prevent segregation. Figure 7-31 illustrates correct and incorrect procedures in handling and transporting concrete.

PLACING CONCRETE

Concrete should be placed as nearly as possible in its final position. It should not be placed in large quantities in one position and allowed to flow or be worked over a long distance in the form. The mortar will tend to flow ahead of the coarser materials, thus causing stone pockets and sloping work planes.

Concrete placed in forms should *never* be allowed to drop freely more than 3 or 4 ft. When drops exceed this, rubber or metal chutes should be used. Figure 7-32 illustrates a metal drop chute to be used in narrow forms. These chutes should be provided in sections which can be coupled so that the length can be adjusted as required. When it is impractical to reach the bottom of narrow forms by means of drop chutes, concrete may be deposited through openings in the sides of forms, known as *windows*. Figure 7-33 illustrates the proper construction of such a "window," which allows the concrete to overflow into the forms rather than allowing it to enter at an angle.

When concrete is placed on a sloping surface, placing should begin at the bottom of the slope. This system will not only improve the compaction of the concrete as placing progresses but will prevent the flowing out of mortar which would occur if pouring had begun at the top of the slope.

Placing of concrete for a slab should begin against

FIGURE 7-32 Using a drop chute in narrow form. (*Courtesy Portland Cement Association.*)

a wall or against previously placed concrete. Dumping away from previously placed concrete allows the coarse aggregate to separate from the mortar. Figure 7-33 illustrates this.

Concrete should be placed in wall forms in relatively thin layers or *lifts,* 12 to 18 in. deep, each layer placed the full length of the form before the next lift is begun. This operation must take place so that each lift is placed in time to integrate easily and completely with the one below. This time factor will be determined by the kind of cement used in making the concrete, by the richness of the mix, by the presence or absence of accelerators, by the temperature of the concrete, and by atmospheric conditions at the time of placing. In addition, the first batches of each lift should be placed at the ends of the form section or in corners, and placing should then proceed toward the center. This is done to prevent the trapping of water at the ends of the sections, in corners, and along form faces. The integration of each lift with the one below is done with the aid of puddling spades or vibrators.

Concrete being placed in columns and walls should be allowed to stand for about 2 hr before placing the concrete for monolithic girders, beams, and slabs. This allows the concrete in the walls or columns to settle and thus prevents cracking due to settlement, which would occur if all members were placed at one time.

The correct placing of concrete includes proper consolidation to ensure that no pockets or spaces remain unfilled and that the face of the formed concrete has been made as smooth as required by specifications. The compacting or consolidation is done by means of *puddling spades* or *vibrators*. A puddling spade is simply a flat piece of metal attached to a handle, which is worked up and down in the freshly placed concrete. Vibrators may be either *internal* (see Figs. 7-34 and 7-35) or *external*.

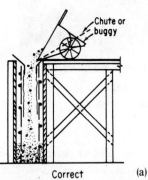
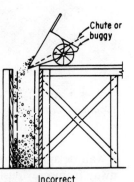
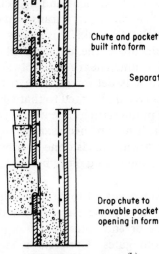
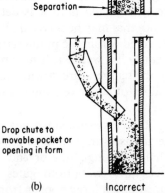

Correct (a)   Incorrect

Separation is avoided by discharging concrete into hopper feeding into drop chute. This arrangement also keeps forms and steel clean until concrete covers them.

Permitting concrete from chute or buggy to strike against form and ricochet on bars and form faces causes separation and honeycomb at the bottom.

Placing in top of narrow form

Correct (b)   Incorrect

Chute and pocket built into form

Separation

Drop chute to movable pocket or opening in form

Drop concrete vertically into outside pocket under each form opening so as to let concrete stop and flow easily over into form without separation.

Permitting rapidly flowing concrete to enter forms on an angle invariably results in separation.

Placing in deep narrow wall through port in form

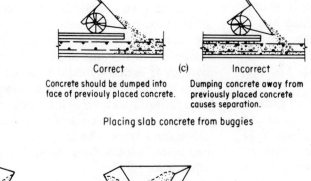

Correct (c)   Incorrect

Concrete should be dumped into face of previouly placed concrete.

Dumping concrete away from previously placed concrete causes separation.

Placing slab concrete from buggies

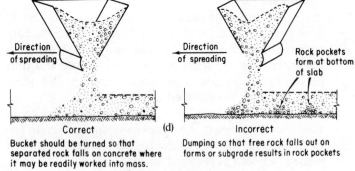
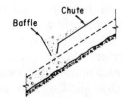
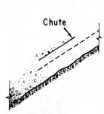

Direction of spreading

Correct (d)   Incorrect

Rock pockets form at bottom of slab

Bucket should be turned so that separated rock falls on concrete where it may be readily worked into mass.

Dumping so that free rock falls out on forms or subgrade results in rock pockets

If separation has not been eliminated in filling placing buckets (A tempory expedient until correction has been made)

Baffle   Chute   Chute

Correct (e)   Incorrect

A baffle and drop at end of chute will avoid separation and concrete remains on slope.

Discharging concrete from free end chute onto a slope causes separation of rock which goes to bottom of slope. Velocity tends to carry concrete down the slope.

Placing concrete on a sloping surface

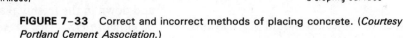

FIGURE 7–33 Correct and incorrect methods of placing concrete. (*Courtesy Portland Cement Association.*)

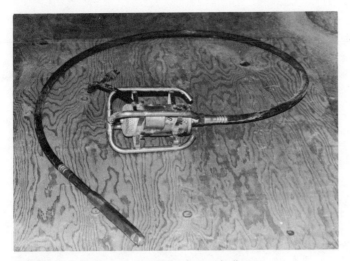

FIGURE 7-34 Electrically driven internal vibrator.

Internal vibrators should always be inserted vertically into the concrete (see Fig. 7-36) and should be used for consolidation only—not to move concrete from one place to another within a form. External vibrators are used against the outside of forms and are most effective in producing smooth surfaces against the form faces.

Puddling or vibrating eliminates stone pockets and air bubbles and consolidates each layer of concrete with the one below. The process should also bring enough mortar to the surface or the form face to ensure the proper finish. It must be remembered that excessive vibration causes segregation because coarse aggregate is forced away from the vibrator and pockets of mortar form around it. It should also be remembered that vibrating tends to increase the pressure of concrete against the forms and that special care must be taken to see that forms are strong and tight enough to withstand the additional pressure.

Underwater Concreting

Placing concrete under water requires special techniques. Of course, it cannot be dropped through the water but must be conducted to its final position. A number of methods are used to place concrete for underwater structures, among them the use of a *tremie,* an *underwater bucket,* a *concrete pump,* or *preplaced aggregate.*

FIGURE 7-35 Gasoline-powered internal vibrator. (*Courtesy Whiteman Mfg. Co.*)

FIGURE 7-36 Correct and incorrect methods of consolidating concrete. (*Courtesy Portland Cement Association.*)

Correct	(a)	Incorrect
Start placing at bottom of slope so that compaction is increased by weight of newly added concrete. Vibration consolidates the concrete.		When placing is begun at top of slope the upper concrete tends to pull apart especially when vibrated below as this starts flow and removes support from concrete above.

When concrete must be placed in a sloping lift

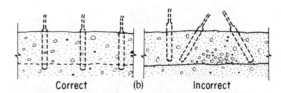

Correct	(b)	Incorrect
Vertical penetration of vibrator a few inches into previous lift (which should not yet be rigid) at systematic regular intervals will give adequate consolidation.		Haphazard random penetration of the vibrator at all angles and spacings without sufficient depth will not assure intimate combination of the two layers.

Systematic vibration of each new lift

Tremie

One of the most widely used methods of placing concrete under water is by the use of a tremie. This is a pipe whose diameter should normally be 8 times the maximum size of the coarse aggregate in the concrete being used. It is commonly made in 10- and 12-in. diameters in 10-ft lengths. A funnel-shaped hopper is bolted to the top, and sections are bolted together with a gasket at each joint, to make a pipe long enough to reach from the surface to the bottom of the underwater structure.

The bottom end is fitted with a watertight valve or a plug of some kind, and the tremie is lowered by hoisting equipment into position. Placement is begun by filling the tremie with concrete while the bottom valve is closed and then opening it to allow concrete to flow out. It flows outward from the bottom of the pipe, pushing the existing concrete surface outward and upward. Concrete is fed to the tremie continuously as long as the valve is open and as long as the flow is smooth so that the concrete surface next to the water is not physically disturbed; high-quality concrete will result. The tremie is slowly raised as concreting progresses, but the end is kept submerged in concrete at all times, as illustrated in Fig. 7–37.

FIGURE 7–37 Diagram of tremie.

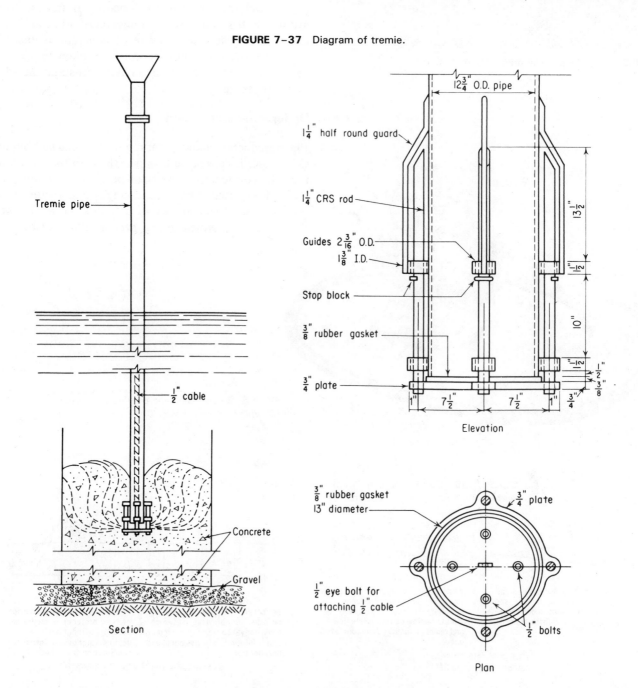

If a plug is used at the bottom of the pipe, it will be pushed out as concrete is introduced into the tremie and the process will continue as above.

Where large areas are involved, several tremies are used together, the concrete from each flowing into a common mass. In most cases, tremies should not be spaced more than 20 to 25 ft apart.

The mix proportions for tremie concrete will be somewhat different from normal structural mixes because the concrete must flow into place and consolidate without any mechanical help. The slump should be from 6 to 9 in., and generally rounded coarse aggregate is preferable to crushed stone, since it will aid flowability. The fine aggregate proportions will usually be from 40 to 50% of the total aggregates, and the water/cement ratio is kept to 0.44 or less by weight. Air-entraining agents and pozzolans are often used to improve the concrete flow characteristics.

Underwater Bucket

An underwater bucket has bottom gates which cannot be opened until the bucket is resting on the bottom or on previously placed concrete. This operation allows depositing the concrete and removing the bucket without disturbing the concrete or unduly agitating the water. A canvas cover prevents the swirling action of water from disturbing the concrete while the bucket is being lowered.

Concrete Pump

Using a concrete pump to place under water is similar to using a tremie. The lower end of the line must be kept submerged in concrete and is then slowly raised to prevent the pressure from overcoming the pump as placing progresses.

Preplaced Aggregate

Using grouted aggregate involves first placing a series of small vertical pipes in the form, reaching from the bottom to a level above that which the concrete is to reach. The forms are then filled with the required coarse aggregate. Finally, structural grout is pumped through the pipes, forcing the water out of the forms and filling the spaces between the particles of aggregate. The pipes are slowly withdrawn as grouting proceeds, but their lower ends must be kept below the level of grout already in the form.

HOT WEATHER CONCRETING

Placing concrete during hot weather presents some special problems, and the success of a concreting project under such conditions depends on their solution.

High temperatures accelerate the set of concrete, and only a short time is available in which finishing may be done. In addition, more mixing water is usually required for the same amount of slump, and if more cement is not added also, to maintain the water/cement ratio, the concrete strength will be reduced. Higher water content also will result in greater drying shrinkage. Figure 7-38 indicates the relationship between concrete temperature and the amount of water required per cubic yard of concrete.

During hot weather, a concrete temperature of 50 to 60°F is desirable, and for massive projects it should be from 40 to 50°F. The most practical way to achieve these temperatures is to control the temperature of the concrete ingredients.

Aggregate stockpiles should be shaded from the sun and kept moist by sprinkling in order to take advantage of the cooling brought about by evaporation. Coarse aggregate may be sprayed with cold water or chilled by blowing cold air over it while it is on the way to the batching plant or mixer.

Water may be cooled by insulating and burying the supply line, by refrigeration, or by adding crushed or flaked ice to the mixing water, provided that the ice is all melted by the time mixing is completed.

The subgrade, forms, and steel should be sprinkled with cool water just before concrete is placed during hot weather. It is also helpful to wet down the entire area since the evaporation will help to cool the surrounding air and increase the relative humidity.

Transporting and placing should be carried out as rapidly as possible. The depth of lifts in walls may have to be reduced to ensure consolidation with the concrete below and the area of slab pours reduced to avoid cold

FIGURE 7-38 Relationship of water content to concrete temperature. (*Courtesy Portland Cement Association.*)

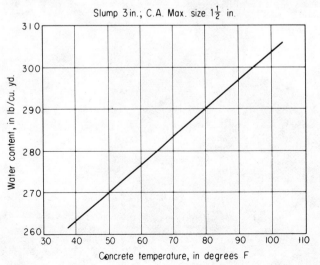

joints. All protection possible should be given from the sun and wind.

Finishing steps should be attended to at the earliest possible moment and the curing and protection procedures carried out promptly. Wood forms should be sprayed with water while still in place and loosened as soon as practicable so that water can be run down inside the forms. On flat surfaces, the curing water should not be much cooler than the concrete, in order to minimize cracking due to stresses caused by temperature change. Preferably after 24 hr of moist curing, curing paper, heat-reflecting plastic sheets, or a curing compound should be applied without delay.

FIGURE 7-39 Polethylene film enclosure. (*Courtesy Portland Cement Association.*)

COLD WEATHER CONCRETING

Experiments in cold weather construction have been carried out by contractors and subcontractors for many years, and much valuable experience has been gained. This experience, together with new techniques and materials which have recently appeared on the market, has led to a marked change in the attitude of owners, builders, architects, and engineers toward cold weather construction.

Most industrial building projects are not interrupted during the winter months, though it has been considered desirable to have the building "closed in" before the advent of cold weather. But "closing in" meant completing exterior walls and roofs, a long job in the case of large buildings. So it has become common practice to enclose a structural frame with large tarpaulins or polyethylene film, as shown in Fig. 7-39. Polyethylene has the added advantage of letting light into the enclosed space.

Within the enclosure, the use of equipment such as space heaters (Fig. 7-40) or automatic oil furnaces (see

FIGURE 7-40 Space heater. (*Courtesy Kelley Machine Div. Wiesner-Rapp Co. Inc.*)

FIGURE 7-41 Countercurrent mixers in precast plant. (*Courtesy Almar Specialty Machines Inc.*)

FIGURE 7–42 Oil-burning heater. (*Courtesy Portland Cement Association.*)

FIGURE 7–43 Enclosed hanging scaffold. (*Courtesy Portland Cement Association.*)

Fig. 7–42) permits temperatures which allow most types of construction work to continue throughout the winter.

In multistory construction, too large to be completely covered, enclosures may be used one floor at a time. This takes the form of a hanging scaffold enclosed in tarpaulin or plastic film which protects masons and masonry while closure walls are being built (Fig. 7–43). When heat is required, warm air may be piped into the enclosure.

In recent years it is becoming more and more common for major projects to be initiated during the winter. In most regions except the northern states, the Prairie Provinces, and northwestern Ontario, frost penetration under snow cover seldom exceeds 1 ft. This presents no problem for modern excavating machinery and, in fact, sometimes facilitates excavation by providing firmer footing for equipment and trucks.

When it is known in advance that excavating is to be done in soil having considerable frost penetration, the depth of frost can be kept to a minimum by leaving the snow undisturbed over the area or by covering the site with a thick layer of straw before the first snowfall. As soon as an excavation has been completed, steps must be taken to prevent frost from getting into the ground. Straw may again be used, or a shelter such as that shown in Fig. 7–44 may be provided for even better protection.

Protection for concrete during placing and curing operations is essential whenever below-freezing temperatures are to be expected. Concrete should be warm when placed and should remain at a reasonable temperature until it has gained sufficient strength to prevent frost damage. Concrete which has developed a strength of 500

FIGURE 7–44 Tarpaulin-enclosed foundation. (*Courtesy Portland Cement Association.*)

psi is usually considered past the danger stage, although it cannot withstand repeated cycles of freezing and thawing. To attain a minimum of 500 psi, the American Concrete Institute recommends that the temperature be maintained at 70 °F for 3 days, or 50 °F for 5 days.

In cold weather, it is sometimes necessary to heat the water and aggregate in order to produce concrete which will have a placing temperature of between 50° and 70 °F. Water is the most practical component to heat, since more heat units can be stored in water than in other materials. However, water should not be heated above 175 °F because of the danger of causing a *flash* set. If the bulk of the aggregate has a temperature appreciably below 45 °F, it must also be heated. Aggregate is usually heated by either steam coils or by the injection of live steam.

A formula has been devised to calculate the temperatures to which water—and aggregate, if necessary—must be heated to produce concrete of a given temperature:

$$X = \frac{Wt + 0.22W't}{W + 0.22W'}$$

where

X = desired concrete temperature

W = weight of water

W' = weight of cement and aggregate

t = temperature of water

t' = temperature of solids

Remember that the moisture in the aggregate will have the same temperature as the aggregate. Remember also that if it is necessary to calculate a temperature to which aggregate must be heated, it is assumed that the added water will also have been heated to a given temperature.

Large exposed areas, such as floor slabs, should be protected on both sides. It is difficult to supply heat to the upper surface of the slab if it is not enclosed, but it should be covered with insulation, as illustrated in Fig. 7–45. Heat can be supplied to the underside of the slab by heaters, as in Fig. 7–46.

Enclosures such as that shown in Fig. 7–47 may be used while erecting concrete or masonry walls. As building progresses, protection may be removed from one location and reused in another (see Fig. 7–48).

It may at times be very difficult to supply heat to isolated structures during their curing period. An alternative to doing so is to build insulated forms (Fig. 7–49) and to use concrete with the maximum allowable placing temperature. The insulation will allow the concrete to retain its heat and to cure properly despite low temperatures.

There are two particular precautions which should be taken when using heaters in enclosed spaces. Heaters in spaces containing freshly placed concrete should be adequately vented. Failure to do so may result in an accumulation of carbon monoxide gas which presents a hazard to health and acts as a retarder to the setting of concrete. The second precaution concerns the use of oil

FIGURE 7–45 Surface of slab insulated against heat loss. (*Courtesy Portland Cement Association.*)

FIGURE 7–46 Supplying heat to underside of slab. (*Courtesy Portland Cement Association.*)

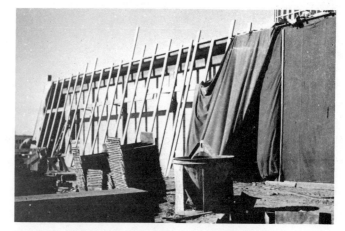

FIGURE 7–48 Removing part of protective enclosure. (*Courtesy Portland Cement Association.*)

FIGURE 7–49 Insulated forms. (*Courtesy Portland Cement Association.*)

FIGURE 7–47 Wall enclosure for winter construction. (*Courtesy Portland Cement Association.*)

burning heaters in rooms which have received their base plaster coat. An oily film may be deposited upon the plaster surface and may very seriously affect the adhesion of the finish plaster coat.

Shelters used in winter construction can be divided into two general groups. The first is the self-supporting type. One example of this is a shelter made from light laminated arches covered with plastic film, now in fairly common use. Another is an air-supported unit made from plastic-coated fabric and often large enough to cover an area the size of a football field. The second type of shelter may use the existing frame of the building for support and encompass all the shelters discussed earlier in this chapter.

Cost is a prime factor in deciding whether or not to build in the winter. It is generally conceded that the average increase in costs for winter construction will vary from 5 to 10%. This includes the cost of providing tarpaulins, heaters, insulation, fuel, and snow removal. Indirect savings may result from higher productivity, uninterrupted schedules, and greater control of temperatures on the job—thus offsetting direct costs.

CONCRETE JOINTS

See Chapter 6.

FINISHING AND CURING CONCRETE

The treatment of exposed concrete surfaces to produce the desired appearance, texture, or wearing qualities is known as *finishing*. The procedure to follow depends primarily on whether the surface is horizontal or vertical. Horizontal surfaces are usually exposed and must be finished before the concrete has hardened. Timing is a very important factor in this operation. Finishing must be done when the concrete is neither too hard to be worked nor so soft that it will fail to retain the desired finish.

The concrete must first be struck off level as soon as it has been placed. This is done by the use of *strike-off bars* worked against the top edge of *screeds* previously set to the proper height (see Fig. 7–50). These bars are operated either by hand or by power, the latter acting as vibrators as well as strike-off bars. Striking off removes all humps and hollows, leaving a true and even surface.

Further finishing is then delayed until the surface is quite stiff but still workable. Any water appearing on the surface should be removed with a broom, burlap, or other convenient means. Do not sprinkle cement or a mixture of cement and fine aggregate on the surface to take up excess water. This forms a layer that will dust and wear easily, or develop a great many fine cracks due to shrinkage. A short wood *float* (see Fig. 7–51) produces an even surface, with all larger particles of aggregate embedded in mortar. This should be followed by a long float (see Fig. 7–52), which removes the marks left by the short one and produces a gritty nonskid surface.

If a scored surface is required, it can be produced by drawing a stiff, coarse broom across the surface, as shown in Fig. 7–53.

When a smooth surface is required, it may be produced by using a steel trowel, illustrated in Fig. 7–54. Finishing can also be done by the use of power floats and power trowels. These are commonly used on large projects and are operated as in Fig. 7–55. Smooth surfaces may also be produced by the use of power grinders, as illustrated in Fig. 7–56.

Trimmed edges and joints are made by using edg-

FIGURE 7–50 Power-operated strike-off bars. (*Courtesy Whiteman Mfg. Co.*)

FIGURE 7–52 Magnesium float. (*Courtesy Portland Cement Association.*)

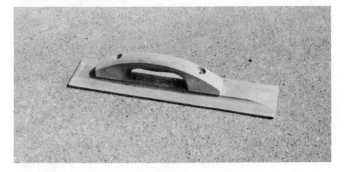

FIGURE 7–51 Short wood float.

FIGURE 7-53 Broomed surface. (*Courtesy Portland Cement Association.*)

FIGURE 7-54 Steel trowel.

FIGURE 7-55 Power trowel in operation. (*Courtesy Whiteman Mfg. Co.*)

FIGURE 7-56 Power concrete grinder in operation.

ing tools such as those in Fig. 7-57. Special hard surfaces are applied and finished as described in Chapter 11.

When vertical surfaces have to be given a smooth finish, a different technique is used. First, the forms must be removed after the concrete is sufficiently strong. The tie holes must then be filled with mortar, or the snap ties must be broken off, and all depressions must be patched. All rough fins and protrusions must be removed. This can be done by rubbing the surface with a flat emery stone. Next, the surface is washed to remove all loosened material and at the same time dampen the concrete.

Grout is then applied to the surface with a stiff brush and rubbed in with a piece of burlap or a cork float.

FIGURE 7-57 (a) Jointing tool; (b) edging tool.

(a)   (b)

This will fill in all the pores and small holes. After the grout has set sufficiently so that it will not rub out of the holes easily, the surface must be rubbed down with clean burlap to remove all excess material.

It has been shown that the strength, watertightness, and durability of concrete improve with age as long as conditions are favorable for continued hydration of the cement. The improvement in strength is rapid at the early stages but continues more slowly for an indefinite period. The conditions required are sufficient moisture and temperatures preferably in the 65 to 70 °F. range.

Fresh concrete contains more than enough water for complete hydration, but under most job conditions, much of it will be lost through evaporation unless certain precautions are taken. In addition, hydration proceeds at a much slower rate when temperatures are low, and practically ceases when they are near freezing. Thus, in order for concrete to develop its full potential, it must be protected against loss of moisture and low temperatures, particularly during the first 7 days.

Concrete can be kept moist in several ways, such as by leaving forms in place, by sprinkling with water, by covering with water or a waterproof cover, and by sealing the surface with a wax coating.

When forms are left on, the top surfaces should be covered to prevent evaporation. In hot weather, it may be necessary to sprinkle water on wooden forms to prevent them from drying out and taking moisture from the concrete.

Where concrete is to be kept moist by sprinkling, care should be taken not to allow the surfaces to dry out between applications. One method of doing this is to cover the concrete with burlap or canvas.

Flat surfaces can be covered with water, damp sand, or waterproof material such as waxed paper or polyethylene film. Sprinkling may also be used, but it must be constant enough to prevent the concrete from drying out. A wax film applied as a spray is commonly used to seal a concrete surface.

HOT CONCRETE

In the foregoing paragraphs, a good deal of emphasis has been placed on the necessity for keeping fresh concrete temperatures in the range of from 50 to 90°F. For concrete that is to be placed in the field, the maintenance of such temperatures is necessary in order to allow time for consolidation, finishing, etc.

But in situations in which concrete is to be mixed in a central plant for use in producing reinforced precast products or for concrete block, it has been found that, under the proper conditions, fresh concrete can be placed successfully at temperatures up to 175°F.

The primary reason for using hot concrete is that it sets and gains early strength much more rapidly than concrete placed at conventional temperatures. For example it has been shown that curing times of 3 hr can produce approximately 60 to 70% of the 28-day design strength. Therefore more intensive use can be made of expensive plant facilities. Molds can be used two or sometimes three times in an 8-hr day, rather than once, as in the case with normal temperature concrete.

Moderate concrete temperatures of up to 90°F can be achieved economically by heating the mixing water and the aggregates. When hotter concrete is required, injection of steam into the mixer is used to produce temperatures of from 120 to 175°F. Temperature is controlled by regulating the amount of steam entering the mixer. Steam pressure of from 30 to 70 psi is used, with steam temperature from 300 to 320°F.

To achieve the best results, the steam should be injected where mixing intensity is high, and for this reason, counter-current mixers, which do produce high-intensity mixing, are often used in plants producing hot concrete (see Fig. 7-41). Normally, steam is injected into the mixer at the same time the aggregates are dumped but just prior to adding cement and water. Water produced in the mix due to the condensation of steam is included in the total amount of water required for strength and slump control. If more water is required for slump adjustment, it should be the same temperature as the concrete.

A fast, efficient transportation system is necessary to convey fresh concrete from the mixer to the forms. In some cases, monorail hoppers, traveling at about 200 ft/min, are used. It has been found that external form vibrators have usually been more effective than internal vibrators.

Cleanup of mixing and casting equipment is very important because hot concrete sets up very quickly. The mixers will have to be cleaned twice a day, conveyors and buckets up to six times a day.

DESIGN OF STRUCTURAL LIGHTWEIGHT CONCRETE

Structural lightweight concrete, mentioned earlier in this chapter, is defined as concrete which will have a 28-day compressive strength of more than 2500 psi and an air-dry unit weight of less than 115 lb/cu ft but normally more than 85 lb/cu ft.

Properties of Structural Lightweight Aggregates

Structural lightweight aggregates, usually made from expanded *clay, shale, slate,* or *blast furnace slag* or from *pelletized fly ash,* normally are produced in two sizes: *fine,* with a maximum size of ⅜ in., and *coarse,* with a usual maximum of ¾ in. They have considerably lower unit weights than normal weight aggregates, ranging from 35 to 70 lb/cu ft. ASTM C330 limits the dry, loose unit weight of lightweight aggregate for structural concrete to a maximum of 70 lb/cu ft for fine, 55 lb/cu ft for coarse, and 65 lb/cu ft for combined fine and coarse.

In contrast to normal weight aggregates, which usually absorb 1 to 2% water by weight of dry aggregates, lightweight aggregates may absorb from 5 to 20% water by weight of dry material, based on 24 hr absorption tests. This can amount to as much as 250 lb of water per cubic yard of concrete, if total absorption is reached. As a result, to help control the uniformity of lightweight mixtures, *prewetted but not saturated* aggregates are generally used. Prewetting should be done 24 hr in advance of mixing to allow the moisture to distribute itself evenly through the aggregate.

The bulk specific gravity for lightweight aggregates is generally between 1.0 and 2.4, and for any specific aggregate, the bulk specific gravity will usually increase as the maximum particle size decreases. It is often difficult to determine accurately the bulk specific gravity of many lightweight aggregates, and consequently the mix design method for structural lightweight concrete is based on the concept of *specific gravity factors,* which take into account the actual moisture conditions of the aggregate, rather than on bulk specific gravity.

In some cases, normal weight fine aggregate may be substituted for part or all of the lightweight fines in lightweight concrete mixtures. This may be done for *economy* or to improve *strength, workability, durability, modulus of elasticity,* or *finishing qualities* and generally decreases the water required for a given slump. However, such a combination will increase unit weight from 10 to 18 lb/cu ft and have a detrimental effect on such properties as fire resistance and thermal insulation.

Proportioning

Proper proportioning of lightweight concrete mixtures may sometimes be achieved with the conventional procedures used for normal weight mixes in a few specific instances. Generally speaking, however, the variations in rate of absorption and total absorption of most lightweight aggregates make it very difficult to establish a water/cement ratio accurately enough to be used as a basis for proportioning. Instead it is found to be more practical to design lightweight aggregate mixes by a series of trial mixes proportioned on a *cement content* basis. Tests are made at various ages to determine what cement content is required to produce a given strength.

Lightweight structural concrete intended for watertight structures should be designed for at least 3750 psi if exposed to fresh water or 4000 psi for salt water exposure. The latter strength should also be used for concrete exposed to sulfate concentrations.

Water Content

Concrete made from normal weight aggregates with ¾-in. maximum size will require from 310 to 360 lb of water per cubic yard. On the other hand, if lightweight aggregates with the same maximum size are used, water requirements may range from 290 to 500 lb/cu yd of concrete. When lightweight fines are replaced by normal weight fine aggregate, the amount of water required will be reduced and entrained air in the mix will usually mean a reduction in water of 2 to 3% for each 1% of entrained air.

Cement Content

The cement content required to produce a given strength will vary widely, depending on the characteristics of the lightweight aggregates. In Table 7-9 an approximation is given of the cement contents to be used for establishing trial mix proportions to obtain a specified strength.

TABLE 7-9 Approximate relationship between strength and cement content

Compressive strength (psi)	Cement content (lb/cu yd)
2500	425-700
3000	475-750
4000	550-850
5000	650-950

Proportion of Fine to Coarse Aggregate

Lightweight concrete mixes require a greater percentage of fines than does normal weight concrete because there are significant differences between the bulk specific gravities of fine and coarse lightweight particles and because most lightweight aggregates are rough-surfaced and angular in shape. Workable mixes of lightweight concrete require 40 to 60% of fines by *volume.* That percentage can be reduced when normal weight fines are substituted for the lightweight.

Usually between 28 and 32 cu ft of combined dry, loose lightweight aggregates are required to make 1 cu yd of concrete. The total volume of aggregates and the proportion of fine to coarse, which depends on *gradation, shape, size,* and *surface texture,* have a great influence on the workability and finishing qualities of the fresh concrete.

Entrained Air

Most lightweight concretes contain 2 to 4% entrapped air, but this has little effect on workability and durability and it is desirable to add entrained air to most lightweight mixes. When the concrete is subject to freezing and thawing cycles, air should not be less than $6(\pm 1\frac{1}{2})\%$ total when maximum aggregate size is $\frac{3}{4}$ in. and $7\frac{1}{2}(\pm 1\frac{1}{2})\%$ when maximum aggregates size is $\frac{3}{8}$ in. Even when freeze-thaw resistance is not required, 4 to 8% air is recommended for workability.

Trial Mixes

It is common practice to determine the cement content required for a specific strength by the use of trial mixes, and these should be made with at least three different cement contents. Each mix should have sufficient entrained air to ensure workability and durability. Each trial batch should produce at least $1\frac{1}{2}$ cu ft of concrete, from which measurements of air content and unit weight will be made and four or five test cylinders molded, on which compressive strength tests will be made. The test results will determine the cement content that produces the required strength. Based on the mix proportions of the trial mixes, the mix proportions for the selected cement content can be accurately estimated. Small trial batches made in the lab may require some further adjustment when enlarged to job mixes.

Example

Design a lightweight aggregate concrete mix which is to test 3000 psi in 28 days. Concrete is to have $6(\pm 1\frac{1}{2})\%$ air entrainment and $4(\pm 1)$ in. of slump. Type 1 or normal portland cement and an air entraining agent are to be used.

Proceed as follows:

1. Determine the loose, dry unit weights of aggregates and their moisture content at time of mixing.

   Loose, dry unit wt. of C.A. = 48 lb/cu ft
   Loose, dry unit wt. of F.A. = 62 lb/cu ft
   Moisture content of C.A.   = 4%
   Moisture content of F.A.   = 7%

2. Determine the specific gravity factors (the relationship of aggregate weight to displaced volume) of the fine and coarse aggregates by the pycnometer method (see PCA literature for procedure).

   Specific gravity factor for C.A. = 1.40
   Specific gravity factor for F.A. = 1.98

3. Estimate quantities. From Table 7–9, the cement content should range from 475 to 750 lb/cu yd. Design trial mixes with cement contents of 500, 600, and 700 lb/cu yd, respectively. Assume the total dry, loose volume of aggregates required to be 30 cu ft, with equal volumes of fine and coarse for the first trial.

4. Calculate the loose weights of the damp fine and coarse aggregate required per cubic yard.

   Loose wt. of damp C.A. = $15 \times 48 \times 1.04 = 749$ lb
   Loose wt. of damp F.A. = $15 \times 62 \times 1.07 = 995$ lb

5. Calculate the displaced volumes of the damp aggregates, using the specific gravity factors at hand.

   $$\text{Displaced vol. of C.A.} = \frac{749}{62.4 \times 1.40} = 8.57 \text{ cu ft}$$

   $$\text{Displaced vol. of F.A.} = \frac{995}{62.4 \times 1.98} = 8.05 \text{ cu ft}$$

6. Calculate the displaced volumes of cement and air.

   $$\text{Displaced vol. of cement} = \frac{500}{62.4 \times 3.15} = 2.54 \text{ cu ft}$$

   Vol. of air = $27 \times 0.06 = 1.62$ cu ft

7. Calculate the volume and weight of water required. The volume of water will be 27 − combined volumes of cement, air, and aggregates.

   Vol. of water = $27 - (8.57 + 8.05 + 2.54 + 1.62) =$ 6.22 cu ft
   Wt. of water = $6.22 \times 62.4 = 388$ lb

8. Calculate the weights of materials required for a $1\frac{1}{2}$-cu ft trial batch. Divide the weights required per cubic yard by 18 (27/1.5).

   Wt. of cement = $500 \div 18 = 28$ lb
   Wt. of water = $388 \div 18 = 21.5$ lb
   Wt. of C.A.   = $749 \div 18 = 41.5$ lb
   Wt. of F.A.   = $995 \div 18 = 55$ lb

9. Make a chart similar to that shown in Table 7–10 and record all calculations made to this point, for ready reference.

10. Mix the materials and test the mixture for *air content, slump, unit weight,* and *workability.* If all are satisfactory, take cylinders for compression tests. If a correction in the amount of aggregate or water is required, adjust mix proportions as shown in steps 11 through 14.

11. If more coarse aggregate may be used and more water is required to produce the desired slump, add the amounts required. Assume 4.0 lb of C.A. and 4.5 lb of water.

12. Calculate the correction required per cubic yard for each.

    Correction for C.A.   = $4 \times 18 = 72$ lb
    Correction for water = $4.5 \times 18 = 81$ lb

13. Calculate the corrected batch proportions.
    (a) Corrected weights:

    Cement (no change)   = 500 lb
    F.A. (no change)   = 995 lb
    C.A.   = $749 + 72 = 821$ lb
    Water   = $388 + 81 = 469$ lb

TABLE 7-10 Initial batch proportions and trial batch adjustments

	1	2	3	4	5	6	7	8	9
	Estimated batch proportions (damp aggregates)		Weights for 1½ cu ft trial batch (lb)	Correction for 1½ cu ft (lb)	Correction for 1 cu yd (lb)	Corrected batch proportions (damp aggregates)		Adjusted batch proportions (damp aggregates)	
	Weight (lb)	Displaced volume (cu ft)				Weight (lb)	Displaced volume (cu ft)	Weight (lb)	Displaced volume (cu ft)
Cement	500	$\dfrac{500}{62.4 \times 3.15} = 2.54$	28.0			500	2.54	$0.925 \times 500 = 462$	$0.925 \times 2.54 = 2.35$
Air		$27 \times 0.06 = 1.62$					1.62		1.62
C.A.	749	$\dfrac{749}{62.4 \times 1.40} = 8.57$	41.5	+4.0	+72	$749 + 72 = 821$	$\dfrac{8.21}{62.4 \times 1.40} = 9.39$	$0.925 \times 821 = 759$	$0.925 \times 9.39 = 8.67$
F.A.	995	$\dfrac{995}{62.4 \times 1.98} = 8.05$	55.0			995	8.05	$0.925 \times 995 = 920$	$0.925 \times 8.05 = 7.44$
Water	$6.22 \times 6.24 = 338$	$27 - 20.78 = 6.22$	21.5	+4.5	+81	$338 + 81 = 469$	$469 \div 62.4 = 7.51$	$0.925 \times 469 = 434$	$0.925 \times 7.51 = 6.92$
Total		27.00							27.00

(b) Corrected displaced volumes:

Cement (no change)	=	2.54 cu ft
F.A. (no change)	=	8.05 cu ft
Air (no change)	=	1.62 cu ft
C.A.	$= \dfrac{821}{62.4 \times 1.40} =$	9.39 cu ft
Water	$= 469 \div 62.4 =$	7.51 cu ft
Total	=	29.11 cu ft

14. Multiply the weights and displaced volumes of materials from step 13 by 0.925 (27/29.11) to adjust the quantities to a yield of 27 cu ft.

(a) Adjusted weights:

Cement	$= 500 \times 0.925 =$	462 lb
F.A.	$= 995 \times 0.925 =$	920 lb
C.A.	$= 821 \times 0.925 =$	759 lb
Water	$= 469 \times 0.925 =$	434 lb

(b) Adjusted displaced volumes:

Cement	$= 2.54 \times 0.925 =$	2.35 cu ft
Air	=	1.62 cu ft
F.A.	$= 8.05 \times 0.925 =$	7.44 cu ft
C.A.	$= 9.39 \times 0.925 =$	8.67 cu ft
Water	$= 7.51 \times 0.925 =$	6.92 cu ft
Total	=	27.00 cu ft

Record all of these calculations on the chart.

The adjusted quantities above are for damp aggregates. For future adjustments of batch proportions, it is necessary to convert the batch weights of materials to dry weights. These will then serve as a basis for the calculation of adjustments for change in *aggregate moisture conditions, cement content, proportions of aggregates, slump,* or *air content,* in order to maintain a yield of 27 cu ft. The procedure for converting batch weights of damp aggregates to batch weights for dry aggregates is as follows:

1. Calculate the dry weights of C.A. and F.A. from the batch weights for damp aggregates. The weight of cement is to remain constant.

   Dry wt. of C.A. $= 759 \div 1.04 = 730$ lb
   Dry wt. of F.A. $= 920 \div 1.07 = 860$ lb

2. Calculate the specific gravity factors for the dry state. Assume that the specific gravity factor for C.A. is 1.37 and F.A., 1.99.

3. Calculate the displaced volumes of dry aggregates, using specific gravity factors.

   Displaced vol. of C.A. $= \dfrac{730}{62.4 \times 1.37} = 8.54$ cu ft

   Displaced vol. of F.A. $= \dfrac{860}{62.4 \times 1.99} = 6.92$ cu ft

4. Maintaining the displaced volumes of cement and air constant, calculate the required displaced volume of adding water and the required weight of added water.

   Displaced vol. of added water $= 27 - (2.35 + 1.62 + 8.54 + 6.92) = 7.57$ cu ft
   Req'd wt. of added water $= 7.57 \times 62.4 = 472$ lb

To record these calculations, make a chart similar to that shown in Table 7–11. Into the first two columns, transfer the figures from columns 8 and 9 of Table 7–10. Into columns 3 and 4 enter the calculations from above.

Aggregates may have different moisture contents from time to time, and it will be necessary, under these circumstances, to adjust the batch weights of materials to maintain a yield of 27 cu ft.

TABLE 7–11   Adjustments for changes in aggregate moisture conditions

	1	2	3	4	5	6
	Batch proportions (damp aggregates)		Converted batch proportions (dry aggregates)		Adjusted proportions (aggregate in new moisture condition)	
	Weight (lb)	Displaced volume (cu ft)	Weight (lb)	Displaced volume (cu ft)	Weight (lb)	Displaced volume (cu ft)
Cement	462	2.35	462	2.35	462	2.35
Air		1.62		1.62		1.62
C.A.	759	8.67	$\dfrac{759}{1.04} = 730$	$\dfrac{730}{62.4 \times 1.37} = 8.54$	$730 \times 1.02 = 745$	$\dfrac{745}{62.4 \times 1.35} = 8.84$
F.A.	920	7.44	$\dfrac{920}{1.07} = 860$	$\dfrac{860}{62.4 \times 1.99} = 6.92$	$860 \times 1.04 = 894$	$\dfrac{894}{62.4 \times 1.97} = 7.27$
Water	434	6.92	$7.57 \times 62.4 = 472$	$27 - 19.43 = 7.57$	$6.92 \times 62.4 = 432$	$27 - 20.08 = 6.92$
Total		27.00				27.00

Assume new moisture conditions of 2% for C.A. and 4% for F.A. Proceed as follows:

1. Calculate the specific gravity factors for the new moisture contents. Assume that of the C.A. to be 1.35 and of the F.A., 1.97.
2. Maintain the weight of cement and the displaced volumes of cement and air constant.
3. Calculate the weights of aggregates for the new moisture conditions.

$$\text{Wt. of moist C.A.} = 730 \times 1.02 = 745 \text{ lb}$$
$$\text{Wt. of moist F.A.} = 860 \times 1.04 = 894 \text{ lb}$$

4. Calculate the displaced volumes of aggregates, using new specific gravity factors.

$$\text{Displaced vol. of C.A.} = \frac{745}{62.4 \times 1.35} = 8.84 \text{ cu ft}$$

$$\text{Displaced vol. of F.A.} = \frac{894}{62.4 \times 1.97} = 7.27 \text{ cu ft}$$

5. Calculate the required displaced volume of added water and, from this, the required weight of added water.

$$\text{Displaced volume of added water} = 27 - (2.35 + 1.62 + 8.84 + 7.27) = 6.92 \text{ cu ft}$$
$$\text{Wt. of added water} = 6.92 \times 62.4 = 432 \text{ lb}$$

6. Enter these results into Table 7–11.

As indicated earlier, two more trial mixes will be made, using cement contents of 600 and 700 lb/cu yd. Much of the trial-and-error work may be simplified by beginning with proportions obtained from the first trial mix. They will have to be adjusted to compensate for the increase in cement content by a corresponding decrease in fine aggregate. Proceed as follows:

1. Maintain the weights and volumes displaced by dry, coarse aggregate, air, and water, shown in columns 3 and 4 of Table 7–11, constant. Transfer these values to columns 1 and 2 of Table 7–12.
2. Calculate the volume displaced by the new cement content, 600 lb/cu yd.

$$\text{Displaced vol. of cement} = \frac{600}{62.4 \times 3.15} = 3.05 \text{ cu ft}$$

3. Calculate the required displaced volume of dry fine aggregate as the difference between 27 cu ft and the sum of the displaced volumes of air, cement, coarse aggregate, and water.

$$\text{Req'd vol. of dry F.A.} = 27 - (3.05 + 1.62 + 8.54 + 7.57) = 6.22 \text{ cu ft}$$

4. Calculate the required weight of dry F.A., using the specific gravity factor for the dry state.

$$\text{Req'd wt. of dry F.A.} = 6.22 \times 62.4 \times 1.99 = 772 \text{ lb}$$

5. Convert the weight of dry F.A. to the weight at the original moisture condition.

$$\text{Wt. of F.A. at original moisture condition} = 772 \times 1.07 = 826 \text{ lb}$$

6. Calculate the weight of the C.A. at the original moisture condition.

$$\text{Wt. of C.A. at original moisture condition} = 730 \times 1.04 = 759 \text{ lb}$$

TABLE 7–12 Adjustment of proportions for change in cement content

	1	2	3	4	5	6
	Values from cols. 3 & 4 of Table 7–11		Adjusted proportions (dry aggregates)		Converted proportions (damp aggregates)	
	Weight (lb)	Displaced volume (cu ft)	Weight (lb)	Displaced volume (cu ft)	Weight (lb)	Displaced volume (cu ft)
Cement	462	2.35	600	$\frac{600}{62.4 \times 3.15} = 3.05$	600	3.05
Air		1.62		1.62		1.62
C.A.	730	8.54	730	8.54	$730 \times 1.04 = 759$	$\frac{759}{62.4 \times 1.35} = 9.01$
F.A.	860	6.92	$6.22 \times 62.4 \times 1.99 = 772$	$27 - 20.78 = 6.22$	$772 \times 1.07 = 826$	$\frac{826}{62.4 \times 1.97} = 6.72$
Water	472	7.57	472	7.57	$6.6 \times 62.4 = 412$	$27 - 20.4 = 6.60$
Total				27.00		27.00

7. Calculate the displaced volumes of damp F.A. and C.A.

$$\text{Displaced vol. of damp C.A.} = \frac{759}{62.4 \times 1.35} = 9.01 \text{ cu ft}$$

$$\text{Displaced vol. of damp F.A.} = \frac{826}{62.4 \times 1.97} = 6.72 \text{ cu ft}$$

8. Calculate the required volume of added water.

   Req'd vol. of added water $= 27 - (3.05 + 1.62 + 9.01 + 6.72) = 6.60$ cu ft

9. Calculate the required weight of added water.

   Req'd wt. of added water $= 6.60 \times 62.4 = 412$ lb

Minor changes in slump, air content, or proportion of fine to total aggregate may be desirable at times. These changes will necessitate corresponding adjustments in mix proportions to maintain yield and other characteristics. The following general rules may be used as a guide.

1. For each 1% increase or decrease in the percentage of fine to total aggregate, increase or decrease water by approximately 3 lb and cement content by 1%/cu yd.
2. For each 1-in. increase or decrease in slump, increase or decrease water by approximately 10 lb and cement content by 3%/cu yd.
3. For each 1% increase or decrease in air content, decrease or increase water by approximately 5 lb/cu yd and increase or decrease cement content by 2%/cu yd.

To calculate the adjustments, use the mix proportions that are based on dry aggregates. Maintaining the coarse aggregate constant, the quantities of cement, air, or water are increased or decreased according to the rules outlined above. These adjustments may result in an increase or decrease in the total displaced volume. To maintain the original volume, the proportion of fine aggregates should be decreased or increased accordingly.

Example

Assume that the proportions given in columns 3 and 4 of Table 7-12 should be adjusted to decrease the slump by 2 in. and increase air content by 2%. Proceed as follows:

1. Transfer the proportions given in columns 3 and 4 of Table 7-12 to columns 1 and 2 of Table 7-13.
2. Decrease of 2 in. in slump requires $2 \times 10 = 20$-lb decrease in water. Decrease in cement $= 2 \times 0.03 \times 600 = 36$ lb/cu yd.
3. Two percent increase in air content requires $2 \times 5 = 10$-lb decrease in water. Increase in cement $= 2 \times 0.02 \times 600 = 24$ lb/cu yd.
4. Total decrease in water $= 20 + 10 = 30$ lb. Total decrease in cement $= 36 - 24 = 12$ lb. Increase in air content $= 0.02 \times 27 = 0.54$ cu ft.
5. Enter these calculations in Table 7-13 and complete in accordance with previous similar steps.

TABLE 7-13 Adjustment of proportions for change in air content and slump

	1	2	3	4	5	6
	Values from cols. 3 & 4 of Table 7-12		Adjusted proportions (dry aggregates)		Converted proportions (damp aggregates)	
	Weight (lb)	Displaced volume (cu ft)	Weight (lb)	Displaced volume (cu ft)	Weight (lb)	Displaced volume (cu ft)
Cement	600	3.05	$600 - 12 = 588$	$\dfrac{588}{62.4 \times 3.15} = 2.99$	588	2.99
Air		1.62		$1.62 + 0.54 = 2.16$		2.16
C.A.	730	8.54	730	8.54	$730 \times 1.04 = 759$	$\dfrac{759}{62.4 \times 1.35} = 9.01$
F.A.	772	6.22	$6.32 \times 62.4 \times 1.99 = 785$	$27 - 20.77 = 6.23$	$785 \times 1.07 = 840$	$\dfrac{840}{62.4 \times 1.97} = 6.83$
Water	472	7.57	$472 - 30 = 442$	$\dfrac{442}{62.4} = 7.08$	$6.01 \times 62.4 = 375$	$27 - 20.99 = 6.01$
Total		27.00		27.00		27.00

REVIEW QUESTIONS

1. Give two reasons why as much aggregate as possible should be used when designing a concrete mix of any given strength.

2. Explain the purpose of each of the following aggregate tests: **(a)** silt test, **(b)** specific gravity test, **(c)** fineness modulus test, and **(d)** abrasion test.

3. Concrete is required for interior reinforced building columns 12″ by 12″, 10′ high. It is to test 4000 psi and have 3 to 4 in. of slump, with $5(\pm 1)\%$ air. The sp. gr. of the coarse aggregate is 2.65, its dry-rodded unit weight is 106 lb, and its maximum size 1 in. The sp. gr. of the fine aggregate is 2.78, and its fineness modulus 2.60. **(a)** Design a trial mix, using 30 lb of cement. **(b)** If there are 24 columns, how much cement is required to do the job?

4. Prepare a seven-point check list to be used when preparing to place concrete.

5. Describe the purpose of each of the following when finishing a concrete slab: **(a)** screed, **(b)** strike-off bar, **(c)** float, **(d)** trowel, and **(e)** drag.

6. Explain briefly **(a)** what is meant by a *flash set* of cement, **(b)** why it may be necessary to heat both water and aggregates for concrete in very cold weather, **(c)** why the placing temperature of concrete is lower for massive structures than it is for ordinary reinforced structures, and **(d)** how rounded coarse aggregate usually produces better flowability in concrete than crushed stone.

7. A lightweight aggregate concrete mix is required which is to test 4000 psi in 28 days. It is to have $5(\pm 1)\%$ air entrainment and $4(\pm 1)$ in. of slump. The coarse aggregate has a dry unit weight of 50 lb/cu ft, moisture content of 3%, maximum size of ¾ in. and a sp. gr. factor of 1.35. Fine aggregate has a dry unit weight of 65 lb/cu ft, free moisture of 6%, and a sp. gr. factor of 1.90. Calculate the weight of materials required for a 2-cu ft trial batch, using middle range cement content.

8. Outline four steps which you would take to ensure that 4000-psi concrete mixed and placed in cold weather would in fact achieve its design strength.

Structural Timber Frame

Timber, as a building frame material, has been used successfully in many forms throughout the world. Timber is one of the most versatile materials used in the construction industry because it is relatively light in weight, has good tension and compression strength, and is very durable when protected from decay and insects, yet it can be easily shaped to accommodate almost any situation. Timber is used for pile foundations, beams, columns, bracing, and trusses. Many old wood-frame buildings are still in good structural condition and are being used today.

Originally, buildings made with a heavy timber frame were structures of two or more storeys intended primarily for industrial and storage purposes. The earliest of these buildings used whole logs of various diameters as structural members. The next step was the use of sawn timbers for the structural frame, and more recently, timber members have been built from a number of small pieces, nailed, bolted, or glued together.

Heavy timber framing has now been expanded from the original industrial purpose to include schools, churches, auditoriums, gymnasiums, apartment buildings, supermarkets, etc.

There are a number of important factors which have contributed to the more efficient use of timber in modern building construction. One is the development and refinement of stress-graded timber, both solid and glue-laminated. This has made it possible to apply precise structural design procedures to heavy timber framing, resulting in a completely engineered structure.

The production of modern types of timber connectors has made it possible to develop the full strength of wood when stresses are being transferred from one member to another. It is no longer necessary to provide much larger sections in order to accommodate the fastenings formerly used to transfer stresses from member to member.

Full recognition of the fire resistivity of large timber sections has helped in the growth of timber building. Extensive tests have been successfully carried out to determine the ability of timber to withstand fire. The use of smooth surfaces and rounded edges has improved the ability of timber to oppose the inroads of fire. The use of fire-retardant chemicals has also proven successful in increasing the fire resistance of timber elements. Chemicals such as water-soluble ammonium salts are impregnated into the wood under pressure and when exposed to flame release a gas that impedes the spread of the fire.

Another important development in wood technology has been the use of preservative materials to help timber withstand the deteriorating effects of moisture, disease, and insects. Pressure treatment with these preservatives—including creosote, Wolman salts, and solutions of copper compounds—has given deep penetration and long-lasting effects.

Probably the most important achievement of all is the development of techniques of producing heavy timbers by glue-laminating. This aspect of timber building is dealt with later in the chapter.

Heavy timber construction means that type of wood construction in which a degree of fire resistance is obtained by limiting the minimum sizes of structural members and the minimum thickness and composition of wood floors and roofs. All wood parts are arranged in heavy, solid masses and smooth, flat surfaces so as to avoid thin sections, sharp projections, and concealed or inaccessible spaces.

The basic components of structural timber frame buildings may be classified as columns, beams, purlins, the connectors that are required for them, and decking or planking. The size of each used in any specific case will depend on the load; the span of girders, beams, and

planks; the unsupported height of columns; and the species and grade of timber being used. Complete data on allowable working stresses for species of timber used in construction and grades within a species are available from such sources as the Canadian Institute of Timber Construction, the American Institute of Timber Construction, the Southern Pine Association, the National Lumber Manufacturers Association, or the West Coast Lumberman's Association.

TIMBER CONNECTORS

Various types of connections have been developed for the joining of members in the timber frame and for the anchoring of the timber frame to its supporting foundation (Fig. 8-1).

FIGURE 8-1 Typical timber connectors.

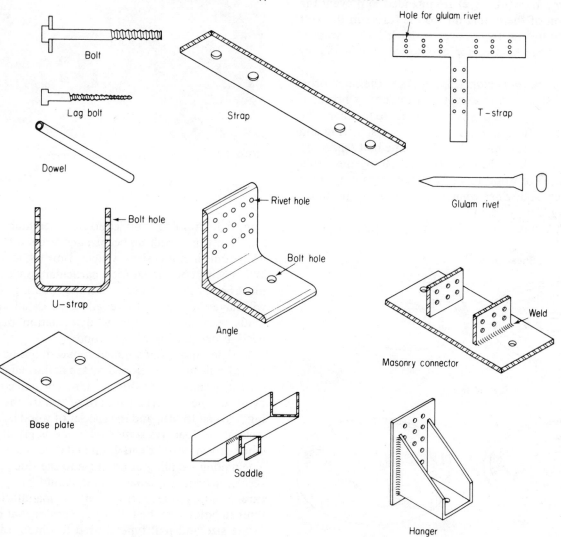

Timber connectors are required to transfer load from one member to another without producing any adverse effects on the connected members; that is, they must be able to distribute the load over a large enough area to prevent member cracking and material deformation in the vicinity of the connection. *T-straps, U-straps, wood or steel side plates, angles and saddles* in conjunction with *bolts, lag screws, glulam rivets, shear plates, and split-ring connectors* are used in various combinations to achieve this transfer of load.

Glulam rivets are special heavy-duty nails made of high-strength steel for use in glue-laminated member connections. At present these rivets are limited to Douglas fir sections only. Further testing of this connection is still being done to extend its use to other species. Glulam rivets are used in conjunction with prepunched steel side plates to achieve load transfer from one section to another. The rivets are driven with a pneumatic hammer as no predrilling of the timber sections is required.

Shear plates (Fig. 8–2) are intended to prevent the deformation of the timber due to the forces in the bolt by distributing the forces over a much larger area of wood. They can be used for steel-to-wood or wood-to-wood connections.

Split-ring connectors (Fig. 8–3) use the same principle as do shear plates except that the bolt is not used in the transfer of load from one member to another. The load is passed from one member into the split ring and on into the other member (Fig. 8–4). The bolt serves only as a clamp to keep the whole connection together. Split rings are used in wood-to-wood connections only. Like shear plates, split rings come in two sizes, 2½ in. in diameter and 4 in. in diameter.

Nails are used extensively in light truss construc-

FIGURE 8–3 Split-ring connectors.

FIGURE 8–4 Connection using split rings.

tion, light framing, application of sheathing and decking, formwork, built-up beams, and stressed-skin panels. They come in many sizes, shapes, finishes, and coatings. When being considered for a particular application, they must be selected to ensure that the resulting connection will have the required strength to resist all anticipated loads without encountering deformation of the connection or splitting of the wood.

The capacity of a nailed connection is based on the lateral resistance developed by the nail when driven into the side grain of the wood section. The nail capacity depends on the nail type (smooth or spiral), the nail coating, the nail length, and the species of wood being nailed. The lateral load resistance of a nail is greatly reduced when driven into the end grain and is not recommended. Toe nailing—nailing at an angle to the side grain—also results in reduced capacity and should be applied with care. Building code specifications and manufacturer's recommendations can be followed to ensure that the appropriate size and nail type is used to obtain satisfactory results.

FIGURE 8–2 Shear plates.

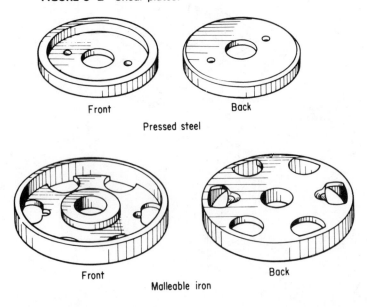

Front   Back

Pressed steel

Front   Back

Malleable iron

WOOD COLUMNS

Wood columns in a structural timber frame are normally solid sawn timber or glue-laminated sections. To achieve a minimum ¾-hr fire rating, columns supporting floors must not be smaller than $8'' \times 8''$ nominal in cross section and columns supporting roofs must not be less than $6'' \times 8''$ nominal. Columns may be continuous throughout the entire building height or spliced at various levels by means of *dowels, wood or metal splice plates, or metal straps* (Fig. 8–5). In some cases, upper-storey columns may bear on the floor beam of the lower storey [see Fig. 8–7(d)].

Column-to-Base Connections

The first consideration in erecting a structural timber frame is the method to be employed in achoring the columns to the foundation. This may be done in several ways, some of which are illustrated in Fig. 8–6. Notice that in Fig. 8–6(f), *shear plates* are set into the column behind the straps. This may be done in any of the connections, when conditions require them.

The dowel connection shown in Fig. 8–6(a) is the simplest type of column-to-base connection and may be

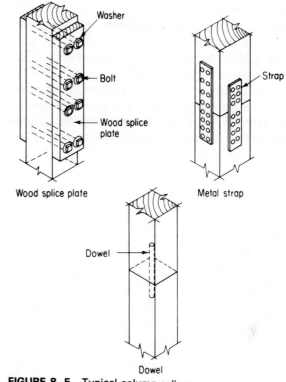

FIGURE 8-5 Typical column splices.

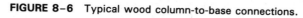

FIGURE 8-6 Typical wood column-to-base connections.

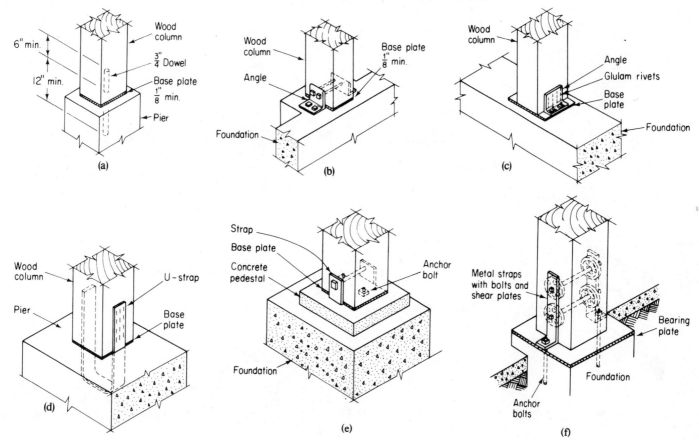

used where uplift and horizontal forces do not have to be considered, but there must be some means of preventing rotation of the column. The dowel is normally ¾ in. in diameter and the base plate a minimum of ⅛ in. thick. The connections in Fig. 8-6(b) and (f) are used where uplift and horizontal forces must be considered. Shear plates may be used if necessary to transmit horizontal forces. The connection in Fig. 8-6(c) is used where the column load must be distributed over a larger area than the end of the column. Plate thickness should not be less than ¼ in. The connection in Fig. 8-6(d) is common in industrial buildings and warehouses. The connection in Fig. 8-6(e) is intended for use with a raised concrete pedestal of limited dimensions, when the column crosssection is large enough to distribute the load adequately. The end of the column is countersunk to accommodate the anchor bolt, nut, and washer. The base plate should be ¼ in. in minimum thickness.

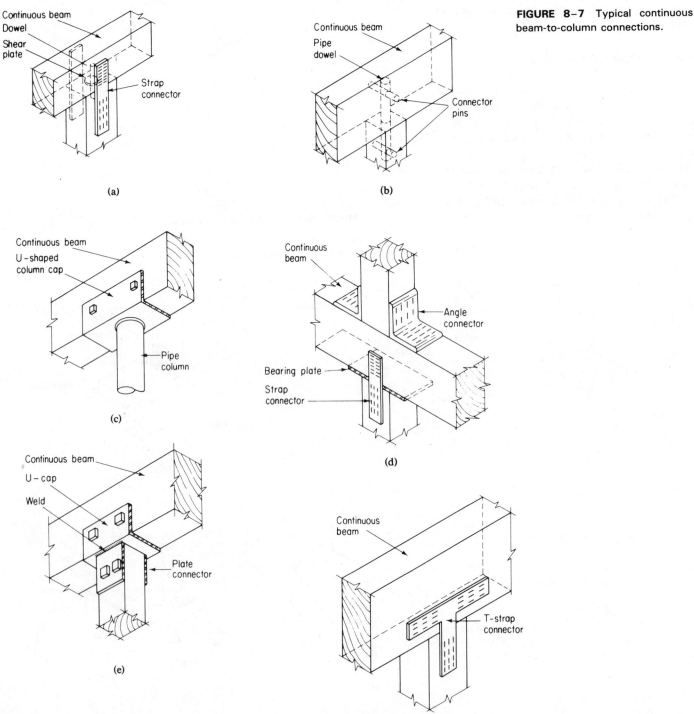

FIGURE 8-7 Typical continuous beam-to-column connections.

WOOD BEAMS

The beams that support floors or roofs in a structural timber frame may be glue-laminated members or they may be of sawn timber. In either case, surfaces should be smooth and edges rounded to reduce the fire hazard. Again, to meet ¾-hr fire rating requirements, floor beams must not be smaller than 6″ × 10″ nominal in depth, and roof beams must be no less than 4″ × 6″ nominal in cross section. Beams may be continuous—extending unbroken over supports (see Fig. 8–7)—or noncontinuous—meeting over a support or fastened to the sides of it (see Fig. 8–8).

Beam-to-Column Connections

Figure 8–7 illustrates a number of ways by which continuous beams are connected to columns, while Fig. 8–8 shows a number of connections for noncontinuous beams at columns. In a number of cases, the connections are interchangeable, perhaps with some modification.

Figure 8–7(a) is a standard beam-to-column connection providing for uplift. The shear plate and dowel are used only when lateral forces make it necessary. A loose bearing plate may be added if the cross section of the column does not provide enough bearing area for the

FIGURE 8–8 Typical noncontinuous beam-to-column connections.

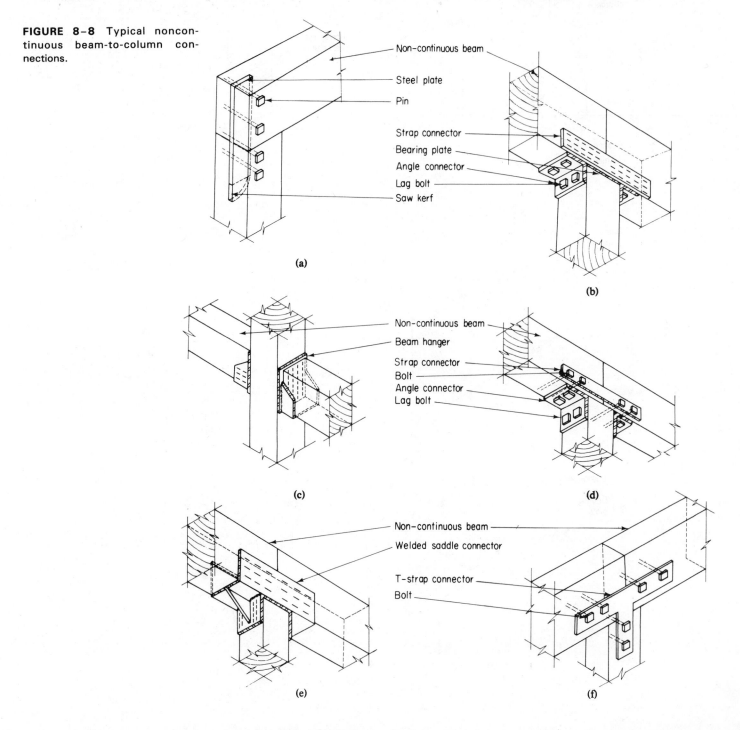

Non-continuous beam

Steel plate

Pin

Strap connector

Bearing plate

Angle connector

Lag bolt

Saw kerf

(a)

(b)

Non-continuous beam

Beam hanger

Strap connector

Bolt

Angle connector

Lag bolt

(c)

(d)

Non-continuous beam

Welded saddle connector

T-strap connector

Bolt

(e)

(f)

beam in compression perpendicular to the grain. Concealed connections are shown in Fig. 8–7(b) and in Fig. 8–8(a). In the latter connection, the bolt heads can also be hidden by countersinking and plugging the holes. Figure 8–7(c) is designed specifically for pipe columns, while Figures 8–7(e) and 8–8(e) are designed for situations in which the beam and the column are not the same width.

Beam-to-Masonry Wall Connections

Buildings may be designed with masonry exterior walls and a structural timber floor and roof frame. In such cases, the framing members must be connected to the masonry wall or to pilasters built into it. Figure 8–9 illustrates several methods for connecting timber beams to a masonry wall.

FIGURE 8–9 Typical beam-to-masonry wall connections.

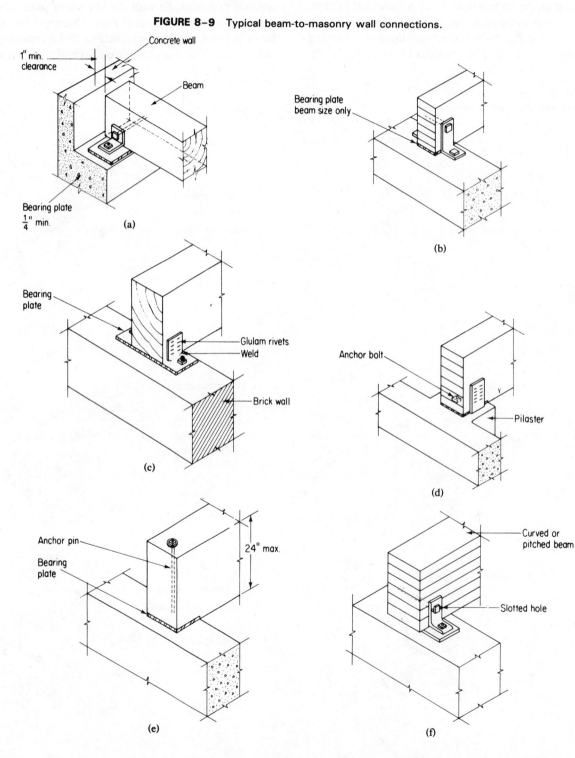

Figure 8–9(a) is a standard anchorage to a masonry wall, providing for uplift and resisting horizontal forces. The bearing plate should be a minimum of ¼ in. thick and the clearance between the end of the beam and the wall at least 1 in. Figure 8–9(d) is used where the pilaster is not wide enough to permit outside anchor bolts. The beam must be countersunk to accommodate the anchor nut and washer. Figure 8–9(e) is intended for beams not exceeding 24 in. in depth. Figure 8–9(f) is used with curved or pitched beams, the slotted hole in the angles

allowing for slight horizontal movement, resulting from vertical deflection.

Roof Beams

The method used to connect roof beams to exterior masonry walls depends on whether there is a parapet extending above the roof or if the roof is a *flush deck*—covers the top of the wall (see Fig. 8–10). In the case of a flush deck, a fascia is required along the top outside

FIGURE 8–10   Roof beam connections to masonry wall.

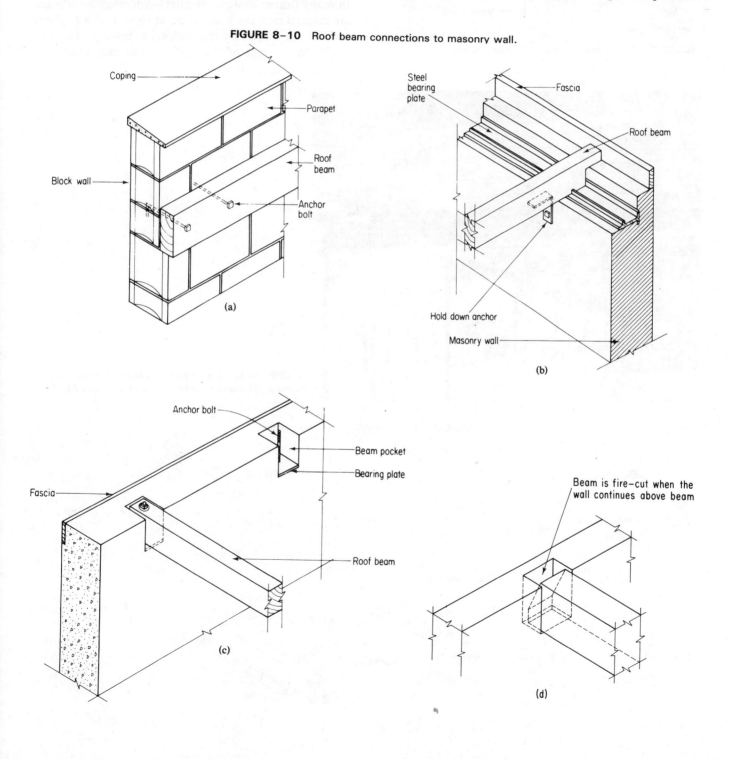

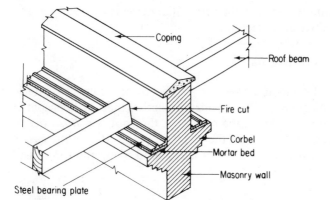

FIGURE 8-11 Timber frame to masonry fire wall.

edge of the wall to provide solid backing for the roof flashing. Notice in Fig. 8-10(c) that the beam pocket is large enough to allow at least ¾-in. clearance at the sides and end of the beam.

When a *party wall* or *fire wall* is involved in structural timber framing, the framing members tie into it from both sides, as shown in Fig. 8-11.

Intermediate Beams and Purlins

In many timber designs, smaller—*intermediate*—beams are framed into the main ones, at right angles to them, and it is to these that the decking is fastened (see Fig. 8-12). In the case of the roof, *purlins* may span from

FIGURE 8-12 Intermediate beams framed with hangers. *(Courtesy Council of Forest Industries of B. C.)*

FIGURE 8-13 Intermediate beam supports.

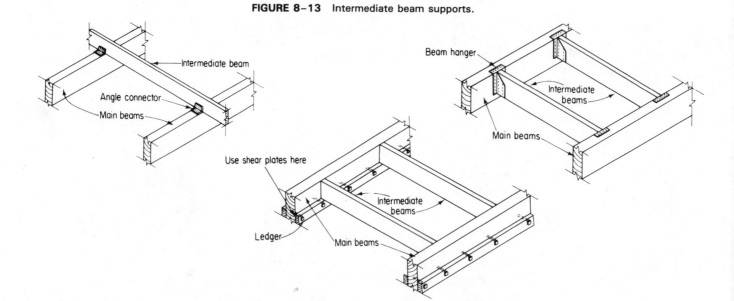

FIGURE 8-14   Decking on main beams. *(Courtesy of Forest Industries of B. C.)*

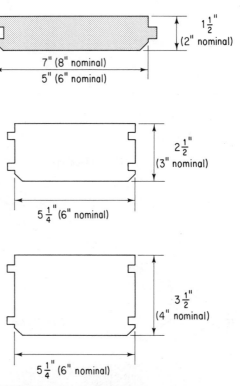

FIGURE 8-15   Types of plank decking.

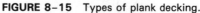
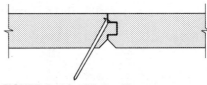

FIGURE 8-16   Blind nailing.

beam to beam and carry the roof deck. These smaller members may be carried *on top* of the main beams, supported on *ledgers* fastened to the sides of the beams or suspended by *hangers* anchored to the beam sides (see Fig. 8–13). Other designs will call for the decking to be applied directly to the main beams (see Fig. 8–14).

FLOOR AND ROOF DECKS

Timber floor and roof decks can be constructed of *plank decking or laminated decking. Plank decking* is used in most applications where extreme wearing of the exposed surface is not a concern. *Laminated decking* is used for bridge decks, loading ramps, and warehouse floors where high wear is anticipated.

Plank decking is tongue-and-groove material that is laid on its wide side over supporting members (Fig. 8–14). Plank decking is normally available in three thicknesses: 2 in., 3 in., and 4 in. nominal. The 2-in. decking has a single tongue and groove, while the 3-in. and 4-in. thicknesses have a double tongue and groove (Fig. 8–15). The 2-in. decking may come in 6- and 8-in. nominal widths while the 3- and 4-in. material comes in 6-in. nominal width only. Plank decking may be plain or have decorative grooves cut into the exposed side for acoustical and aesthetic purposes. Floor decking is usually *blind-nailed;* that is, the nails are driven through the tongue of the decking into the supporting member. The nail head is concealed by the groove of the next plank (Fig. 8–16).

Laminated decking is usually 2-in. thick nominal material that can vary in width from 4 to 12 in. nominal. Laminated decking is placed on edge over supporting

members and spiked into place to provide a load-carrying surface (Fig. 8–17).

Wood species used for plank decking include Red cedar, White pine, Douglas fir, Western hemlock, and the Spruce-Pine-Fir group. Douglas fir, Spruce, or Jack pine are recommended for floor decks because they wear well. Cedar is used in exposed areas because of its durability against decay and its attractive appearance. There are two grades of plank decking: *select* and *commercial.* The select grade is used where strength and appearance are both important (churches, chapels, etc.) and commercial grade is used where appearance is not a major consideration. Commercial grade has a lower strength; however, it is less costly then the select grade. Almost any species of wood can be used for laminated decking. The strength and durability required for its anticipated use will determine the species chosen.

Three basic patterns are used for laying decking: *simple span, two-span continuous,* and *controlled random spliced* (see Fig. 8–18). Since most decking is supplied in random lengths, the controlled random pattern is the most popular because it makes the best use of the

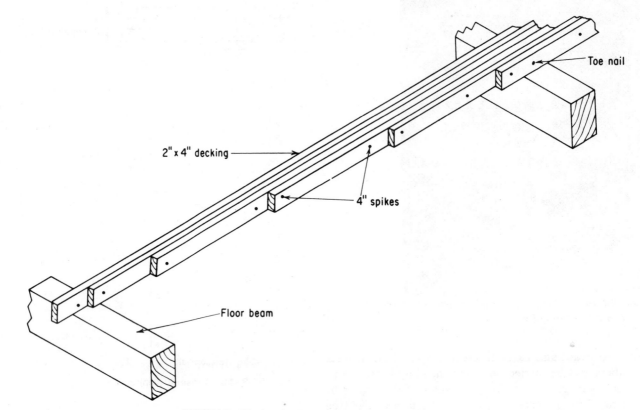

FIGURE 8-17 Laminated timber decking.

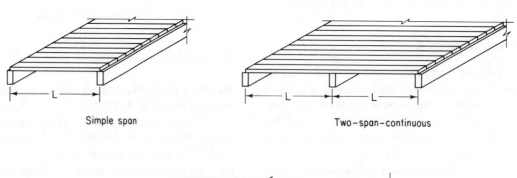

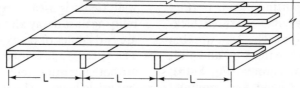

FIGURE 8-18 Deck laying patterns.

material supplied. A deck using this arrangement must extend over at least three spans, have the end joints staggered at least 24 in. in adjacent courses and 6 in. or more in every other course. Each plank must bear on at least one support and end joints in the first half of the end spans are not allowed. To obtain continuity, nailing requirements must be followed carefully.

LAMINATED TIMBER

Bonding wood with adhesives is an old art—examples of it have been found in Egyptian tombs. It was first used in Europe for constructional purposes in the early twentieth century, where it gained ready acceptance, but not until the early 1930s was glue-laminated structural timber introduced in America. World War II and its heavy demand for military and industrial construction provided the impetus which made glue-laminated construction commercially important.

Architects and builders are finding glue-laminated timber the answer to some of their most pressing problems. Such timbers have proved to be extremely versatile and have allowed architects to develop shapes and sizes which would be difficult to obtain with solid timber, concrete, or steel.

MATERIALS USED

Laminated timbers are widely recognized as being among the finest permanent structural materials. They are made from kiln-dried stock, mostly *Douglas fir* and *Southern pine*, with some *Western larch* and *Western hemlock*. Both fir and pine are extremely strong, straight-grained, tough, resilient, and durable woods. The larch and hemlock are not quite so strong but match the fir and pine quite closely in color and texture and are used in the portions of the laminated timber where stresses will be lowest.

Tests have shown that adhesives can be made to develop the full strength of wood and that time alone does not affect the stength of the joint when the proper adhesive is used for given conditions. For laminated members which are protected from appreciable amounts of moisture and relatively high humidity, *casein glue* is satisfactory. When laminated members are continuously immersed in water or subjected to intermittent wetting and drying, such as exterior exposures or in buildings where high humidities are encountered for long periods of time, highly water-resistant adhesives such as *phenol*, *resorcinol*, or *melamine* resin glues should be used.

The manufacture of glue-laminated timber begins with the selection and grading of the laminating lumber. The finished product will be only as good as the pieces which went into it, so the laminating stock must be of good quality. Of the materials available to the laminator, the highest quality is almost free of natural defects which restrict strength and is used almost exclusively for face laminations where appearance, as well as strength, is a factor. The lower grades of laminating stock are almost equivalent to select structural, construction, and standard grades of joists and planks, with special additional limitations which qualify them for laminating purposes. One of these limitations relates to cross grain. Limitations in this aspect of material selection are more severe than commonly imposed by grading rules. They apply to only the outer 10% of the depth of a member stressed principally in bending but apply to all laminations of members under tension or compression. Working unit stresses must be modified to the slope of grain, as specified in Table 8–1.

TABLE 8–1 Strength ratios corresponding to slope of grain

Slope of grain	Maximum strength ratio	
	Fiber stress in bending or tension	Stress in compression parallel to grain
1 in 6	—	0.56
1 in 8	0.53	0.66
1 in 10	0.61	0.74
1 in 12	0.69	0.82
1 in 14	0.74	0.87
1 in 15	0.76	1.00
1 in 16	0.80	—
1 in 18	0.85	—
1 in 20	1.00	—

PREPARATION OF MATERIAL

The laminating stock must be dried to a moisture content as close as practicable to that which it will attain in service. It must, in any case, not be less than 7% or more than 16%. There are two reasons for these limits. One is to provide the moisture conditions under which the best possible glue bonds may be obtained, and the other is to assure that there will be as little dimensional change as possible once the member has been placed in service (Fig. 8–19).

Laminated timbers are often required to be longer than commercially available stock, and the individual pieces must accordingly be spliced end to end to make full-length laminations. This is done in several ways, including *plain scarf joints, stepped scarf joints, finger joints,* and *butt* joints. Figure 8–20 illustrates all four.

Plain scarf joints are relatively easy to make and are used extensively in important structural members. When they occur in tension members or in the tension portion of a member subjected to bending, a reduction in allowable stress is necessary. The scarf is made by a machine which produces a smooth, sloping surface at one or both ends of each board.

A stepped scarf joint employs a mechanical as well as a glue bond, but is more difficult to make. In addition, when stepped scarf joints are used, the portion of the thickness of the lamination occupied by the step is disregarded in calculating the moment of inertia and the net effective area.

Finger joints are cut with special saws and are often used where the laminating stock is ¾-in. material. Notice in Fig. 8–20 that each finger has a square end and that the bottom of each V has a flat surface. This eliminates the likelihood of the fingers splitting the end of the matching piece as the two halves of the joint are fitted together.

Butt joints are not recommended, especially in important structural members, but are sometimes used in

FIGURE 8-19 Measuring moisture content of laminating stock with electric moisture meter. *(Courtesy Canadian Institute of Timber Construction.)*

members or portions of members subjected to compression only. If butt joints are used in the compression portion of a member subjected to bending, the cross-sectional area of all laminations containing butt joints at a particular cross section should be disregarded in computing the effective moment of inertia for that cross section. Butt joints in adjacent laminations should not be spaced closer than 10 times the thickness of the lamination.

Every piece comprising a single lamination (as well as the completed end joint) must be of equal thickness

FIGURE 8-20 End joints for laminating stock.

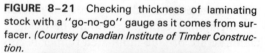

Plain scarf

Stepped scarf

Butt joint

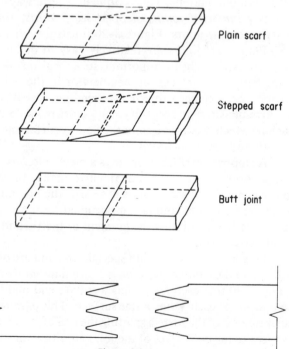

Finger joint

FIGURE 8-21 Checking thickness of laminating stock with a "go-no-go" gauge as it comes from surfacer. *(Courtesy Canadian Institute of Timber Construction.)*

to avoid gaps in the glue lines which will affect strength. For practical reasons, laminating plants maintain a constant thickness of all laminating stock. The standard thicknesses used by most manufacturers are 0.75 and 1.75 in., with a permissible tolerance of ± 0.0075 in. Figure 8–21 shows laminating stock being checked for thickness as it emerges from a surfacer.

ASSEMBLY OF MATERIAL

The first step in the assembly of a glue-laminated timber is the positioning and gluing of end joints. This must be done in such a way that the completed joint, after bonding, will be the same thickness as the lamination. End jointing may be a separate operation, but many manufacturers have devised systems of indexing scarf joints for gluing during the assembly operation. One is to bore a small hole at the time the scarf is made, so a dowel can be inserted during assembly to hold the two pieces in the right position. When used, the stepped scarf helps to position the two pieces at a joint. Other methods of marking meeting points automatically—including hand pinning—are also employed. Finger joints do not require indexing prior to assembly.

Laminations are coated with adhesive by passing them through a mechanical glue spreader which controls the amount of adhesive applied. Scarf joints are coated and pinned during this operation, when the single operation assembly system is used. Finger joints are sprayed with adhesive, fitted together, and set electronically in a single operation. The entire lamination is then laid on edge in a jig or press made of steel frames, according to the shape of the member being formed. The position of each lamination is previously determined according to grade, either by assembling in a trial run or by showing the position of every piece on a drawing. Successive laminations are laid on edge next to the first until the full depth of the member has been reached (see Fig. 8–22).

Clamping begins as soon as the final laminations have been placed in the jig. Steel bolts with wood or metal clamp blocks are commonly used to press the laminations together. Clamping begins at the middle and progresses toward the ends, tightening the nuts with an impact wrench set to produce a pressure of from 10 to 200 psi on the glue lines (Fig. 8–23). A torque wrench is used to check the tension on the clamp nuts (Fig. 8–24).

Some adhesives require a minimum glue line temperature of about 75°F, and the temperature of the laminations may be checked prior to assembly. When the adhesive has set, the clamps are removed and the member is passed through a large double surfacer to remove irregularities (such as squeezed-out beads of glue) and to bring the member to its finished width. Irregular buildups are sawed to the required shape, ends are trimmed, and daps or other fabrication completed. Surface irregularities are corrected as necessary for the required appearance grade, and identification marks are placed on the ends of the member. Finally, it is given a coat of transparent moisture seal.

If necessary, finished members are wrapped before shipment to protect their appearance. Some manufacturers use a wrap of double-creped, impregnated paper, while others use an inner wrap of polyethylene covered with jute.

FIGURE 8–22 Assembling a large glue-laminated beam. *(Courtesy Canadian Institute of Timber Construction.)*

FIGURE 8–23 Tightening clamp bolts with an air-operated impact wrench. *(Courtesy Canadian Institute of Timber Construction.)*

FIGURE 8–24 Checking tension on clamp bolts with a calibrated torque wrench. *(Courtesy Canadian Institute of Timber Construction.)*

When dictated by service conditions, glue-laminated timbers may be pressure-treated for protection against decay or insect damage. Straight, or almost straight, members to be accommodated in a pressure-treating cylinder may be treated after laminating. Creosote or Wolman salts are generally used for this purpose. If the shape of the member prevents it from being placed in the cylinder after fabrication, the laminations may be pressure-treated beforehand. Wolman salts are used in this case, since this preservative does not affect subsequent glue bonding.

PROPERTIES OF GLUE-LAMINATED SECTIONS

The Laminated Timber Institute of Canada has adopted the SI metric system for the sizing of glue-laminated timber sections. Table 8–2 indicates the standard widths adopted. Nominal sizes are not used in the metric system, so all member sizes are designated by their actual dimensions. Lamination widths greater than 10¾ in. (275 mm) are composed of two boards with joints staggered in successive layers (Fig. 8–25). The interior laminations normally are not edge-glued unless loading is to be applied parallel to the width of the laminations.

Properties of sections using ¾-in. (19-mm) laminations and 1½-in. (38-mm) laminations are given in Table 8–3. They include the cross-sectional area, section modulus, moment of inertia, and approximate linear density.

TABLE 8–2 Nominal and finished widths

Initial width of glulam stock		Finished width of glulam member	
Nominal width (in.)	Actual width (mm)	Width (in.)	Width (mm)
4	89	3	80
6	140	5	130
8	184	6¾	175
10	235	8¾	275
12	286	10¾	275
4 + 10, 6 + 8	89 + 235, 140 + 184	12¼	315
6 + 10, 8 + 8	140 + 285, 184 + 184	14¼	365

FIGURE 8–25 Edge joints staggered.

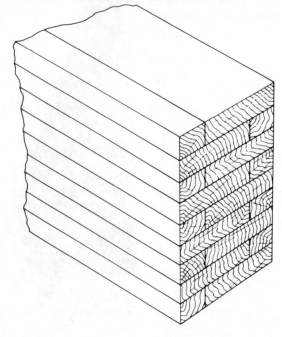

TABLE 8–3 Properties of sections

80 mm Wide Laminations

Size	Number of Lams		Area	Axis X-X			Axis Y-Y		Approx Linear Density kg/m @
b and d			bd	S $bd^2/6$	I $bd^3/12$	$\sqrt{\dfrac{d}{b^2}}$	S $db^2/6$	I $db^3/12$	
Millimetres (mm)	38mm Lams	19mm Lams	$mm^2 \times 10^3$	$mm^3 \times 10^3$	$mm^4 \times 10^6$		$mm^3 \times 10^3$	$mm^4 \times 10^6$	545 kg/m^3
80 x 114	3	6	9.12	173.28	9.88	0.13	121.60	4.86	5.0
133	—	7	10.64	235.85	15.68	0.14	141.87	5.67	5.8
152	4	8	12.16	308.05	23.41	0.15	162.13	6.49	6.6
171	—	9	13.68	389.88	33.33	0.16	182.40	7.30	7.5
190	5	10	15.20	481.33	45.73	0.17	202.67	8.11	8.3
209	—	11	16.72	582.41	60.86	0.18	222.93	8.92	9.1
80 x 228	6	12	18.24	693.12	79.02	0.19	243.20	9.73	9.9
247	—	13	19.76	813.45	100.46	0.20	263.47	10.54	10.8
266	7	14	21.28	943.41	125.47	0.20	283.73	11.35	11.6
285	—	15	22.80	1083.00	154.33	0.21	304.00	12.16	12.4
304	8	16	24.32	1232.21	187.30	0.22	324.27	12.97	13.2
323	—	17	25.84	1391.05	224.66	0.22	344.53	13.78	14.1
80 x 342	9	18	27.36	1559.52	266.68	0.23	364.80	14.59	14.9
361	—	19	28.88	1737.61	313.64	0.24	385.07	15.40	15.7
380	10	20	30.40	1925.33	365.81	0.24	405.33	16.21	16.6
399	—	21	31.92	2122.68	423.47	0.25	425.60	17.02	17.4
418	11	22	33.44	2329.65	486.90	0.26	445.87	17.83	18.2
437	—	23	34.96	2546.25	556.36	0.26	466.13	18.65	19.0
80 x 456	12	24	36.48	2772.48	632.13	0.27	486.40	19.46	19.9
475	—	25	38.00	3008.33	714.48	0.27	506.67	20.27	20.7
494	13	26	39.52	3253.81	803.69	0.28	526.93	21.08	21.5
513	—	27	41.04	3508.92	900.04	0.28	547.20	21.89	22.4
532	14	28	42.56	3773.65	1003.79	0.29	567.47	22.70	23.2
551	—	29	44.08	4048.01	1115.23	0.29	587.73	23.51	24.0
80 x 570	15	30	45.60	4332.00	1234.62	0.30	608.00	24.32	24.8
589	—	31	47.12	4625.61	1362.24	0.30	628.27	25.13	25.7
608	16	32	48.64	4928.85	1498.37	0.31	648.53	25.94	26.5

Source: Reproduced by permission of the Canadian Wood Council.

TABLE 8-3 Properties of sections (continued)

130 mm Wide Laminations

Size b and d	Number of Lams		Area	Axis X-X			Axis Y-Y		Approx Linear Density kg/m @ 545 kg/m³
			bd	S $bd^2/6$	I $bd^3/12$	$\sqrt{\dfrac{d}{b^2}}$	S $db^2/6$	I $db^3/12$	
Millimetres (mm)	38mm Lams	19mm Lams	mm²x10³	mm³x10³	mm⁴x10⁶		mm³x10³	mm⁴x10⁶	
130 x 114	3	6	14.82	281.58	16.05	0.08	321.10	20.87	8.1
133	—	7	17.29	383.26	25.49	0.09	374.62	24.35	9.4
152	4	8	19.76	500.59	38.04	0.09	428.13	27.83	10.8
171	—	9	22.23	633.55	54.17	0.10	481.65	31.31	12.1
190	5	10	24.70	782.17	74.31	0.11	535.17	34.79	13.5
209	—	11	27.17	946.42	98.90	0.11	588.68	38.26	14.8
130 x 228	6	12	29.64	1126.32	128.40	0.12	642.20	41.74	16.1
247	—	13	32.11	1321.86	163.25	0.12	695.72	45.22	17.5
266	7	14	34.58	1533.05	203.90	0.13	749.23	48.70	18.8
285	—	15	37.05	1759.88	250.78	0.13	802.75	52.18	20.2
304	8	16	39.52	2002.35	304.36	0.13	856.27	55.66	21.5
323	—	17	41.99	2260.46	365.06	0.14	909.78	59.14	22.9
130 x 342	9	18	44.46	2534.22	433.35	0.14	963.30	62.61	24.2
361	—	19	46.93	2823.62	509.66	0.15	1016.82	66.09	25.6
380	10	20	49.40	3128.67	594.45	0.15	1070.33	69.57	26.9
399	—	21	51.87	3449.35	688.15	0.15	1123.85	73.05	28.2
418	11	22	54.34	3785.69	791.21	0.16	1177.37	76.53	29.6
437	—	23	56.81	4137.66	904.08	0.16	1230.88	80.01	30.9
130 x 456	12	24	59.28	4505.28	1027.20	0.16	1284.40	83.49	32.3
475	—	25	61.75	4888.54	1161.03	0.17	1337.92	86.96	33.6
494	13	26	64.22	5287.45	1306.00	0.17	1391.43	90.44	35.0
513	—	27	66.69	5702.00	1462.56	0.17	1444.95	93.92	36.3
532	14	28	69.16	6132.19	1631.16	0.18	1498.47	97.40	37.7
551	—	29	71.63	6578.02	1812.24	0.18	1551.98	100.88	39.0
130 x 570	15	30	74.10	7039.50	2006.26	0.18	1605.50	104.36	40.4
589	—	31	76.57	7516.62	2213.65	0.19	1659.02	107.84	41.7
608	16	32	79.04	8009.39	2434.85	0.19	1712.53	111.31	43.0
627	—	33	81.51	8517.79	2670.33	0.19	1766.05	114.79	44.4
646	17	34	83.98	9041.85	2920.52	0.20	1819.57	118.27	45.7
665	—	35	86.45	9581.54	3185.86	0.20	1873.08	121.75	47.1
130 x 684	18	36	88.92	10136.8	3466.81	0.20	1926.60	125.23	48.4
703	—	37	91.39	10707.8	3763.81	0.20	1980.12	128.71	49.8
722	19	38	93.86	11294.4	4077.31	0.21	2033.63	132.19	51.1
741	—	39	96.33	11896.7	4407.75	0.21	2087.15	135.66	52.5
760	20	40	98.80	12514.6	4755.57	0.21	2140.67	139.14	53.8
779	—	41	101.27	13148.2	5121.23	0.21	2194.18	142.62	55.2
130 x 798	21	42	103.74	13797.4	5505.17	0.22	2247.70	146.10	56.5
817	—	43	106.21	14462.2	5907.83	0.22	2301.22	149.58	57.8
836	22	44	108.68	15142.7	6329.67	0.22	2354.73	153.06	59.2
855	—	45	111.15	15838.8	6771.12	0.22	2408.25	156.54	60.5
874	23	46	113.62	16550.6	7232.63	0.23	2461.77	160.01	61.9
893	—	47	116.09	17278.0	7714.65	0.23	2515.28	163.49	63.2
130 x 912	24	48	118.56	18021.1	8217.63	0.23	2568.80	166.97	64.6
931	—	49	121.03	18779.8	8742.01	0.23	2622.32	170.45	65.9
950	25	50	123.50	19554.1	9288.23	0.24	2675.83	173.93	67.3
969	—	51	125.97	20344.1	9856.74	0.24	2729.35	177.41	68.6
988	26	52	128.44	21149.7	10448.0	0.24	2782.87	180.89	70.0
1007	—	53	130.91	21971.0	11062.4	0.24	2836.38	184.36	71.3
130 x 1026	27	54	133.38	22807.9	11700.4	0.25	2889.90	187.84	72.6

Source: Reproduced by permission of the Canadian Wood Council.

TABLE 8–3 Properties of sections (continued)

175 mm Wide Laminations

Size	Number of Lams		Area	Axis X-X			Axis Y-Y		Approx Linear Density kg/m @ 545 kg/m³
b and d			bd	S $bd^2/6$	I $bd^3/12$	$\sqrt{\dfrac{d}{b^2}}$	S $db^2/6$	I $db^3/12$	
Millimetres (mm)	38mm Lams	19mm Lams	$mm^2 \times 10^3$	$mm^3 \times 10^3$	$mm^4 \times 10^6$		$mm^3 \times 10^3$	$mm^4 \times 10^6$	
175 x 152	4	8	26.60	673.87	51.21	0.07	775.83	67.89	14.5
171	—	9	29.92	852.86	72.92	0.07	872.81	76.37	16.3
190	5	10	33.25	1052.92	100.03	0.08	969.79	84.86	18.1
209	—	11	36.58	1274.03	133.14	0.08	1066.77	93.34	19.9
228	6	12	39.90	1516.20	172.85	0.09	1163.75	101.83	21.7
247	—	13	43.22	1779.43	219.76	0.09	1260.73	110.31	23.5
175 x 266	7	14	46.55	2063.72	274.47	0.09	1357.71	118.80	25.4
285	—	15	49.88	2369.06	337.59	0.10	1454.69	127.29	27.2
304	8	16	53.20	2695.47	409.71	0.10	1551.67	135.77	29.0
323	—	17	56.53	3042.93	491.43	0.10	1648.65	144.26	30.8
342	9	18	59.85	3411.45	583.36	0.11	1745.63	152.74	32.6
361	—	19	63.17	3801.03	686.09	0.11	1842.60	161.23	34.4
175 x 380	10	20	66.50	4211.67	800.22	0.11	1939.58	169.71	36.2
399	—	21	69.82	4643.36	926.35	0.11	2036.56	178.20	38.0
418	11	22	73.15	5096.12	1065.09	0.12	2133.54	186.68	39.8
437	—	23	76.47	5569.93	1217.03	0.12	2230.52	195.17	41.7
456	12	24	79.80	6064.80	1382.77	0.12	2327.50	203.66	43.5
475	—	25	83.13	6580.73	1562.92	0.12	2424.48	212.14	45.3
175 x 494	13	26	86.45	7117.72	1758.08	0.13	2521.46	220.63	47.1
513	—	27	89.78	7675.76	1968.83	0.13	2618.44	229.11	48.9
532	14	28	93.10	8254.87	2195.79	0.13	2715.42	237.60	50.7
551	—	29	96.43	8855.03	2439.56	0.13	2812.40	246.08	52.5
570	15	30	99.75	9476.25	2700.73	0.14	2909.38	254.57	54.3
589	—	31	103.07	10118.5	2979.91	0.14	3006.35	263.06	56.1
175 x 608	16	32	106.40	10781.8	3277.69	0.14	3103.33	271.54	57.9
627	—	33	109.72	11466.2	3594.67	0.14	3200.31	280.03	59.8
646	17	34	113.05	12171.7	3931.46	0.15	3297.29	288.51	61.6
665	—	35	116.38	12898.2	4288.66	0.15	3394.27	297.00	63.4
684	18	36	119.70	13645.8	4666.86	0.15	3491.25	305.48	65.2
703	—	37	123.03	14414.4	5066.67	0.15	3588.23	313.97	67.0
175 x 722	19	38	126.35	15204.1	5488.69	0.15	3685.21	322.46	68.8
741	—	39	129.68	16014.8	5933.51	0.16	3782.19	330.94	70.6
760	20	40	133.00	16846.6	6401.73	0.16	3879.17	339.43	72.4
779	—	41	136.32	17699.5	6893.97	0.16	3976.15	347.91	74.2
798	21	42	139.65	18573.4	7410.81	0.16	4073.13	356.40	76.1
817	—	43	142.98	19468.4	7952.85	0.16	4170.10	364.88	77.9
175 x 836	22	44	146.30	20384.4	8520.71	0.17	4267.08	373.37	79.7
855	—	45	149.63	21321.5	9114.97	0.17	4364.06	381.86	81.5
874	23	46	152.95	22279.7	9736.24	0.17	4461.04	390.34	83.3
893	—	47	156.27	23258.9	10385.1	0.17	4558.02	398.83	85.1
912	24	48	159.60	24259.2	11062.2	0.17	4655.00	407.31	86.9
931	—	49	162.93	25280.5	11768.0	0.17	4751.98	415.80	88.7
175 x 950	25	50	166.25	26322.9	12503.3	0.18	4848.96	424.28	90.5
969	—	51	169.57	27386.3	13268.6	0.18	4945.94	432.77	92.4
988	26	52	172.90	28470.8	14064.6	0.18	5042.92	441.26	94.2
1007	—	53	176.23	29576.4	14891.7	0.18	5139.90	449.74	96.0
1026	27	54	179.55	30703.0	15750.6	0.18	5236.88	458.23	97.8
1045	—	55	182.88	31850.7	16642.0	0.18	5333.85	466.71	99.6
175 x 1064	28	56	186.20	33019.4	17566.3	0.19	5430.83	475.20	101.
1083	—	57	189.52	34209.2	18524.3	0.19	5527.81	483.68	103.
1102	29	58	192.85	35420.1	19516.4	0.19	5624.79	492.17	105.
1121	—	59	196.18	36652.0	20543.4	0.19	5721.77	500.65	107.
1140	30	60	199.50	37905.0	21605.8	0.19	5818.75	509.14	109.
1159	—	61	202.82	39179.0	22704.2	0.19	5915.73	517.63	110.
175 x 1178	31	62	206.15	40474.1	23839.2	0.20	6012.71	526.11	112.
1197	—	63	209.48	41790.2	25011.4	0.20	6109.69	534.60	114.
1216	32	64	212.80	43127.4	26221.5	0.20	6206.67	543.08	116.
1235	—	65	216.13	44485.7	27469.9	0.20	6303.65	551.57	118.
1254	33	66	219.45	45865.0	28757.3	0.20	6400.63	560.05	120.
1273	—	67	222.77	47265.4	30084.4	0.20	6497.60	568.54	121.
175 x 1292	34	68	226.10	48686.8	31451.7	0.21	6594.58	577.03	123.
1311	—	69	229.43	50129.3	32859.8	0.21	6691.56	585.51	125.
1330	35	70	232.75	51592.9	34309.2	0.21	6788.54	594.00	127.
1349	—	71	236.07	53077.5	35800.7	0.21	6885.52	602.48	129.
1368	36	72	239.40	54583.2	37334.9	0.21	6982.50	610.97	130.

Source: Reproduced by permission of the Canadian Wood Council.

TABLE 8-3 Properties of sections (continued)

b

d

y

x —— x

y

225 mm Wide Laminations

Size	Number of Lams		Area	Axis X-X			Axis Y-Y		Approx Linear Density kg/m @
b and d			bd	S bd²/6	I bd³/12	$\sqrt{\dfrac{d}{b^2}}$	S db²/6	I db³/12	
Millimetres (mm)	38mm Lams	19mm Lams	mm²x10³	mm³x10³	mm⁴x10⁶		mm³x10³	mm⁴x10⁶	545 kg/m³
225 x 228	6	12	51.30	1949.40	222.23	0.07	1923.75	216.42	27.9
247	—	13	55.58	2287.84	282.55	0.07	2084.06	234.46	30.3
266	7	14	59.85	2653.35	352.90	0.07	2244.38	252.49	32.6
285	—	15	64.13	3045.94	434.05	0.08	2404.69	270.53	34.9
304	8	16	68.40	3465.60	526.77	0.08	2565.00	288.56	37.3
323	—	17	72.68	3912.34	631.84	0.08	2725.31	306.60	39.6
225 x 342	9	18	76.95	4386.15	750.03	0.08	2885.63	324.63	41.9
361	—	19	81.22	4887.04	882.11	0.08	3045.94	342.67	44.2
380	10	20	85.50	5415.00	1028.85	0.09	3206.25	360.70	46.6
399	—	21	89.78	5970.04	1191.02	0.09	3366.56	378.74	48.9
418	11	22	94.05	6552.15	1369.40	0.09	3526.88	396.77	51.2
437	—	23	98.32	7161.34	1564.75	0.09	3687.19	414.81	53.6
225 x 456	12	24	102.60	7797.60	1777.85	0.09	3847.50	432.84	55.9
475	—	25	106.88	8460.94	2009.47	0.10	4007.81	450.88	58.2
494	13	26	111.15	9151.35	2260.38	0.10	4168.13	468.91	60.5
513	—	27	115.43	9868.84	2531.36	0.10	4328.44	486.95	62.9
532	14	28	119.70	10613.4	2823.16	0.10	4488.75	504.98	65.2
551	—	29	123.97	11385.0	3136.58	0.10	4649.06	523.02	67.5
225 x 570	15	30	128.25	12183.7	3472.37	0.11	4809.38	541.05	69.8
589	—	31	132.52	13009.5	3831.31	0.11	4969.69	559.09	72.2
608	16	32	136.80	13862.4	4214.17	0.11	5130.00	577.13	74.5
627	—	33	141.07	14742.3	4621.72	0.11	5290.31	595.16	76.8
646	17	34	145.35	15649.3	5054.74	0.11	5450.63	613.20	79.2
665	—	35	149.63	16583.4	5513.99	0.11	5610.94	631.23	81.5
225 x 684	18	36	153.90	17544.6	6000.25	0.12	5771.25	649.27	83.8
703	—	37	158.18	18532.8	6514.29	0.12	5931.56	667.30	86.1
722	19	38	162.45	19548.1	7056.88	0.12	6091.88	685.34	88.5
741	—	39	166.73	20590.5	7628.79	0.12	6252.19	703.37	90.8
760	20	40	171.00	21660.0	8230.80	0.12	6412.50	721.41	93.1
779	—	41	175.27	22756.5	8863.67	0.12	6572.81	739.44	95.5
225 x 798	21	42	179.55	23880.1	9528.18	0.13	6733.13	757.48	97.8
817	—	43	183.82	25030.8	10225.1	0.13	6893.44	775.51	100.
836	22	44	188.10	26208.6	10955.2	0.13	7053.75	793.55	102.
855	—	45	192.38	27413.4	11719.2	0.13	7214.06	811.58	105.
874	23	46	196.65	28645.3	12518.0	0.13	7374.38	829.62	107.
893	—	47	200.93	29904.3	13352.2	0.13	7534.69	847.65	109.
225 x 912	24	48	205.20	31190.4	14222.8	0.13	7695.00	865.69	112.
931	—	49	209.48	32503.5	15130.4	0.14	7855.31	883.72	114.
950	25	50	213.75	33843.7	16075.7	0.14	8015.63	901.76	116.
969	—	51	218.02	35211.0	17059.7	0.14	8175.94	919.79	119.
988	26	52	222.30	36605.4	18083.0	0.14	8336.25	937.83	121.
1007	—	53	226.57	38026.8	19146.5	0.14	8496.56	955.86	123.

TABLE 8-3 Properties of sections (continued)

225 mm Wide Laminations

| Size | Number of Lams | | Area | Axis X-X | | | Axis Y-Y | | Approx Linear Density kg/m @ |
| b and d | | | bd | S bd²/6 | I bd³/12 | $\sqrt{\dfrac{d}{b^2}}$ | S db²/6 | I db³/12 | |
Millimetres (mm)	38mm Lams	19mm Lams	mm²x10³	mm³x10³	mm⁴x10⁶		mm³x10³	mm⁴x10⁶	545 kg/m³
225 x 1026	27	54	230.85	39475.3	20250.8	0.14	8656.88	973.90	126.
1045	—	55	235.13	40950.9	21396.8	0.14	8817.19	991.93	128.
1064	28	56	239.40	42453.6	22585.3	0.14	8977.50	1009.97	130.
1083	—	57	243.68	43983.3	23816.9	0.15	9137.81	1028.00	133.
1102	29	58	247.95	45540.1	25092.6	0.15	9298.13	1046.04	135.
1121	—	59	252.23	47124.0	26413.0	0.15	9458.44	1064.07	137.
225 x 1140	30	60	256.50	48735.0	27778.9	0.15	9618.75	1082.11	140.
1159	—	61	260.77	50373.0	29191.1	0.15	9779.06	1100.14	142.
1178	31	62	265.05	52038.1	30650.4	0.15	9939.38	1118.18	144.
1197	—	63	269.33	53730.3	32157.6	0.15	10099.6	1136.21	147.
1216	32	64	273.60	55449.6	33713.3	0.15	10260.0	1154.25	149.
1235	—	65	277.88	57195.9	35318.4	0.16	10420.3	1172.29	151.
225 x 1254	33	66	282.15	58969.3	36973.7	0.16	10580.6	1190.32	154.
1273	—	67	286.42	60769.8	38680.0	0.16	10740.9	1208.36	156.
1292	34	68	290.70	62597.4	40437.9	0.16	10901.2	1226.39	158.
1311	—	69	294.98	64452.0	42248.3	0.16	11061.5	1244.43	161.
1330	35	70	299.25	66333.7	44111.9	0.16	11221.8	1262.46	163.
1349	—	71	303.52	68242.5	46029.5	0.16	11382.1	1280.50	165.
225 x 1368	36	72	307.80	70178.4	48002.0	0.16	11542.5	1298.53	168.
1387	—	73	312.08	72141.3	50030.0	0.17	11702.8	1316.57	170.
1406	37	74	316.35	74131.3	52114.3	0.17	11863.1	1334.60	172.
1425	—	75	320.63	76148.4	54255.7	0.17	12023.4	1352.64	175.
1444	38	76	324.90	78192.6	56455.0	0.17	12183.7	1370.67	177.
1463	—	77	329.17	80263.8	58713.0	0.17	12344.0	1388.71	179.
225 x 1482	39	78	333.45	82362.1	61030.3	0.17	12504.3	1406.74	182.
1501	—	79	337.73	84487.5	63407.9	0.17	12664.6	1424.78	184.
1520	40	80	342.00	86640.0	65846.4	0.17	12825.0	1442.81	186.
1539	—	81	346.27	88819.5	68346.6	0.17	12985.3	1460.85	189.
1558	41	82	350.55	91026.1	70909.3	0.18	13145.6	1478.88	191.
1577	—	83	354.83	93259.8	73535.3	0.18	13305.9	1496.92	193.
225 x 1596	42	84	359.10	95520.6	76225.4	0.18	13466.2	1514.95	196.
1615	—	85	363.38	97808.4	78980.3	0.18	13626.5	1532.99	198.
1634	43	86	367.65	100123.	81800.7	0.18	13786.8	1551.02	200.
1653	—	87	371.92	102465.	84687.6	0.18	13947.1	1569.06	203.
1672	44	88	376.20	104834.	87641.5	0.18	14107.5	1587.09	205.
1691	—	89	380.48	107230.	90663.4	0.18	14267.8	1605.13	207.
225 x 1710	45	90	384.75	109653.	93753.9	0.18	14428.1	1623.16	210.
1729	—	91	389.02	112104.	96913.9	0.18	14588.4	1641.20	212.
1748	46	92	393.30	114581.	100144.	0.19	14748.7	1659.23	214.
1767	—	93	397.58	117085.	103445.	0.19	14909.0	1677.27	217.
1786	47	94	401.85	119617.	106818.	0.19	15069.3	1695.30	219.

Source: Reproduced by permission of the Canadian Wood Council.

TABLE 8-3 Properties of sections (continued)

275 mm Wide Laminations

Size	Number of Lams		Area	Axis X-X			Axis Y-Y		Approx Linear Density kg/m @
b and d			bd	S bd²/6	I bd³/12	$\sqrt{\dfrac{d}{b^2}}$	S db²/6	I db³/12	545 kg/m³
Millimetres (mm)	38mm Lams	19mm Lams	mm²x10³	mm³x10³	mm⁴x10⁶		mm³x10³	mm⁴x10⁶	
275 x 304	8	16	83.60	4235.73	643.83	0.06	3831.67	526.85	45.5
323	—	17	88.82	4781.75	772.25	0.07	4071.15	559.78	48.4
342	9	18	94.05	5360.85	916.71	0.07	4310.63	592.71	51.2
361	—	19	99.28	5973.05	1078.13	0.07	4550.10	625.64	54.1
380	10	20	104.50	6618.33	1257.48	0.07	4789.58	658.57	56.9
399	—	21	109.72	7296.71	1455.69	0.07	5029.06	691.50	59.8
275 x 418	11	22	114.95	8008.18	1673.71	0.07	5268.54	724.42	62.6
437	—	23	120.18	8752.75	1912.47	0.08	5508.02	757.35	65.5
456	12	24	125.40	9530.40	2172.93	0.08	5747.50	790.28	68.3
475	—	25	130.63	10341.1	2456.02	0.08	5986.98	823.21	71.1
494	13	26	135.85	11184.9	2762.69	0.08	6226.46	856.14	74.0
513	—	27	141.07	12061.9	3093.88	0.08	6465.94	889.07	76.8
275 x 532	14	28	146.30	12971.9	3450.53	0.08	6705.42	921.99	79.7
551	—	29	151.52	13915.0	3833.59	0.09	6944.90	954.92	82.5
570	15	30	156.75	14891.2	4244.01	0.09	7184.38	987.85	85.4
589	—	31	161.98	15900.5	4682.71	0.09	7423.85	1020.78	88.2
608	16	32	167.20	16942.9	5150.65	0.09	7663.33	1053.71	91.1
627	—	33	172.43	18018.4	5648.77	0.09	7902.81	1086.64	93.9
275 x 646	17	34	177.65	19126.9	6178.02	0.09	8142.29	1119.57	96.8
665	—	35	182.88	20268.6	6739.32	0.09	8381.77	1152.49	99.6
684	18	36	188.10	21443.4	7333.64	0.10	8621.25	1185.42	102.
703	—	37	193.32	22651.2	7961.91	0.10	8860.73	1218.35	105.
722	19	38	198.55	23892.1	8625.08	0.10	9100.21	1251.28	108.
741	—	39	203.77	25166.2	9324.08	0.10	9339.69	1284.21	111.
275 x 760	20	40	209.00	26473.3	10059.8	0.10	9579.17	1317.14	114.
779	—	41	214.23	27813.5	10833.3	0.10	9818.65	1350.06	117.
798	21	42	219.45	29186.8	11645.5	0.10	10058.1	1382.99	120.
817	—	43	224.68	30593.2	12497.3	0.10	10297.6	1415.92	122.
836	22	44	229.90	32032.7	13389.6	0.11	10537.0	1448.85	125.
855	—	45	235.13	33505.3	14323.5	0.11	10776.5	1481.78	128.
275 x 874	23	46	240.35	35010.9	15299.8	0.11	11016.0	1514.71	131.
893	—	47	245.57	36549.7	16319.4	0.11	11255.5	1547.63	134.
912	24	48	250.80	38121.6	17383.4	0.11	11495.0	1580.56	137.
931	—	49	256.02	39726.5	18492.7	0.11	11734.4	1613.49	139.
950	25	50	261.25	41364.5	19648.1	0.11	11973.9	1646.42	142.
969	—	51	266.48	43035.7	20850.8	0.11	12213.4	1679.35	145.
275 x 988	26	52	271.70	44739.9	22101.5	0.11	12452.9	1712.28	148.
1007	—	53	276.92	46477.2	23401.2	0.12	12692.4	1745.20	151.
1026	27	54	282.15	48247.6	24751.0	0.12	12931.8	1778.13	154.
1045	—	55	287.38	50051.1	26151.7	0.12	13171.3	1811.06	157.
1064	28	56	292.60	51887.7	27604.2	0.12	13410.8	1843.99	159.
1083	—	57	297.83	53757.4	29109.6	0.12	13650.3	1876.92	162.
275 x 1102	29	58	303.05	55660.1	30668.7	0.12	13889.7	1909.85	165.
1121	—	59	308.27	57596.0	32282.5	0.12	14129.2	1942.77	168.
1140	30	60	313.50	59565.0	33952.0	0.12	14368.7	1975.70	171.
1159	—	61	318.73	61567.0	35678.1	0.12	14608.2	2008.63	174.
1178	31	62	323.95	63602.1	37461.6	0.12	14847.7	2041.56	176.
1197	—	63	329.17	65670.4	39303.7	0.13	15087.1	2074.49	179.

Source: Reproduced by permission of the Canadian Wood Council.

TABLE 8-3 Properties of sections (continued)

275 mm Wide Laminations

Size b and d	Number of Lams		Area	Axis X-X			Axis Y-Y		Approx Linear Density kg/m @
			bd	S bd²/6	I bd³/12	$\sqrt{\dfrac{d}{b^2}}$	S db²/6	I db³/12	
Millimetres (mm)	38mm Lams	19mm Lams	mm²x10³	mm³x10³	mm⁴x10⁶		mm³x10³	mm⁴x10⁶	545 kg/m³
275 x 1216	32	64	334.40	67771.7	41205.2	0.13	15326.6	2107.42	182.
1235	—	65	339.63	69906.1	43167.0	0.13	15566.1	2140.35	185.
1254	33	66	344.85	72073.6	45190.1	0.13	15805.6	2173.27	188.
1273	—	67	350.08	74274.2	47275.5	0.13	16045.1	2206.20	191.
1292	34	68	355.30	76507.9	49424.1	0.13	16284.5	2239.13	194.
1311	—	69	360.52	78774.7	51636.8	0.13	16524.0	2272.06	196.
275 x 1330	35	70	365.75	81074.5	53914.5	0.13	16763.5	2304.99	199.
1349	—	71	370.98	83407.5	56258.3	0.13	17003.0	2337.92	202.
1368	36	72	376.20	85773.6	58669.1	0.13	17242.5	2370.84	205.
1387	—	73	381.42	88172.7	61147.8	0.14	17481.9	2403.77	208.
1406	37	74	386.65	90604.9	63695.3	0.14	17721.4	2436.70	211.
1425	—	75	391.88	93070.3	66312.6	0.14	17960.9	2469.63	213.
275 x 1444	38	76	397.10	95568.7	69000.6	0.14	18200.4	2502.56	216.
1463	—	77	402.33	98100.2	71760.3	0.14	18439.9	2535.49	219.
1482	39	78	407.55	100664.	74592.6	0.14	18679.3	2568.41	222.
1501	—	79	412.77	103262.	77498.5	0.14	18918.8	2601.34	225.
1520	40	80	418.00	105893.	80478.9	0.14	19158.3	2634.27	228.
1539	—	81	423.23	108557.	83534.7	0.14	19397.8	2667.20	231.
275 x 1558	41	82	428.45	111254.	86667.0	0.14	19637.2	2700.13	233.
1577	—	83	433.67	113984.	89876.5	0.14	19876.7	2733.06	236.
1596	42	84	438.90	116747.	93164.4	0.15	20116.2	2765.98	239.
1615	—	85	444.13	119543.	96531.4	0.15	20355.7	2798.91	242.
1634	43	86	449.35	122372.	99978.7	0.15	20595.2	2831.84	245.
1653	—	87	454.58	125235.	103507.	0.15	20834.6	2864.77	248.
275 x 1672	44	88	459.80	128130.	107117.	0.15	21074.1	2897.70	250.
1691	—	89	465.02	131059.	110810.	0.15	21313.6	2930.63	253.
1710	45	90	470.25	134021.	114588.	0.15	21553.1	2963.56	256.
1729	—	91	475.48	137016.	118450.	0.15	21792.6	2996.48	259.
1748	46	92	480.70	140043.	122398.	0.15	22032.0	3029.41	262.
1767	—	93	485.92	143104.	126433.	0.15	22271.5	3062.34	265.
275 x 1786	47	94	491.15	146198.	130555.	0.15	22511.0	3095.27	267.
1805	—	95	496.38	149326.	134766.	0.15	22750.5	3128.20	270.
1824	48	96	501.60	152486.	139067.	0.16	22990.0	3161.13	273.
1843	—	97	506.83	155679.	143458.	0.16	23229.4	3194.05	276.
1862	49	98	512.05	158906.	147941.	0.16	23468.9	3226.98	279.
1881	—	99	517.28	162165.	152516.	0.16	23708.4	3259.91	282.
275 x 1900	50	100	522.50	165458.	157185.	0.16	23947.9	3292.84	285.
1919	—	101	527.72	168784.	161948.	0.16	24187.3	3325.77	287.
1938	51	102	532.95	172142.	166806.	0.16	24426.8	3358.70	290.
1957	—	103	538.17	175534.	171760.	0.16	24666.3	3391.62	293.
1976	52	104	543.40	178959.	176812.	0.16	24905.8	3424.55	296.
1995	—	105	548.63	182417.	181961.	0.16	25145.3	3457.48	299.
275 x 2014	53	106	553.85	185908.	187210.	0.16	25384.7	3490.41	302.
2033	—.	107	559.08	189433.	192558.	0.16	25624.2	3523.34	304.
2052	54	108	564.30	192990.	198008.	0.16	25863.7	3556.27	307.
2071	—	109	569.53	196581.	203559.	0.17	26103.2	3589.19	310.
2090	55	110	574.75	200204.	209213.	0.17	26342.7	3622.12	313.
2109	—	111	579.97	203861.	214971.	0.17	26582.1	3655.05	316.
275 x 2128	56	112	585.20	207550.	220834.	0.17	26821.6	3687.98	319.

Source: Reproduced by permission of the Canadian Wood Council.

TABLE 8–3 Properties of sections (continued)

315 mm Wide Laminations

Size b and d Millimetres (mm)	Number of Lams 38mm Lams	Area bd $mm^2 \times 10^3$	Axis X-X S $bd^2/6$ $mm^3 \times 10^3$	Axis X-X I $bd^3/12$ $mm^4 \times 10^6$	$\sqrt{\dfrac{d}{b^2}}$	Axis Y-Y S $db^2/6$ $mm^3 \times 10^3$	Axis Y-Y I $db^3/12$ $mm^4 \times 10^6$	Approx Linear Density kg/m @ 545 kg/m³
315 x 304	8	95.76	4851.84	737.48	0.06	5027.40	791.82	52.2
342	9	107.73	6140.61	1050.04	0.06	5655.83	890.79	58.7
380	10	119.70	7581.00	1440.39	0.06	6284.25	989.77	65.2
418	11	131.67	9173.01	1917.16	0.06	6912.68	1088.75	71.7
456	12	143.64	10916.6	2488.99	0.07	7541.10	1187.72	78.2
494	13	155.61	12811.8	3164.54	0.07	8169.53	1286.70	84.7
315 x 532	14	167.58	14858.7	3952.43	0.07	8797.95	1385.68	91.3
570	15	179.55	17057.2	4861.32	0.08	9426.38	1484.65	97.8
608	16	191.52	19407.3	5899.84	0.08	10054.8	1583.63	104.
646	17	203.49	21909.0	7076.64	0.08	10683.2	1682.61	111.
684	18	215.46	24562.4	8400.35	0.08	11311.6	1781.58	117.
722	19	227.43	27367.4	9879.63	0.09	11940.0	1880.56	124.
315 x 760	20	239.40	30324.0	11523.1	0.09	12568.5	1979.54	130.
798	21	251.37	33432.2	13339.4	0.09	13196.9	2078.52	137.
836	22	263.34	36692.0	15337.2	0.09	13825.3	2177.49	143.
874	23	275.31	40103.4	17525.2	0.09	14453.7	2276.47	150.
912	24	287.28	43666.5	19911.9	0.10	15082.2	2375.45	156.
950	25	299.25	47381.2	22506.0	0.10	15710.6	2474.42	163.
315 x 988	26	311.22	51247.5	25316.2	0.10	16339.0	2573.40	169.
1026	27	323.19	55265.4	28351.2	0.10	16967.4	2672.38	176.
1064	28	335.16	59435.0	31619.4	0.10	17595.9	2771.35	183.
1102	29	347.13	63756.2	35129.6	0.11	18224.3	2870.33	189.
1140	30	359.10	68229.0	38890.5	0.11	18852.7	2969.31	196.
1178	31	371.07	72853.4	42910.6	0.11	19481.1	3068.29	202.
315 x 1216	32	383.04	77629.4	47198.7	0.11	20109.6	3167.26	209.
1254	33	395.01	82557.0	51763.3	0.11	20738.0	3266.24	215.
1292	34	406.98	87636.3	56613.0	0.11	21366.4	3365.22	222.
1330	35	418.95	92867.2	61756.7	0.12	21994.8	3464.19	228.
1368	36	430.92	98249.7	67202.8	0.12	22623.3	3563.17	235.
1406	37	442.89	103783.	72960.0	0.12	23251.7	3662.15	241.
315 x 1444	38	454.86	109469.	79037.0	0.12	23880.1	3761.12	248.
1482	39	466.83	115307.	85442.4	0.12	24508.5	3860.10	254.
1520	40	478.80	121296.	92184.9	0.12	25137.0	3959.08	261.
1558	41	490.77	127436.	99273.1	0.13	25765.4	4058.05	267.
1596	42	502.74	133728.	106715.	0.13	26393.8	4157.03	274.
1634	43	514.71	140172.	114521.	0.13	27022.2	4256.01	280.
315 x 1672	44	526.68	146768.	122698.	0.13	27650.7	4354.99	287.
1710	45	538.65	153515.	131255.	0.13	28279.1	4453.96	293.
1748	46	550.62	160413.	140201.	0.13	28907.5	4552.94	300.
1786	47	562.59	167464.	149545.	0.13	29535.9	4651.92	306.
1824	48	574.56	174666.	159295.	0.14	30164.4	4750.89	313.
1862	49	586.53	182019.	169460.	0.14	30792.8	4849.87	319.
315 x 1900	50	598.50	189524.	180048.	0.14	31421.2	4948.85	325.
1938	51	610.47	197181.	191069.	0.14	32049.6	5047.82	332.
1976	52	622.44	204990.	202530.	0.14	32678.1	5146.80	339.
2014	53	634.41	212950.	214440.	0.14	33306.5	5245.78	346.
2052	54	646.38	221061.	226809.	0.14	33934.9	5344.75	352.
2090	55	658.35	229325.	239644.	0.15	34563.3	5443.73	359.
2128	56	670.32	237740.	252955.	0.15	35191.8	5542.71	365.

Source: Reproduced by permission of the Canadian Wood Council.

TABLE 8-3 Properties of sections (continued)

365 mm Wide Laminations

Size b and d	Number of Lams	Area	Axis X-X			Axis Y-Y		Approx Linear Density kg/m @
		bd	S bd^2/6	I bd^3/12	$\sqrt{\dfrac{d}{b^2}}$	S db^2/6	I db^3/12	
Millimetres (mm)	38mm Lams	mm^{2}x10^3	mm^{3}x10^3	mm^{4}x10^6		mm^{3}x10^3	mm^{4}x10^6	545 kg/m^3
365 x 342	9	124.83	7115.31	1216.72	0.05	7593.83	1385.87	68.0
380	10	138.70	8784.33	1669.02	0.05	8437.58	1539.86	75.5
418	11	152.57	10629.0	2221.47	0.06	9281.34	1693.84	83.1
456	12	166.44	12649.4	2884.07	0.06	10125.1	1847.83	90.6
494	13	180.31	14845.5	3666.84	0.06	10968.8	2001.82	98.2
532	14	194.18	17217.2	4579.80	0.06	11812.6	2155.80	106.
365 x 570	15	208.05	19764.7	5632.95	0.07	12656.3	2309.79	113.
608	16	221.92	22487.8	6836.32	0.07	13500.1	2463.77	121.
646	17	235.79	25386.7	8199.91	0.07	14343.8	2617.76	128.
684	18	249.66	28461.2	9733.74	0.07	15187.6	2771.75	136.
722	19	263.53	31711.4	11447.8	0.07	16031.4	2925.73	144.
760	20	277.40	35137.3	13352.1	0.08	16875.1	3079.72	151.
365 x 798	21	291.27	38738.9	15456.8	0.08	17718.9	3233.70	159.
836	22	305.14	42516.1	17771.7	0.08	18562.6	3387.69	166.
874	23	319.01	46469.1	20307.0	0.08	19406.4	3541.68	174.
912	24	332.88	50597.7	23072.5	0.08	20250.2	3695.66	181.
950	25	346.75	54902.0	26078.4	0.08	21093.9	3849.65	189.
988	26	360.62	59382.1	29334.7	0.09	21937.7	4003.63	196.
365 x 1026	27	374.49	64037.7	32851.3	0.09	22781.4	4157.62	204.
1064	28	388.36	68869.1	36638.4	0.09	23625.2	4311.60	212.
1102	29	402.23	73876.2	40705.8	0.09	24468.9	4465.59	219.
1140	30	416.10	79059.0	45063.6	0.09	25312.7	4619.58	227.
1178	31	429.97	84417.4	49721.8	0.09	26156.5	4773.56	234.
1216	32	443.84	89951.5	54690.5	0.10	27000.2	4927.55	242.
365 x 1254	33	457.71	95661.3	59979.7	0.10	27844.0	5081.53	249.
1292	34	471.58	101546.	65599.3	0.10	28687.7	5235.52	257.
1330	35	485.45	107608.	71559.3	0.10	29531.5	5389.51	264.
1368	36	499.32	113844.	77869.9	0.10	30375.3	5543.49	272.
1406	37	513.19	120257.	84541.0	0.10	31219.0	5697.48	279.
1444	38	527.06	126845.	91582.6	0.10	32062.8	5851.46	287.
365 x 1482	39	540.93	133609.	99004.8	0.11	32906.5	6005.45	295.
1520	40	554.80	140549.	106817.	0.11	33750.3	6159.44	302.
1558	41	568.67	147664.	115030.	0.11	34594.0	6313.42	310.
1596	42	582.54	154955.	123654.	0.11	35437.8	6467.41	317.
1634	43	596.41	162422.	132699.	0.11	36281.6	6621.39	325.
1672	44	610.28	170064.	142174.	0.11	37125.3	6775.38	332.
365 x 1710	45	624.15	177882.	152089.	0.11	37969.1	6929.37	340.
1748	46	638.02	185876.	162456.	0.11	38812.8	7083.35	347.
1786	47	651.89	194045.	173283.	0.12	39656.6	7237.34	355.
1824	48	665.76	202391.	184580.	0.12	40500.4	7391.32	363.
1862	49	679.63	210911.	196358.	0.12	41344.1	7545.31	370.
1900	50	693.50	219608.	208627.	0.12	42187.9	7699.29	378.
365 x 1938	51	707.37	228480.	221397.	0.12	43031.6	7853.28	385.
1976	52	721.24	237528.	234678.	0.12	43875.4	8007.27	393.
2014	53	735.11	246751.	248479.	0.12	44719.1	8161.25	400.
2052	54	748.98	256151.	262811.	0.12	45562.9	8315.24	408.
2090	55	762.85	265726.	277683.	0.13	46406.7	8469.22	415.
2128	56	776.72	275476.	293107.	0.13	47250.4	8623.21	423.

Source: Reproduced by permission of the Canadian Wood Council.

Douglas fir laminations, 1½ in. thick, may be bent to a minimum radius of 27 ft 6 in. when the ends are tangent, and 35 ft 6 in. when the member is curved throughout its length. Limiting radii for ¾-in. laminations are 9 ft 4 in. and 12 ft 6 in., respectively. When smaller curves are necessary, the laminations must be reduced in thickness, resulting in higher costs due to additional material handling and waste.

TYPES OF UNITS

Glue-laminated members may be used independently or in conjunction with other materials in the construction of structural units. For example, trusses may be built with glue-laminated chords and sawn timber webs (see Fig. 8-26). Units such as folded plate or other stressed skin panels may be combinations of glue-laminated timber and plywood (see Fig. 8-27).

The greatest use of glue-laminated lumber is in the fabrication of beams and arches. Three basic types of beams are made: *straight, tapered, and curved* (see Fig. 8-28). Straight and tapered beams (Fig. 8-29) are usual-

FIGURE 8-26 Glue-laminated and sawn timber used together in a truss. *(Courtesy Bondwood Structures, Alberta Ltd.)*

ly made with a slight camber to allow for normal deflection and to improve appearance. Straight beams are common in the type of construction shown in Fig. 8-30.

The *three-hinged arch* (most commonly used) may

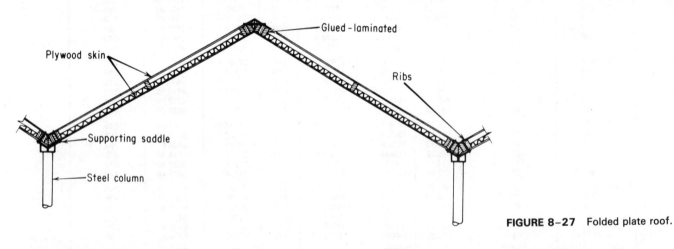

FIGURE 8-27 Folded plate roof.

FIGURE 8-28 Basic beam shapes.

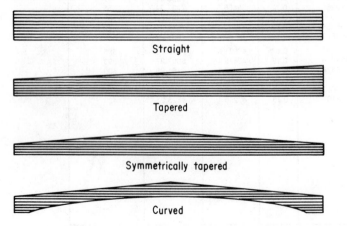

FIGURE 8-29 Symmetrically tapered beam.

FIGURE 8-30 Straight glue-laminated beams. *(Courtesy Canadian Wood Council.)*

have legs which are *straight, circular,* or of *varying curvature.* Straight-legged arches are normally either *high* V or *low* V (see Fig. 8-31), as the V arch allows maximum utilization of enclosed space. The building shown in Fig. 8-32 is an example of high V arches.

Circular arches are actually *segments* of a circle, as may be seen in Fig. 8-33. Notice the purlins, spanning from arch to arch, which will be used to carry the sheathing. Arches of varying curvature may be parabolic or gothic in shape or may be a combination of curve and straight line. In Fig. 8-34, gothic arches form the main supporting structure, but additional framing has been added to produce a straight-line, peaked roof. Figures 8-35 and 8-36 illustrate the variation in design possible using glulam structural members.

The simplest types of glue-laminated arches are those made for light construction work—normally from 2-in.-wide laminations. They are also available in a variety of shapes, the basic ones being the *gothic arch,* the *utility arch,* and the *shed rafter.* In the case of the gothic arch, the member is curved throughout. The utility arch has a short, straight portion, which maintains a 4 in 12 slope at the top, while the shed rafter is partly curved and partly straight, with the straight portion tangent to the curve. Light arches (ribs) are commonly used in spans of up to 80 ft, for 2-ft. o.c. spacing. The number of laminations will vary from 4 to 12, depending on the span and the load.

FIGURE 8-31 Arch shapes.

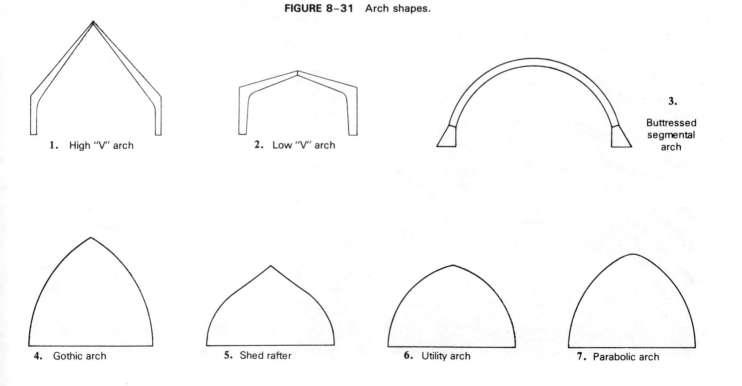

1. High "V" arch

2. Low "V" arch

3. Buttressed segmental arch

4. Gothic arch

5. Shed rafter

6. Utility arch

7. Parabolic arch

FIGURE 8-32 High V arches.

FIGURE 8-34 Glue-laminated gothic arches. *(Courtesy Canadian Wood Council.)*

FIGURE 8-33 Glue-laminated segmental arch frame. *(Courtesy TPL Industries Ltd.)*

FIGURE 8–35 Arches with straight top end. *(Courtesy American Institute of Timber Construction.)*

FIGURE 8–36 Reversed curve glulam arches. *(Courtesy Canadian Wood Council.)*

CONNECTIONS FOR GLULAM BEAMS AND ARCHES

Glulam Beams

The only connection which will be different from those described in Figs. 8-6, 8-7, 8-8, and 8-9 is the one required when a beam is cantilevered over a support. To produce the required length in such a case, two members may be connected between supports. Three such connections are illustrated in Fig. 8-37. Figure 8-37(a) is commonly used for this purpose when large beams are involved, designed so that the bolt is in tension, with the plates providing the required bearing perpendicular to the grain. To design the shape of cut so that the full depth of the beam is available for shear, draw lines at 60° from the ends of both top and bottom plates. The horizontal cut should be at the center of the beam and the sloping cuts within the boundaries of the 60° lines.

Figures 8-37(b) and (c) show *saddle* connections, with square-cut beam ends. Figure 8-37(b) may be used for moderate loads and Fig. 8-37(c) for light loads.

Glulam Arches

Arch connections are of two general types: *upper hinge* and *lower hinge*. Upper hinge connections may be made by *gussets, bolts,* or *bolted plates* (see Fig. 8-38). Light arch legs are usually connected at the top by gussets, while the other two methods are used with heavier arches. The bolted connection is used for arches with a steep pitch,

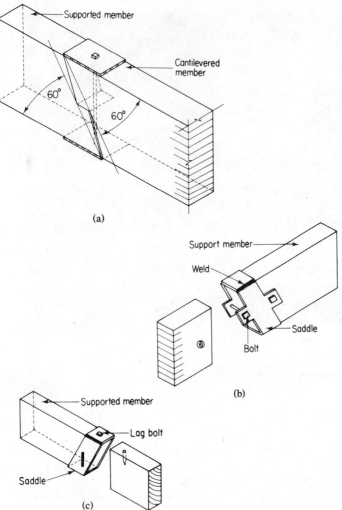

FIGURE 8-37 Connection for cantilevered beam.

FIGURE 8-38 Arch upper hinge connections.

with double shear plates at the center. When the slope of the arch is such that a bolt length becomes excessive, the bolted plate connection is used. In both, a dowel and double shear plates are added for vertical shear.

The bottom—*lower hinge*—of light arches is generally one of the connections shown in Fig. 8-39. Heavier arches may be connected at the base by any of the methods illustrated in Fig. 8-40. Figure 8-40(a) is the most common connection used for this purpose, where the foundation has been designed to carry the horizontal thrust. If the foundation is not so designed, the type of connection shown in Fig. 8-40(b) will be used, which includes a tie rod. Figures 8-40(c) and (d) show connections designed for wood and steel bases, respectively, while Fig. 8-40(e) is intended for smaller arches which may not require a true hinge. A true *hinge* connection, shown in Fig. 8-41, is used for long-span, deep section arches, such as those in Fig. 8-33.

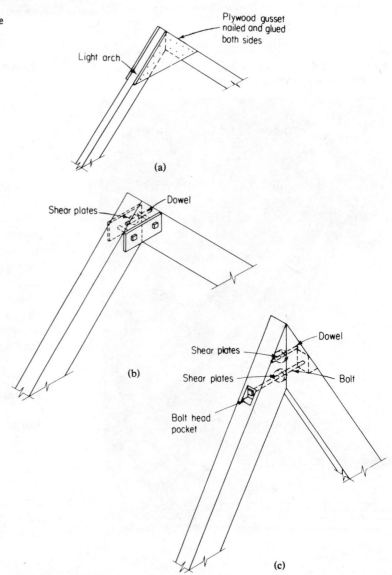

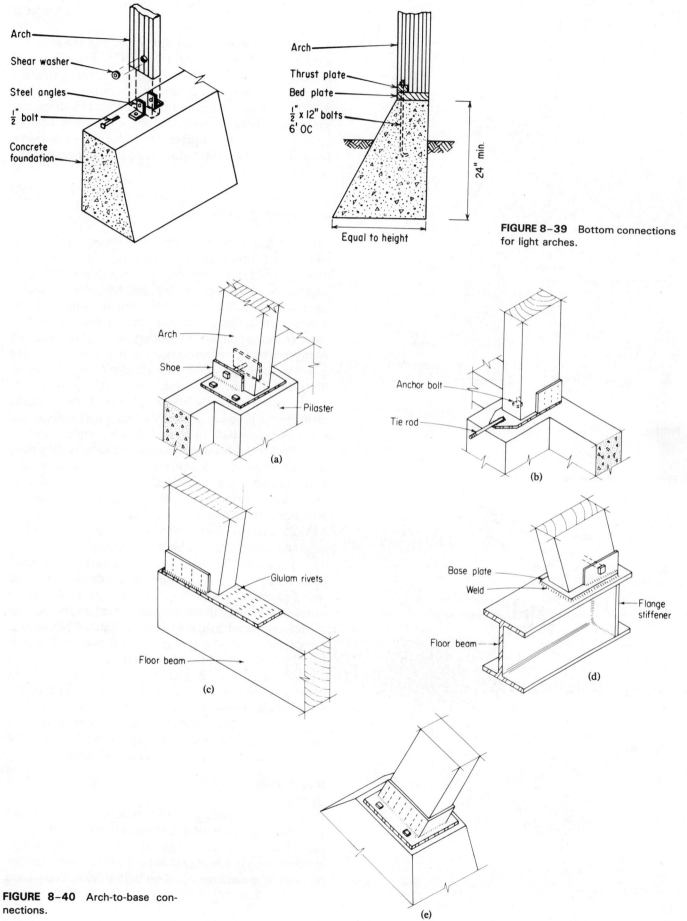

FIGURE 8-39 Bottom connections for light arches.

FIGURE 8-40 Arch-to-base connections.

211

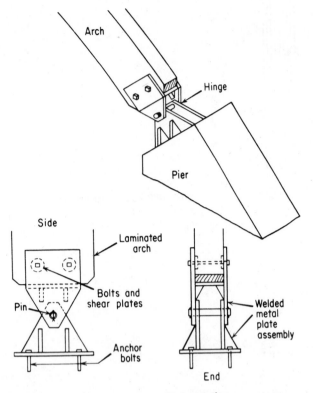

FIGURE 8-41 True hinge arch connection.

FIGURE 8-42 Moment connections.

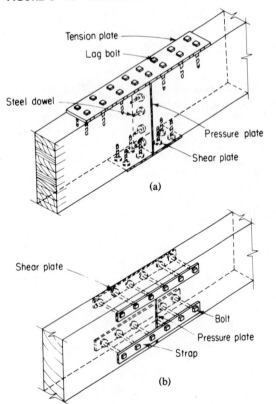

Moment connections are sometimes required for splicing long beams or arches, and Fig. 8-42 illustrates such a connection. In Fig. 8-42(a), the bending moment is resisted by steel straps top and bottom, and in Fig. 8-42(b) the moment connection is provided by straps and shear plates bolted to the sides of the members. In both connections a pressure plate is included and shear transfer is provided by short dowels and shear plates in the end faces.

PLYWOOD STRUCTURES

Plywood is recognized as one of the most versatile and widely used products in the building construction industry. In addition to the ordinary uses to which it is put, such as for sheathing, decking, and formwork, it is also used in the construction of such structural units as *box beams, web beams,* and *stressed skin panels.* Just as glue plays an important part in the makeup of glue-laminated members, so it has an important role not only in the manufacture of plywood itself but also in the fabrication of plywood beams and panels. In all these, plywood is combined with sawn lumber or glue-laminated framing members to produce a variety of strucural units which are strong, relatively light, and usually simple to install.

Plywood sheathing is manufactured by placing three or more wood plies or veneers with their grains at right angles to each other to produce a built-up section of some predetermined thickness. The plies are bonded together with waterproof glue that is set with heat and pressure. The result is a piece of material that is uniform in thickness and very strong in both directions.

Plywood sheathing is manufactured in various grades, depending on the type and condition of the two outside veneers. Unsanded plywood is normally graded as select grade or sheathing grade based on the number of defects found in the face plies. Sanded plywood is designated as good one side or good two sides, again, based on the condition of the face plies.

Normal plywood thicknesses range from ¼ to ¾ in., depending on the use. For example, plywood used for formwork is usually manufactured in three thicknesses; ⅝ in., ¹¹⁄₁₆ in., and ¾ in. Because of its good quality, strength, and water resistance, plywood is a material whose versatility makes its uses almost limitless.

Box Beams

In wood frame buildings in which long, deep beams are required, the weight of a solid glulam beam might be excessive and a properly designed *box beam* may be substituted. This is a beam made by sheathing a framework of sawn or glulam members on both sides with plywood (see Fig. 8–43), the thickness of which depends on the load to which the beam will be subjected.

The strength of such a beam is largely dependent on the glue bond between plywood and frame but is also improved by using plywood sheets which are as long as possible. Long sheets can be produced by *end scarfing*— joining end to end—regular sheets, as shown in Fig. 8–44.

Glue may be applied with a brush or by gun (see Fig. 8–45), at the temperature recommended by the manufacturer. Facilities for applying pressure to the glue joints until the glue has set are also important.

Plywood Web Beams

A plywood *web beam* consists of one or more plywood webs to which sawn lumber or glue-laminated timber flanges are attached along the top and bottom edges. Notice that in the beam shown in Fig. 8–45 the top and bottom flanges are laminated and the webs are sawn lumber. At intervals along the beam, lumber stiffeners are used to separate the flanges and control web buckling. Such beams may be fabricated with nails alone, with nails and glue, or by pressure gluing. The strength of the

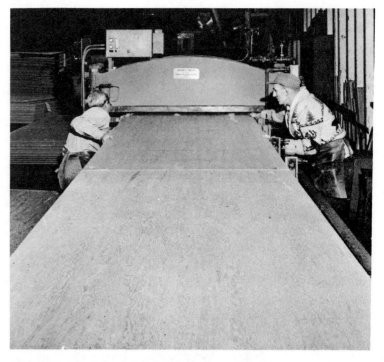

FIGURE 8–44 Joining plywood sheets by end scarfing. *(Courtesy Plywood Mfg. Association of B. C.)*

beam will vary somewhat, depending on which method of fabrication is used. Figure 8–46 illustrates typical plywood web beam designs.

FIGURE 8–45 Large plywood web beam under construction. *(Courtesy Plywood Mfg. Association of B. C.)*

FIGURE 8–43 Large box beam under construction. *(Courtesy Plywood Mfg. Association of B. C.)*

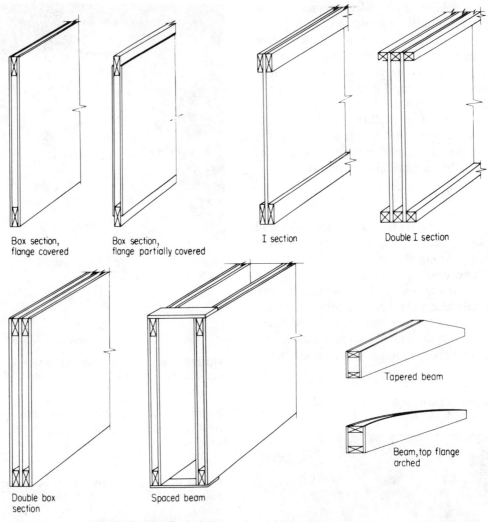

Box section,   Box section,   I section   Double I section
flange covered   flange partially covered

Double box   Spaced beam
section

Tapered beam

Beam, top flange
arched

FIGURE 8–46 Typical plywood web beam designs.

There is no definite limit to the span which may be bridged with plywood web beams. Shallow beams spanning from 10 to 50 ft and deep beams spanning up to 100 ft have been found to be economical. Usually the limiting factor is the 48-in. width, which is the standard used in plywood manufacture. If sheets are edge-spliced to increase width, greater spans can be achieved.

Stressed Skin Panels

A stressed skin panel consists of longitudinal framing members of wood, covered on one or both sides of plywood skins (see Fig. 8–47). The plywood skin, in addition to providing a surface covering, acts with the framing members to contribute to the strength of the unit as a whole. The plywood skins must be attached to the frame with glue in order to develop the full potential strength of the panel.

Spacing of the longitudinal members depends on

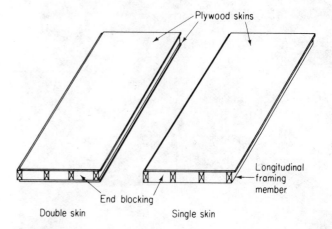

Plywood skins

Longitudinal framing member

End blocking

Double skin   Single skin

FIGURE 8–47 Stressed skin panels.

the thickness of the plywood skin and the direction of the face grain in relation to the long dimension of the longitudinal members. Table 8–4 gives basic spacing with various plywood grades and thicknesses.

TABLE 8-4 Basic spacing of panel framing members

Plywood thickness (in.)	Basic spacing (in.)	
	Face grain parallel to longitudinal members	Face grain perpendicular to longitudinal members
¼ (sanded)	10.1	12.1
⁵⁄₁₆ (unsanded)	11.0	15.6
⅜ (sanded)	17.5	15.5
⅜ (unsanded)	15.3	16.6
½ (sanded)	25.6	25.3
½ (unsanded)	23.0	28.2
⅝ (sanded)	33.1	30.7
⅝ (unsanded)	26.8	38.8
¾ (sanded)	40.3	36.3
¾ (unsanded)	36.3	40.2

Stressed skin panels are used for roof and floor decking (see Fig. 12–29) in the standard rectanglar shapes but may be fabricated in almost any desired shape to fit a particular design. For example, in Fig. 8–48, rectangular panels are used to form a folded plate roof, and in Fig. 8–49 triangular panels make up segments of a roof for an octagonal-shaped building.

A special kind of stressed skin panel is involved when plywood is applied to a curved frame to make skin stressed panel segments for a vaulted roof. Figure 8–50 shows a curved, skin stressed panel, and Fig. 8–51 shows a completed plywood vaulted roof.

FIGURE 8-48 Folded plate roof of stressed skin panels.

FIGURE 8-49 Stressed skin roof panel. *(Courtesy Plywood Mfg. Association of B. C.)*

FIGURE 8-50 Curved panel for vaulted roof. *(Courtesy Plywood Mfg. Association of B. C.)*

FIGURE 8-51 Plywood vaulted roof. *(Courtesy Plywood Mfg. Association of B. C.)*

TERMITE CONTROL FOR WOOD BUILDINGS

In some areas and climates, wood which is in contact with or close to the ground is subject to attack by insects called *termites*. They bore into wood and, working from inside, may almost completely destroy it, leaving only an outer shell. Where wooden buildings or portions of them are subject to termite attack, protection against such infestation must be provided.

One method of providing protection is by pressure-treating the wood likely to be affected with creosote or salt. Creosote is usually used on wood that is not going to be exposed to view, and such treatment, properly done, is considered to provide protection for a 25-yr period. Wood which will be exposed to view may be pressure-treated with strong salt solution. The protection period is about the same as that provided by creosote.

Another method of protecting against attack by termites is to provide barrier shields under wood members at any point where there is a possibility of termites reaching them from the ground. Wooden members should be kept at least 18 in. above the ground and a shield provided between the member and the ground. For members like floor joists, sills, plates, or window and door sills, resting on a masonry or concrete wall, the shield can be

cast into the foundation wall or set in mortar on the top of it, as illustrated in Fig. 8-52. For individual members, such as posts or columns resting on a footing or pier, a shield should be placed between the member and its footing.

Shields are made preferably of copper sheet, 16 oz, minimum, and, in addition to covering the surface on which the wood rests, should project a minimum of 2 in., at a 45° angle, from the edges of the wall or post footing.

REVIEW QUESTIONS

1. Outline five factors which have contributed to the more efficient use of timber in modern building construction.

2. List six advantages in the use of glue-laminated members for timber construction.

3. Answer briefly: **(a)** Why are there severe limitations to the use of cross-grained material in laminated members? **(b)** Why is the moisture content of laminating stock so closely controlled? **(c)** Why is there relatively small tolerance in the thickness of laminating stock?

4. Exlain briefly the purpose of **(a)** a bearing plate, **(b)** a shear plate, and **(c)** a lateral tie.

5. Explain the difference in action between a *shear plate* and a *split ring*.

6. Explain what is meant by **(a)** a three-hinged arch, **(b)** an arch rib, and **(c)** a true hinge connection.

7. Explain **(a)** why a ½-in. space is left between the ends of wood decking and the face of a masonry wall, **(b)** how the above-mentioned space is concealed, and **(c)** the purpose of a *spline*.

8. Show by a scale diagram how you would suspend an object such as an overhead heater from a wooden roof beam.

FIGURE 8-52 Termite shields.

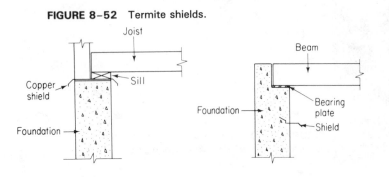

Reinforced Concrete Frame

We have seen that concrete is a basic material in the construction of footings, foundations, and bearing walls. It is also widely used for making the structural frames of large buildings—the columns, the girders, and the beams. Since these members are subjected to tensile stresses and concrete is weak in resisting tension, steel is added to the member—hence the term "reinforced concrete."

A reinforced concrete building frame may be erected by either of two methods. The members may be *cast-in-place* or the frame may be assembled from *precast* members. In the first case, forms are built and erected to form the shape of the frame and concrete is placed on the site. Precast members, on the other hand, are formed, cast, and cured in a plant and are subsequently brought to the building site ready for assembly.

CAST-IN-PLACE REINFORCED CONCRETE STRUCTURAL FRAME

There are two basic considerations in the erection of a cast-in-place structural frame. One of these is the actual building and erecting of forms of the size, shape, and strength required, and the other is the proper placing of the steel rods needed for reinforcement.

The size and shape of the members to be formed are indicated on the building plans, while the size and amount of material necessary to make the forms strong enough are based on the loads and pressures involved. The size, shape, and amount of steel required are also indicated on the plans in the form of column, girder, and beam *schedules*.

REINFORCING STEEL

Reinforcing bars are available in various sizes and grades. The grade designates the yield strength of the material used in the manufacture of the bars. The three most common grades are Grade 40, 50, and 60. For example, a Grade 40 bar has a yield strength of 40,000 psi. Higher grades are available but are subject to a premium.

The size of a bar is designated by a number that represents the number of eighths of an inch that make up the bar diameter (see Table 9–1). A #4 bar has a diameter of $4 \times \frac{1}{8} = \frac{1}{2}$ in. In Canada, metric bar sizes have been adopted using a similar approach for bar designation (Table 9–1). In this instance, the number of the bar approximates the bar diameter in millimeters.

Reinforcing bars may be smooth or deformed—the deformations being standard indentations rolled into the surface of the bar during the manufacturing process.

TABLE 9-1 Reinforcing bar properties

| | IMPERIAL | | | | METRIC | | |
Size	Dia (in.)	Area (in.²)	Weight (lb/ft)	Size	Dia (mm)	Area (mm²)	Mass (kg/m)
#3	0.375	0.11	0.376	#10	11.3	100	0.785
#4	0.500	0.20	0.668	#15	16.0	200	1.570
#5	0.625	0.31	1.043	#20	19.5	300	2.355
#6	0.750	0.44	1.502	#25	25.2	500	3.925
#7	0.875	0.60	2.044	#30	29.9	700	5.495
#8	1.000	0.79	2.670	#35	35.7	1000	7.850
#9	1.128	1.00	3.400	#45	43.7	1500	11.775
#10	1.270	1.27	4.303	#55	56.4	2500	19.625
#11	1.410	1.56	5.313				
#14	1.693	2.25	7.650				
#18	2.257	4.00	13.600				

COLUMN FORMS

Column forms are often subjected to much greater lateral pressure than wall forms because of their comparatively small cross section and relatively high rates of placement. It is therefore necessary to provide tight joints and strong tie support. Some means of accurately locating column forms, anchoring them at their base, and keeping them in a vertical position are also prime considerations. Wherever possible, a *cleanout* opening should be provided at the bottom of the form so that debris may be removed before placing begins (see Fig. 9-3). *Windows* are often built into one side of tall column forms to allow the placing of concrete in the bottom half of the form without having to drop it from the top.

Columns may be square, rectangular, round, or irregular and forms may be of wood, steel, or fiberboard.

Wood—boards or plywood—is commonly used for square or rectangular forms and may also be used for irregular or round ones. However, most round columns are formed by the use of steel or fiberboard tubes of various diameters (see Fig. 9-1). Steel forms are also available for square or rectanglar columns.

There are many ways to build square or rectangular wood column forms, depending on column size and height and on the type of tying equipment to be used. Figure 9-2 illustrates a simple method of making a form for a light column, up to about 12 in. by 12 in. The sheathing is generally plywood, with wood battens and steel rods used as a tying system.

As column sizes increase, either the thickness of the sheathing must be increased or vertical stiffeners must be added to prevent sheathing deflection. In Fig. 9-3 vertical stiffening is used. Also notice the adjustable metal clamps which tie the form together. Several methods of tying column forms are available, and some are illustrated in Fig. 9-4. Ties of this type are generally referred to as *yokes*.

Large columns require heavy yokes and a strong tying system. In Fig. 9-5, the yokes are made from 2″ × 4″s on edge, and the tighteners are patented U clamps and wedges.

Another method of tying columns involves the use of steel strapping (normally made in three widths: ¾, 1¼, and 2 in.) for column banding. The ¾-in. band is made in two thicknesses—0.028 and 0.035 in., with approximate breaking strengths of 2300 and 3100 lb, respectively. The 1¼-in. width is made in thicknesses of 0.035 and 0.050 in., with approximate breaking strengths of 5000 and 7000 lb. The 2-in. band is 0.050 in. thick and has an approximate breaking strength of 11,000 lb.

FIGURE 9-1 Fiberboard columns: (a) in place; (b) removed.

(a)   (b)

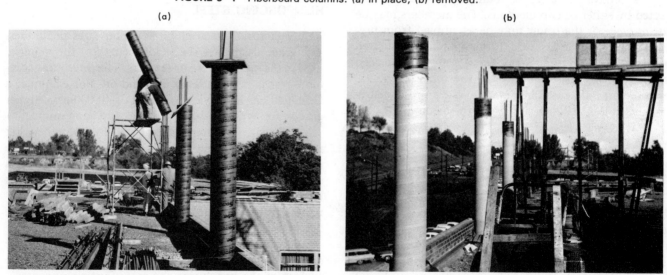

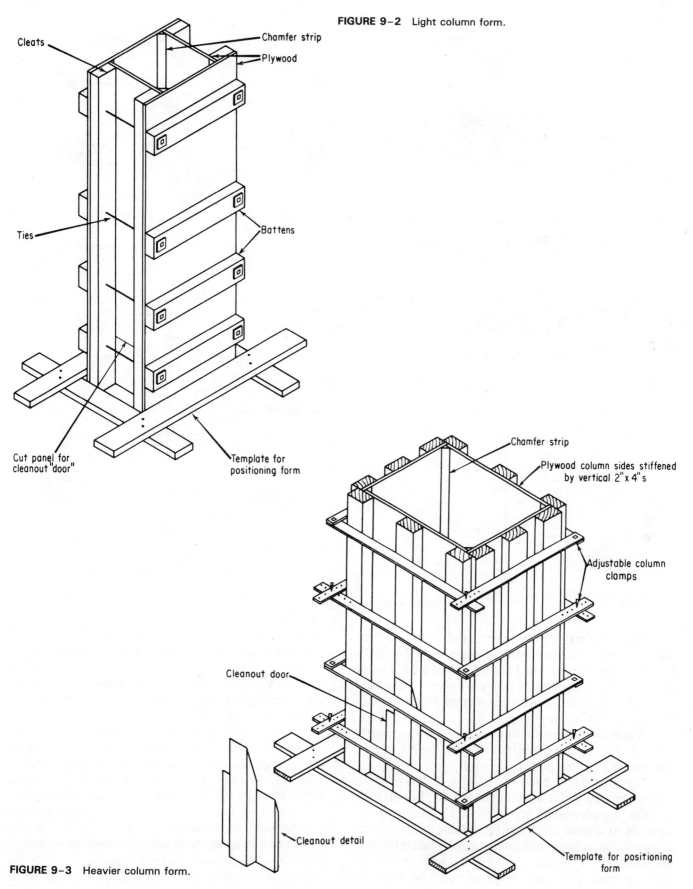

FIGURE 9-2 Light column form.

Cleats

Chamfer strip

Plywood

Ties

Battens

Cut panel for cleanout "door"

Template for positioning form

Chamfer strip

Plywood column sides stiffened by vertical 2"x 4"s

Adjustable column clamps

Cleanout door

Template for positioning form

Cleanout detail

FIGURE 9-3 Heavier column form.

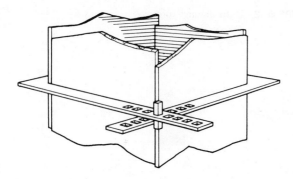

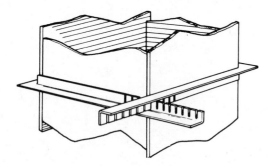

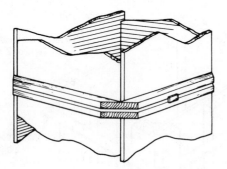

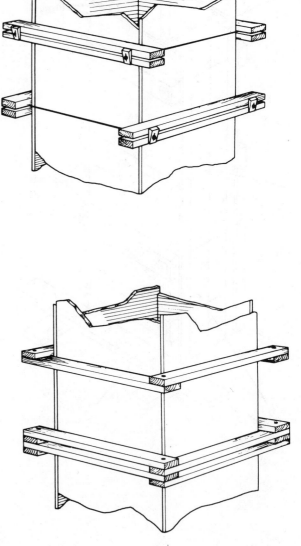

FIGURE 9-4 Light column ties.

Strapping may be used either with horizontal yokes (see Fig. 9–6) or with vertical stiffeners, as in Fig. 9–7. In either case, bands should not be bent at a 90° angle. The metal has a tendency to fracture under those conditions, and there may be danger of form failure caused by cracked bands (see Fig. 9–6).

The band is tightened by a special tool, after which a metal clip is placed over the ends of the band and clamped in place by a crimping tool.

The first step in building a column form is to select the proper type and thickness of sheathing. Boards may be used, but plywood is more frequently employed because of its greater resistance to splitting and warping. It may be ½, ⅝, ¾, or 1 in. thick, depending on the size and height of the column. Table 6–7 may be used to determine the spacing of yokes, based on lateral pressures in-

dicated in Table 6–3. It will be assumed that the first tie will be as close to the bottom of the form as is practical—within 6 or 8 in.—and that the top tie will be at or near the top. Yokes or bands must be checked to ensure that they withstand bending and shear and that deflection will not exceed ¹⁄₁₆ in.

One pair of form sides can be cut the exact width of one column dimension. The other pair must be made wider, depending on the method of form assembly. In Fig. 9–3, for example, the second pair of sides will be wider by twice the thickness of the sheathing. On the other hand, in Fig. 9–2, extra allowance must be made for the width of the cleats.

Lay out, on the wide sides, the position of chamfer strips by drawing lines parallel to the edge of the board. The distance between these lines will be the second dimen-

FIGURE 9-5 Horizontal column form stiffeners. *(Courtesy Jahn Concrete Forming.)*

FIGURE 9-6 Banding a plywood column form. *(Courtesy Plywood Mfg. Association of B. C.)*

FIGURE 9-7 Steel-banded column form. *(Courtesy Acme Steel Co.)*

sion of the column (see Fig. 9-8). Fasten the chamfer strips in place with nails and glue.

If yokes or bands are to be used to tie the column, mark the location of each on the side of the form. If vertical stiffeners are used, tack them in place on the column sides and mark the position of the clamps or bands on a stiffener. When bands are used to tie the column, the yoke pieces can be tacked in place on each side before assembly. With other tying systems, all possible parts should be tacked in place before the form is assembled.

The form can now be assembled on a temporary basis. All four sides can sometimes be put together before erection, while in other situations it is preferable to put three sides together (see Fig. 9-9), set the partially completed form in place, and add the fourth side later. This would probably be done in setting column forms for a job like that shown in Fig. 9-11, where the reinforcement is already in position.

Round wood columns are usually made of 2-in. *cribbing,* as shown in Fig. 9-10. The edges of each piece of cribbing must be beveled so that they will form a cir-

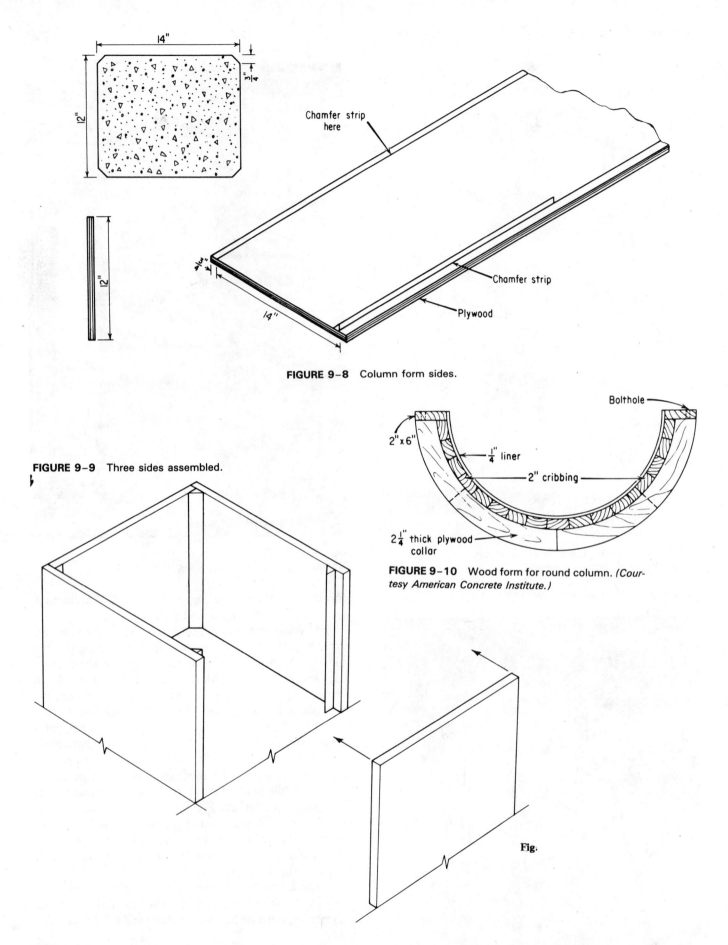

14"

12"

3"/4

12"

1/4"

14"

Chamfer strip here

Chamfer strip

Plywood

FIGURE 9-8 Column form sides.

FIGURE 9-9 Three sides assembled.

Bolthole

2"x6"

1/4" liner

2" cribbing

2 1/4" thick plywood collar

FIGURE 9-10 Wood form for round column. *(Courtesy American Concrete Institute.)*

Fig.

FIGURE 9-11 Column form templates and column reinforcement in place. *(Courtesy Bethlehem Steel Co.)*

cle when fitted together. The amount of bevel on each one will depend on the diameter of the column and the number of pieces used to make the form.

To locate column forms accurately, templates such as the one illustrated in Fig. 9-2 must be made. These are carefully located by chalk line or other convenient means and anchored in position (see Fig. 9-11).

The length of the column form is determined by subtracting the thickness of the bottom of the girder form which the column is to carry from the column height indicated on the plans or in the column schedule (see below). Once the column forms are in place, they must be plumbed, braced, and made ready to support the ends of the girder and beam forms which will be built to them.

The type, amount, and size of reinforcing to be used

in each column will be indicated in a column schedule. It may also list the column sizes and heights. A typical column schedule is outlined in Table 9-2.

The column reinforcing is usually assembled as a unit on the job site, using the material indicated in the schedule. Wire is used to hold the vertical bars and lateral ties together. The unit is placed in position and fastened in place by tying to the dowels which project from the base surface (see Fig. 9-11). Notice that the vertical bars are cut long enough for their upper ends to act as dowels for the columns on the floor above.

Columns may be reinforced with spiral reinforcement as well as vertical bars. Spiral reinforcement consists of a continuous rod formed into a spiral of a given diameter (see Fig. 9-12). Rods of various sizes are used

TABLE 9-2 Typical column schedule

		Reinforced concrete column schedule			
Column no.	Column size (in.)	Vertical reinforcement	Lateral ties	Top elevation (ft)	Bottom elevation (ft)
1, 5, 20	20 × 18	4-#6	#3 @ 11 in. o.c.	98.67	87.33
2	34 × 18	4-#8	#3 @ 12 in. o.c.	97.50	87.33
3	26 × 18	4-#7	#3 @ 12 in. o.c.	97.50	87.33
4	24 × 16	4-#7	#3 @ 12 in. o.c.	98.00	87.33
6	18 × 17	4-#6	#3 @ 11 in. o.c.	98.67	87.33
7, 8, 9	18 × 18	4-#6	#3 @ 11 in. o.c.	96.50	87.67
10	20 × 20	4-#7	#3 @ 12 in. o.c.	96.00	87.33
11, 12, 13	36 × 18	4-#8	#3 @ 12 in. o.c.	97.00	87.33
14, 15, 16, 17	20 × 20	4-#7	#3 @ 12 in. o.c.	98.67	87.33
18	36 × 18	6-#6	#3 @ 12 in. o.c.	97.67	87.67
19	36 × 18	6-#6	#3 @ 12 in. o.c.	97.67	85.50
20	20 × 18	4-#6	#3 @ 11 in. o.c.	98.00	85.50

FIGURE 9–12 Spiral reinforcement.

for this purpose, depending on the diameter of the spiral. The loops of the spiral are kept properly spaced—maintained at the proper *pitch*—by spacer bars, usually three per spiral. Vertical bars are generally used in addition to spirals, spaced evenly around their inside circumference as specified by the schedule.

GIRDER AND BEAM FORMS

Columns, girders, and beams on each floor of a reinforced concrete frame structure are usually placed monolithically (see Fig. 9–13). This means that all the forms must frame into one another. It is also important to remember that some parts of these forms may be removed before others. For example, the beam and girder sides may be removed first, followed later by the column forms, and finally by the beam and girder bottoms. It is therefore necessary to construct the girder and beam forms so that each part can be removed without disturbing the remainder of the form.

The first step in building a girder or beam form is to set the form bottom. Make its width exactly that of the member being formed and its length equal to the distance between columns. The two ends of the form bottom are cut at a 45° angle, as shown in Fig. 9–14, to produce a chamfer at the junction of the girder bottom and column. Chamfer strips with beveled ends are nailed to the girder bottom, flush with the outside edge. Notice that the girder bottom rests on the narrow side of the column form.

The girder must be supported between columns by some type of shoring. T-head shores are a frequent choice, (see Fig. 9–15) their number depending on the load and carrying capacity of each shore. Stringers supported by adjustable metal posts or metal scaffolding (see Fig. 9–16) may also be used to support girder bottoms.

Girder sides overlap the bottom and rest on the

FIGURE 9–13 Concrete structural frame. *(Courtesy American Concrete Institute.)*

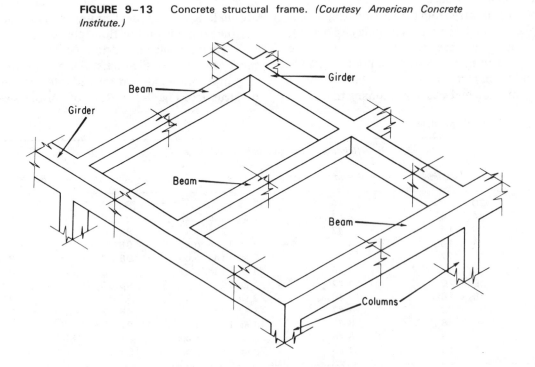

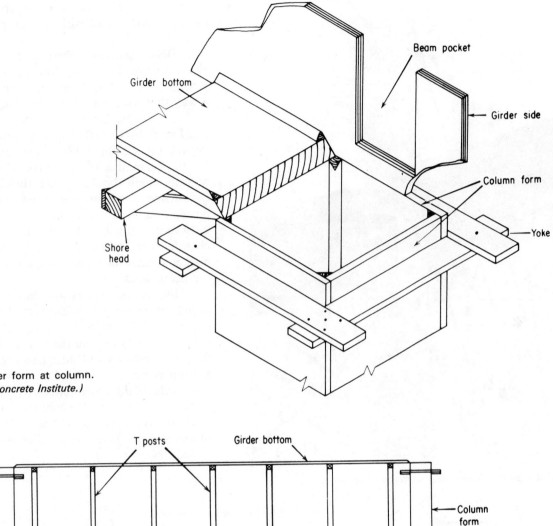

FIGURE 9–14 Girder form at column.
(Courtesy American Concrete Institute.)

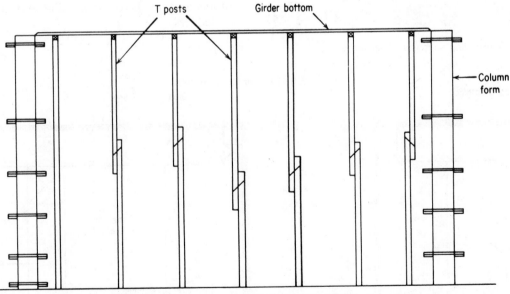

FIGURE 9–15 Girder bottom supported by T posts. *(Courtesy American Concrete Institute.)*

shore heads and the sides of the column form (see Fig. 9–14). They are held in place by ledger strips nailed to the shore heads with double-headed nails. By removing these ledger strips and the *kickers* (see Fig. 9–17), girder form sides may be removed without disturbing the bottom. The bottom and its supports must remain in place until the beam concrete has gained a major portion of its design strength. Full concrete strength will not be at-

FIGURE 9-16 Scaffolding supports girder form.

tained for 28 days. For larger girders, the side forms will have to be provided with vertical stiffeners to prevent buckling.

Beam forms are constructed in the same manner as girder forms. They may frame into either columns or girders (see Fig. 9-13) and they must be framed in such a way that form removal is as simple as possible.

A beam pocket is cut into the girder or column form (see Fig. 9-14) of sufficient size to receive the end of the beam form (see Fig. 9-18). Cut the ends of the side forms for the beam at a 45° angle (see Fig. 9-20). The beam bottom length should be such that its end is flush with the outside of the girder or column form. The end is supported by a block resting on girder ledger strips (see Figs. 9-19 and 9-20). Cut 45° beveled ends on a length of chamfer strip and nail it to the bottom edge of the beam pocket so that the ends fit against the beveled edges of the beam sides.

Forms for girders and beams which frame into round columns require some special attention. Fiberboard column forms have relatively thin shells, and a yoke or collar is needed around the top of the form to support the girder bottom and sides (see Fig. 9-21). The girder bottom is cut the width of the column diameter and a half-circle of the same diameter is cut in the ends. The girder bottom rests on the column collar, as shown in Fig. 9-22, and is supported by shores. Beam sides rest on the shores and on the column collar and are held in place by ledgers.

FIGURE 9-17 Girder form details.

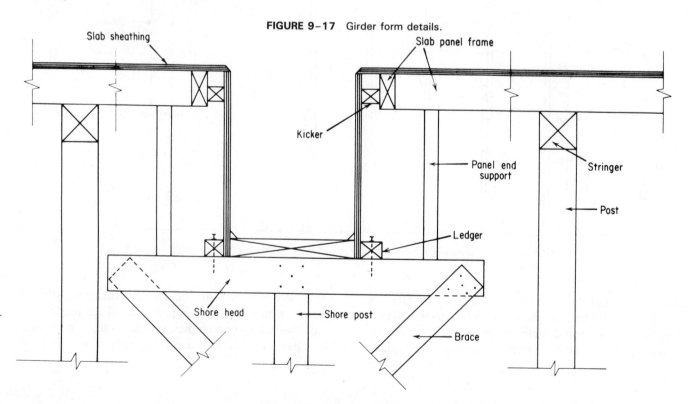

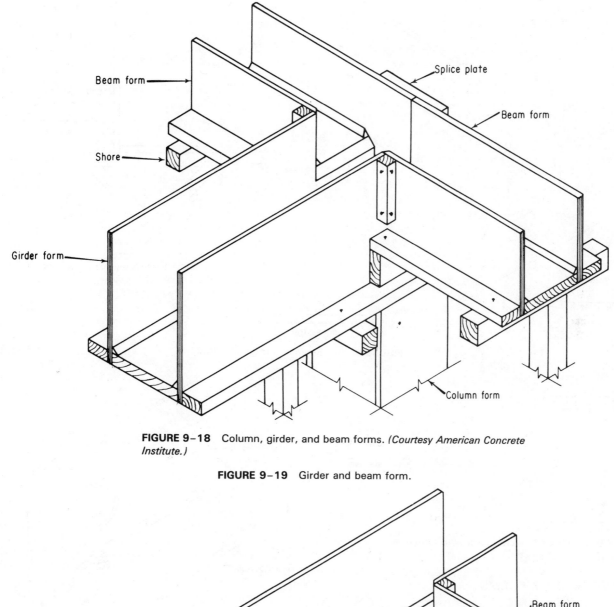

FIGURE 9–18   Column, girder, and beam forms. *(Courtesy American Concrete Institute.)*

FIGURE 9–19   Girder and beam form.

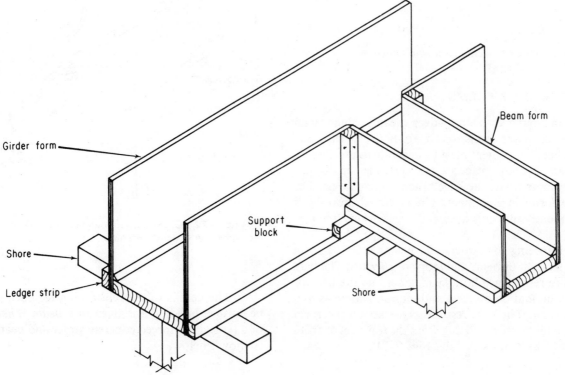

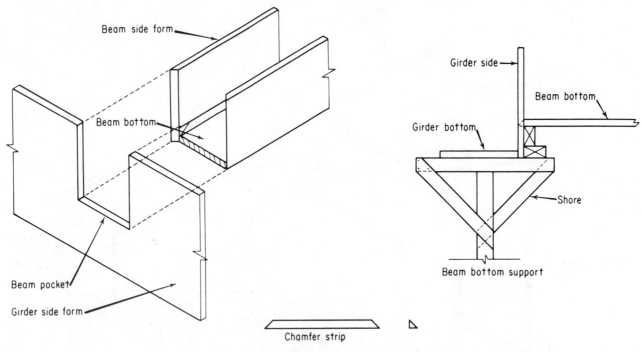

FIGURE 9-20 Beam end to girder form.

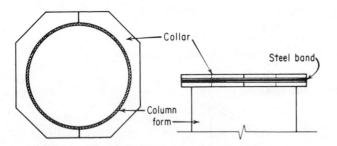

FIGURE 9-21 Round column form with top collar.

SPANDREL BEAM FORMS

Forms for spandrel beams—deep beams which span openings in outer walls—need to be very carefully formed. Form alignment must be accurate to produce an attractive wall. Shore heads are often extended on the outside to accommodate the knee braces which are used to keep the forms in alignment. The extended shore head also frequently supports a catwalk for workers (see Fig. 9-23).

Reinforcing for girders and beams is in the form of bars, straight or bent, and loops or *stirrups*. The bars are used to resist tension in the member and the stirrups resist a combination of shear and tension known as *diagonal tension*. The bars rest on *chairs* set on the form bottom, and the stirrups hang from the top steel or from *loop bars* (see beam schedule, Fig. 9-24).

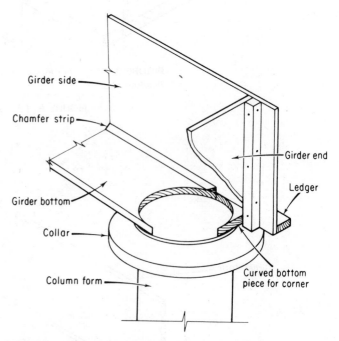

FIGURE 9-22 Girder form to round column. (Courtesy American Concrete Institute.)

The details of the reinforcing for each girder and beam in a building are given in a *beam schedule*. Part of a typical reinforced concrete girder and beam schedule is shown below.

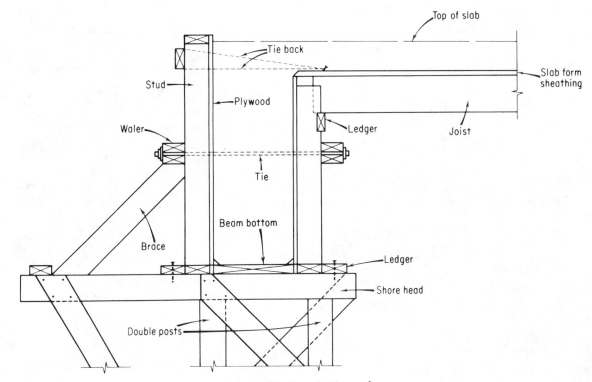

FIGURE 9-23 Spandrel beam form.

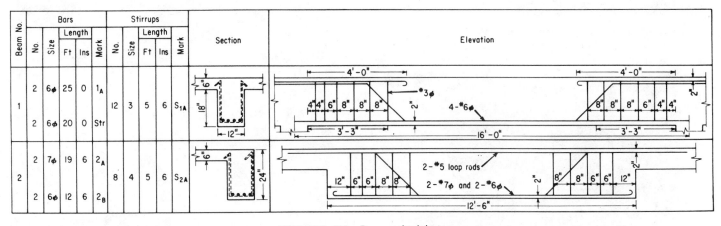

Beam No.	Bars					Stirrups					Section	Elevation
	No.	Size	Length Ft	Length Ins	Mark	No.	Size	Length Ft	Length Ins	Mark		
1	2	6φ	25	0	1ᴀ	12	3	5	6	S₁ᴀ		
	2	6φ	20	0	Str							
2	2	7φ	19	6	2ᴀ	8	4	5	6	S₂ᴀ		
	2	6φ	12	6	2ʙ							

FIGURE 9-24 Beam schedule.

PRECAST CONCRETE STRUCTURAL FRAMES

Precast structural members are of three types. They may be *normally reinforced, prestressed pretensioned,* or *prestressed post-tensioned.* Reinforced precast members are normally designed according to common reinforced concrete practice. However, as a result of factory control, high-strength concrete is produced whether required or not, for it is actually more economical to design reinforced concrete for higher strengths in the range of 5000 psi or better.

The members of a precast structural frame are often prestressed; that is, the tensile reinforcement has been placed under tension before the member is erected. Pretensioning is a technique in which the reinforcement is placed in the forms and stretched against fixed abutments; concrete is then poured around the wires or cables. Pretensioned members are made several at a time in a long bed (see Fig. 9-25), and the major requirements are therefore simplicity and quantity.

Post-tensioning involves first forming and casting the member with ducts through its length. After the con-

FIGURE 9-25 Pretensioned, precast beams.
(Courtesy Portland Cement Association.)

FIGURE 9-27 Prestressed precast concrete structural frame. *(Courtesy Portland Cement Association.)*

FIGURE 9-26 Hydraulic tensioning jack. *(Courtesy Portland Cement Association.)*

FIGURE 9-28 Prestressed columns and roof slabs. *(Courtesy Portland Cement Association.)*

crete is cured, reinforcement is placed in the ducts and anchored at one end. It is stretched against the concrete by means of jacks such as the one illustrated in Fig. 9-26. The *tendons* are gripped and tension is maintained by various patented gripping devices which fit into the end of the member. Post-tensioning is economical if the member is long or large. Both weight and cost may

generally be reduced by post-tensioning units longer than 45 ft or weighing over 7 tons.

Normally reinforced members (which include nearly all columns) are individually made, and this makes possible the production of special shapes to suit a particular situation. In Fig. 9-27, the columns include seats to support the girders. Figure 9-28 illustrates two other column

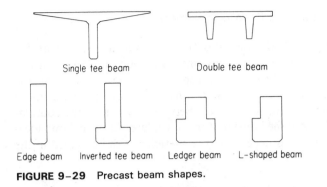

FIGURE 9–29 Precast beam shapes.

Single tee beam

Double tee beam

Edge beam Inverted tee beam Ledger beam L-shaped beam

FIGURE 9–30 Single T prestressed girders. *(Courtesy Portland Cement Association.)*

shapes just as readily produced by precast members. In Fig. 9–30 still another specialized column shape is shown.

Precast girder and beam shapes have become standardized to a considerable degree, but it is possible to produce a special shape if required. In Fig. 9–29, a number of beam shapes in common use are shown. One great advantage in the use of these precast members is the possibility of erecting long free spans. In Fig. 9–30, for example, the single T girders shown have a span of 75 ft, and much greater spans are quite common.

Precast concrete joists are widely used to support concrete floor and roof slabs in many types of buildings. They are small, prestressed units, described in Chapter 12 and illustrated in Fig. 12–10.

Column Erection

The first step in the erection of a precast structural frame is the preparation of the column footings. Since the columns must be mechanically attached to them, anchor

bolts have to be set very accurately into the footing (see the section in Chapter 6 on dowel templates). Each column foot has anchor plates (Fig. 9–31) cast into it so that the column can be bolted to the footing.

FIGURE 9–31 Column-to-footing connection. *(Courtesy Portland Cement Association.)*

A nut is turned onto each anchor bolt before the column is set over them. After erection, the column is plumbed—using guy wires if necessary (see Fig. 9–27), and the nuts are adjusted until the column rests firmly over them. Top nuts are then put on and tightened to hold the column in position. Finally, the space between the top of the footing and the column is filled with grout—often made with metallic aggregate—to provide even pressure over the entire area.

Column-to-Base Connections

Figure 9–32 illustrates five common arrangements for connecting precast column to a foundation pier, a pile cap, a wall footing, or a spread footing. In each one, the *double nut system* and *nonshrink grout* are used and a 2- to 2½-in. space is left between the top of the foundation and the bottom of the base plate. At the column locations, it is necessary that sufficient ties be placed in the

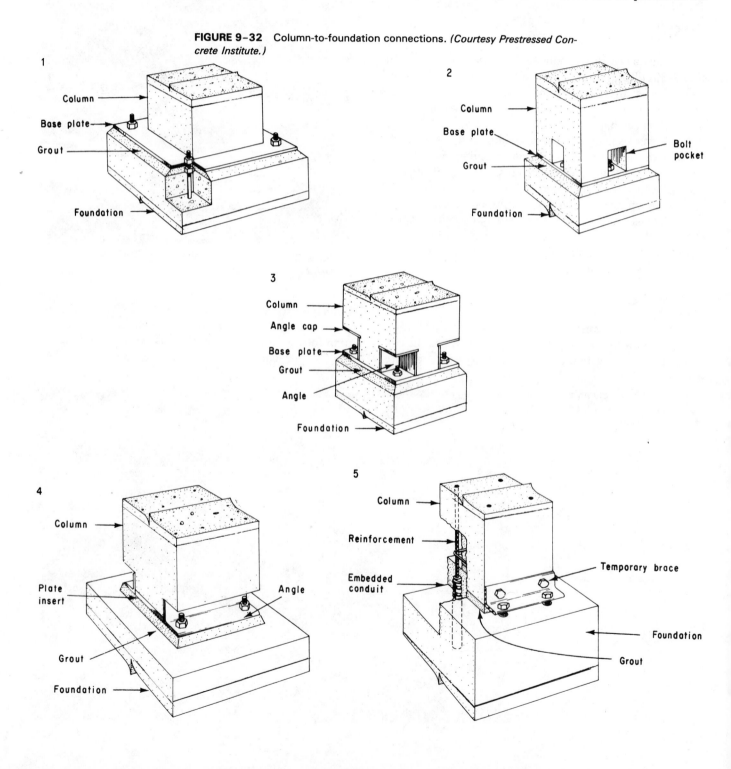

FIGURE 9–32 Column-to-foundation connections. *(Courtesy Prestressed Concrete Institute.)*

top of the pier or wall to secure the anchor bolts. Each of the connection types shown has some unique features, as follows:

1. The base plate is *larger* in dimension than the cross section of the column, and the anchor bolts may be either at the corners or at the middle of the sides of the plate, depending on erection requirements. Column reinforcement is welded to the base plate.

2. The *internal* base plate is either the *same size* or *smaller* than the column cross section, and so there must be *anchor bolt pockets* in the base of the column. After erection, the pockets are usually grouted at the same time as the underbase grout is placed.

3. The base plate is restricted to not more than column cross-section size and a section of ½-in. thick angle, with a cap on one end, is cast into the corners of the column. The base plate is welded to the angles. Reinforcement may be welded to the base plate or to the angles.

4. A full base plate is not used. Instead, angles are attached to the bottom of the column by welding to the main reinforcement or to *dowel bars* which lap the main reinforcement. Often a plate is inserted at the base of the column between the angles. To prevent rotation, welded studs can be attached to the vertical legs of the angles.

5. The column reinforcing bars project from the bottom of the column and are inserted into grout-filled metallic conduit embedded in the foundation. Temporary bracing is required for the column until the grout has gained the desired strength. The angles shown are one means of providing temporary bracing.

Beam Erection

Beams are raised into position by crane, and some type of temporary connection is usually provided until the permanent ones have been completed. In the structure shown in Fig. 9–27, pins cast into the column seats hold the girders in place during erection. Angle iron clips bolted through holes in the columns restrain the girders until the final connections can be made.

Connections may be either *rigid frame* or *simply supported*. Rigid frame connections are made by welding, and simply supported connections are made by inserting pins from one member to another. In Fig. 9–25, the beam shown has a steel plate cast into its ends so that a welded connection may be made between it and a plate or bar cast into the column. In simply supported connections, pins cast into the column seats project into holes in the end of the girder or beam. The space around the pin is grouted after erection.

Beam-to-Column Connections

In Fig. 9–33, seven different methods are shown for connecting beams to columns. These are by no means the only types of connections used for this purpose, but they serve to illustrate some of the basic techniques. Although all the beams shown are rectangular in cross section, any of the other types of beam commonly used could be connected by the same methods.

1. This is a welded connection which conceals a column haunch without using a *dapped-end* beam (see connection 3). As shown, it is for a *simple span* condition, but it can be used as a *moment connection* by using nonshrink grout between the end of the beam and the column and by providing for transfer of tension at the top of the beam.

2. This is a welded connection, with the beam resting on a *corbel* projecting from the side of the column (see Fig. 9–27). It is also for a simple span condition but can be converted into a moment connection. Elastomeric bearing pads are shown, but they are optional.

3. This is called a *dapped-end* connection, also welded. Its development to a moment connection requires grout at two interfaces, which is a difficult field procedure.

4. This welded connection is often used when it is required that it be hidden for architectural reasons. An H shape section is shown projecting from the column, but other structural steel shapes may be used instead.

5. This is a doweled connection, with bars projecting from the column into steel tubes cast into the beam. Tubes are later filled with grout. The connection can be made continuous by welding tension reinforcement as shown in connection 6. Bearing pads are used at the bearing surfaces.

6. This is a welded connection in which the tension bars are welded to small angles. If future extension of the column is required, anchor bolts can be set into the cast-in-place concrete between the beam ends.

7. In this type of connection, a post-tensioning bar is tensioned after the nonshrink grout has been placed between the column and beam end.

Erection of Additional Storeys

Additional storeys will be added to a precast frame building by splicing upper floor columns to those already erected. Again, there are many ways of doing this, and Fig. 9–34 illustrates five of the basic ones. In many ways they are similar to column-to-base connections. Most column splices require nonshrink grout between the members to provide for variations in dimensions. Provision should be made for erection and alignment, such as by using the double nut system.

1. The main column reinforcement is welded to the base plate. Pockets may be placed at the sides of the column or in the corners.

2. This connection also uses an inserted plate between the angles. The angle recess is grouted after erection is complete.

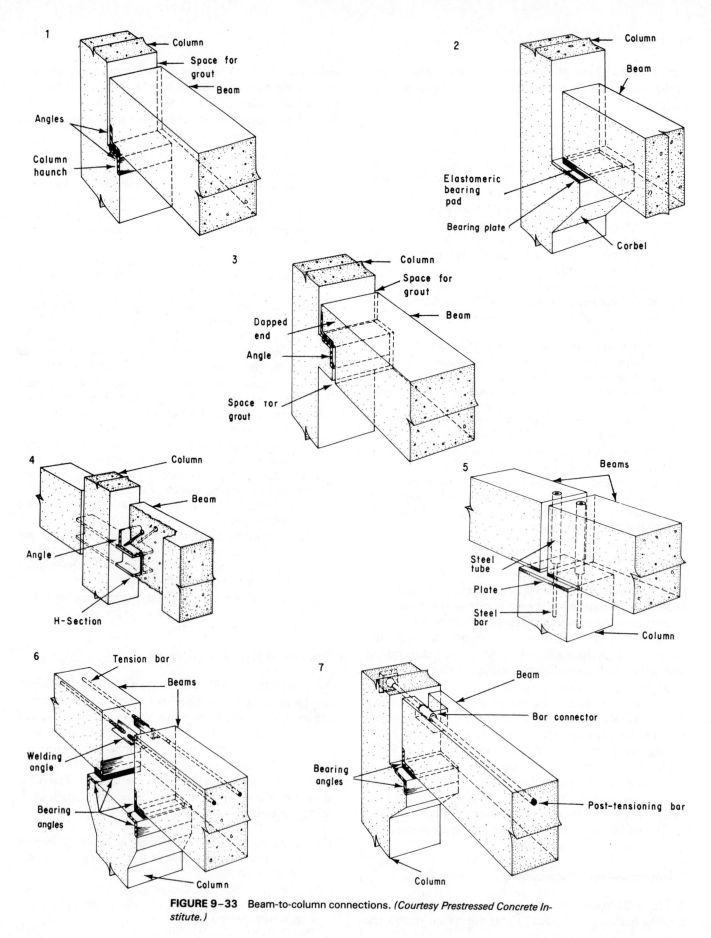

FIGURE 9-33 Beam-to-column connections. *(Courtesy Prestressed Concrete Institute.)*

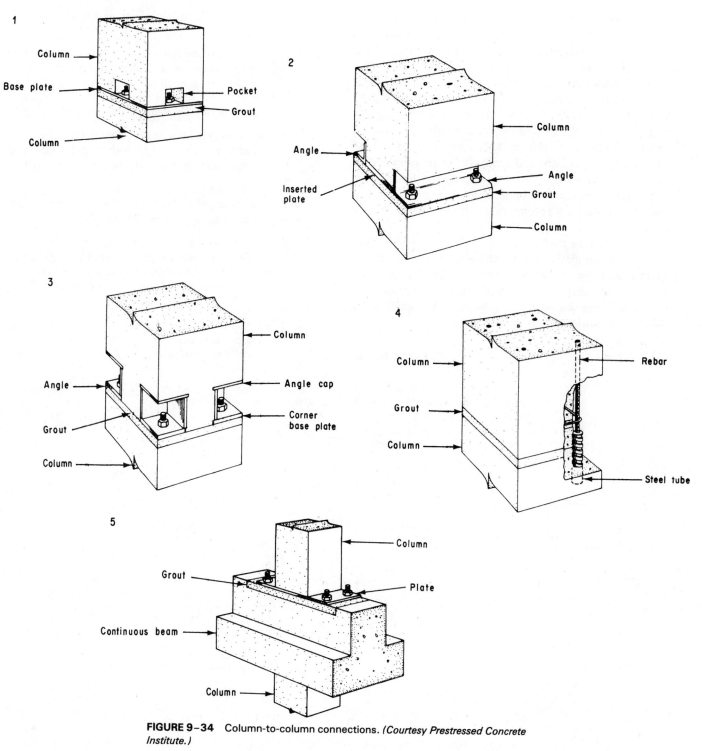

FIGURE 9–34 Column-to-column connections. *(Courtesy Prestressed Concrete Institute.)*

3. Four small corner base plates are used here rather than one full-sized one under the angles.

4. Reinforcing bars projecting from the upper column extend into steel tubes cast into the lower one. Provisions for temporary bracing must be provided.

5. Here, the column is spliced through a continuous beam and reinforcement is required within the beam to transfer the load from column to column.

Floor and Roof Slab Erection

Finally, in most precast structural frame designs, prestressed floor and roof slabs span from beam to beam, from beam to wall, or from wall to wall to complete the structure. In most cases the slabs will be set edge to edge, but they may be spaced and the space between them filled with cast-in-place, reinforced concrete (see Fig. 9-30).

Slab-to-beam connections should take into consideration the transfer of horizontal forces from slab to beam if it is assumed that the floor or roof is acting as a diaphragm. Where cast-in-place topping is used on floors, additional reinforcement in the form of welded wire mesh should be placed *across* the beam to minimize cracking. Figure 9-35 illustrates a couple of typical slab-to-beam connections. In connection 1, the ends of tee legs may be dapped to accommodate units with a greater depth.

When slabs are supported by bearing walls, types of connections different from those shown above are required, and Fig. 9-36 shows some of the possible connections. In most designs, some degree of continuity is required at the slab-to-wall connection, but generally a fully fixed connection is not desirable.

1. A bond beam is provided under the slab ends, and the

joint between the meeting ends of the slabs is fully grouted.

2. *Hairpin anchors* cast into the bond beam are embedded in the grout between slab ends to provide positive anchorage. Bars may be grouted into the keyways between adjacent slabs to tie the slabs together lengthwise and help to prevent roofing problems at the joint.

3. The pocket in the wall must be sufficiently large to receive the extended stem of the single tee slab without difficulty. Elastomeric bearings pads may be used if required.

4. Double tee slabs rest on a bond beam cast on a block wall. Precast *fill-ins* between the tee stems may be used, as forms for concrete between the slab ends.

5. Single or double tee units used as wall panels may have corbels cast with them to support floor or roof slabs. Vertically slotted bolted connections may be used to tie the slabs to the load-bearing wall panels or the welded connections shown may be used.

6. Connecting a slab to a parallel wall requires a connection that can accommodate vertical movements. Slotted angles with *low-friction washers* permit this movement.

FIGURE 9-35 Slab-to-beam connections. *(Courtesy Prestressed Concrete Institute.)*

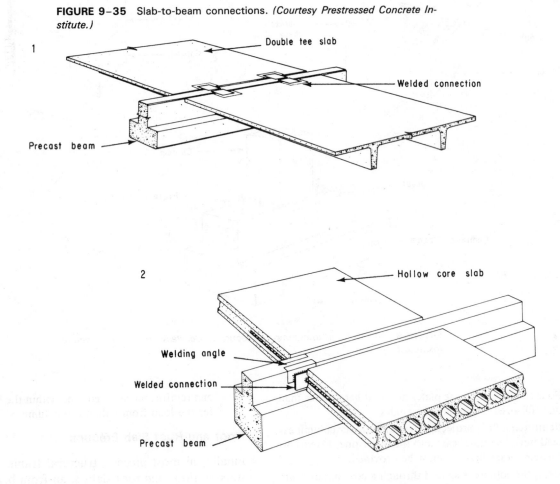

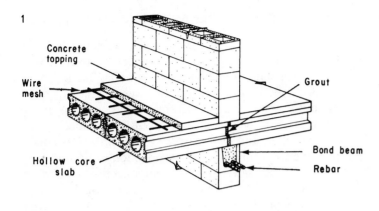

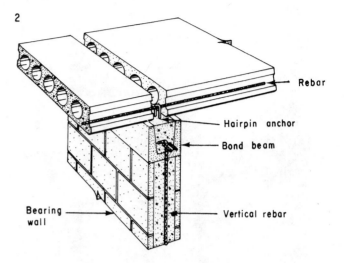

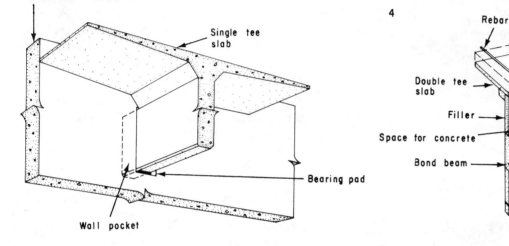

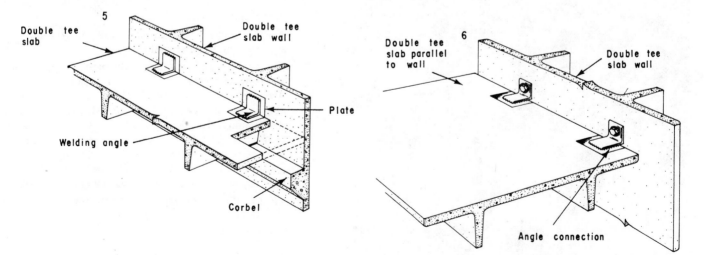

FIGURE 9-36 Slab-to-wall connections. *(Courtesy Prestressed Concrete Institute.)*

ARCHITECTURAL PRECAST CONCRETE

A relatively new advance in the precast concrete industry has been the development of architectural precast concrete. Originally, precast structural units, like those described in the foregoing section, were intended, in the main, to form a structural frame, in the same way wood or steel might be used. More recently, advances in *design, control,* and *manufacturing techniques* have made possible and economical the production of precast units which are themselves finished components of a building.

There are two principal methods by which this type of unit is produced. One is known as *tilt-up* construction, in which the units are cast on the job site and tilted up into position. The other is an independent, *factory-type* industry, geared to the production of a variety of precast products, in which all the facilities and economies of a *production line* type of operation may be utilized.

In most cases, the units produced by the tilt-up method are intended strictly as curtain wall components and, as such, are described in Chapter 14 on curtain wall construction.

Factory-Produced Precast Concrete Units

There are a number of distinct advantages in the factory method of producing precast units. They include:

1. The design freedom which is possible. The styles and designs of buildings may be as diversified as the skill and ingenuity of those in charge of production permit (see Figs. 9–37 and 9–38).
2. A very high degree of quality control of the ingredients in the units is possible.
3. A wide variety of shapes, sizes, and configurations for any one building is possible.
4. The possibility of including in the precast units some or all of the services which are normally installed after erection (see Figs. 9–39 and 9–40). This can result in considerable saving of time and labor following the erection of the building. The precast wall panels for the building shown in Fig. 9–39 were shipped to the job site complete with *glazing, mechanical services,* and *interior finish.* Only a minimum of mechanical connections and interior painting were required following panel installation.
5. The use of precast components makes possible the rapid closure of the building. This allows for earlier access by the service trades.

Among the types of units produced by this method, perhaps the major one is the *precast wall panel* (see Fig. 9–41). Such panels may be used as *curtain walls,* as *loading bearing members,* as *wall-supporting units,* as *formwork for exterior walls,* or as *shear walls.*

FIGURE 9–37 Toronto City Hall in architectural precast concrete. *(Courtesy Prestressed Concrete Institute.)*

FIGURE 9–38 Fine example of precast concrete architecture. *(Courtesy Prestressed Concrete Institute.)*

FIGURE 9–39   Panels may come complete with glazing, mechanical services, and interior finish. *(Courtesy Prestressed Concrete Institute.)*

Curtain wall panels are units which carry no loads except wind load. They are set into or hung from the structural frame to act as closure units and to provide the architectural effect desired.

Load bearing-panels *do* carry the loads of floors and roof. In Fig. 9–42, for example, the load-bearing panels were designed in widths to suit the standard width of floor and roof slabs which they support. Another way in which load-bearing units are utilized is in the segmental construction of a precast frame or of large wall panels. This technique involves the post-tensioning together of a number of segments to form a complete unit. In the building shown in Fig. 9–43, precast segments are post-tensioned together to form a complete building frame.

Precast wall-supporting units are those which support a part of the wall but do not carry any loads from floor or roof. Such units are stacked one above the other so that they carry the weight of the units above them.

FIGURE 9–40   Preglazed, precast wall panel. *(Courtesy Prestressed Concrete Institute.)*

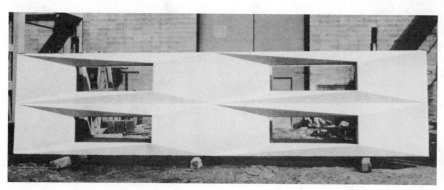

FIGURE 9–41   Large architectural precast wall panel. *(Courtesy Prestressed Concrete Institute.)*

FIGURE 9–42 Erecting load-bearing wall panel. (Courtesy Prestressed Concrete Institute.)

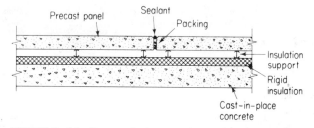

FIGURE 9–44 Precast facing panels used as formwork.

FIGURE 9–43 Frame made from post-tensioned precast segments. (Courtesy Prestressed Concrete Institute.)

FIGURE 9–45 Precast facing panels used as formwork for cast-in-place structural wall. (Courtesy Prestressed Concrete Institute.)

In some cases, precast panels may be intended for exterior finish only, and, in such situations, they may be used as forms for the cast-in-place structural wall behind them. Figure 9–44 indicates how this is done, while Fig. 9–45 shows an example of this technique in operation. Panels may also be used as formwork for cast-in-place columns (see Fig. 9–46). Other types of precast units may also be used for this purpose (see Fig. 9–47).

Walls which are utilized for the transfer of lateral forces are known as shear walls (see Fig. 9–48), and precast wall panels are ideally suited for such a purpose.

In addition to its use in panels and other structural members, architectural precast concrete has a number of other uses, such as for *fences* and *screens* (see Fig. 9–49), *planters, bases, light standards, street furniture* (Fig. 9–50), and *sculpture*.

The key to the success of architectural precast concrete projects will probably rest in the planning which goes into them. One of the major items in that planning

should be the concept that the cost of making the molds should be spread over as many units as possible. In other words, the more units which can be made in one mold, the less the unit cost will be. This leads to the concept of a *master mold*—one from which a maximum number

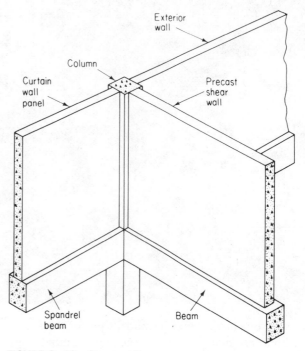

FIGURE 9–48 Shear wall.

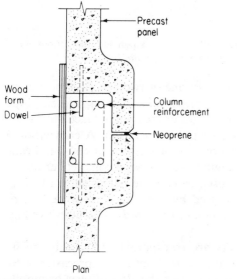

FIGURE 9–46 Precast panels as formwork for column.

FIGURE 9–47 Precast column covers used as formwork. *(Courtesy Prestressed Concrete Institute.)*

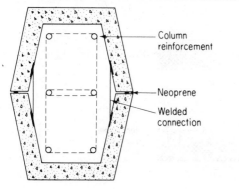

FIGURE 9–49 Precast sun screen. *(Courtesy Prestressed Concrete Institute.)*

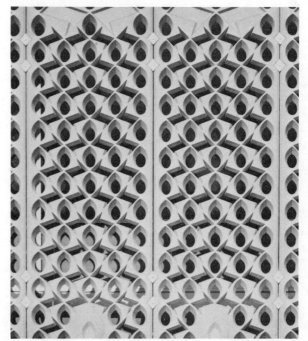

FIGURE 9-50 Precast street furniture. *(Courtesy Prestressed Concrete Institute.)*

of units for one project may be made, some of them perhaps with minor modifications in the original mold. Figure 9-51 shows an office building in which all the panels were cast from one master mold.

FIGURE 9-51 Walls built from one master mold. *(Courtesy Prestressed Concrete Institute.)*

Molds are usually one of two general types: *complete envelope* or *conventional*. A complete envelope mold is one in which all sides remain in place during the entire casting and stripping operation. A conventional mold has one or more removable side or end bulkheads (see Fig. 9-52). Figure 9-53 shows a complete envelope mold which was used to cast the panels shown in Fig. 9-41. Panels produced by these molds may be *closed, open,* or a *combination of open and closed* units (see Fig. 9-54).

Precast units are transported to the job site by truck, and erection is carried out by the precaster or by the general contractor, with the aid of hoists or mobile cranes. In any case, one of the most common methods

FIGURE 9-52 Mold types. *(Courtesy Prestressed Concrete Institute.)*

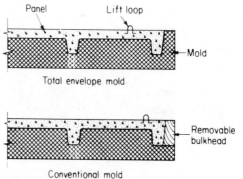

FIGURE 9-53 Total envelope panel mold. *(Courtesy Prestressed Concrete Institute.)*

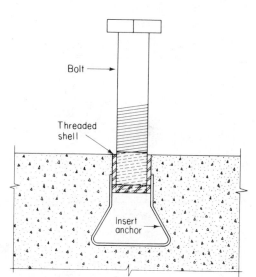

FIGURE 9-55 Concrete insert.

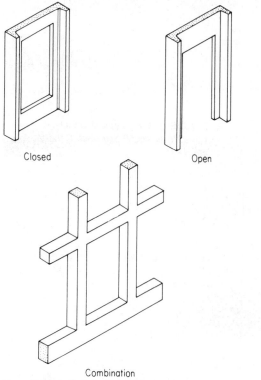

FIGURE 9-54 Precast panel shapes.

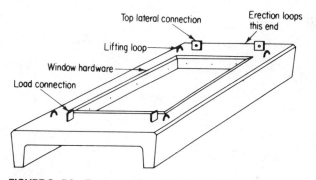

FIGURE 9-56 Typical wall panel hardware. *(Courtesy Prestressed Concrete Institute.)*

of connecting to the unit in order to lift it is by means of inserts and bolts (see Fig. 9–55). Inserts are properly placed during the casting process and should be protected by plastic caps to prevent dirt or ice from getting into them until they are ready to be used.

Units are held in position in the building by connection hardware, part of which is cast into the units at the plant (see Fig. 9–56), while the corresponding parts

are attached to the structural frame. Connections are usually either bolted or welded and are designed to suit the particular project. Figure 9–57 illustrates a few typical connections, but a number of others may be used as well, depending on the circumstances in any particular case.

The joints between precast concrete panels are an important part of the overall project and will have been designed taking into account the *structural requirements*, and *degree of watertightness required,* the *appearance,* and *cost.* They will usually be one of two basic types: *one-stage* or *two-stage* joints.

One-stage joints are those which are simply sealed against water penetration by the use of a sealant near the outer surface (see Fig. 9–58). Two-stage joints are those which have a sealant near the outer surface and an air-seal usually close to the inner face of the panels. Between the two is a chamber which must be vented and drained to the outside. For example, the outer sealant in a two-stage horizontal joint must be broken at intervals to allow any moisture which collects behind it to drain to the outside (see Fig. 9–59).

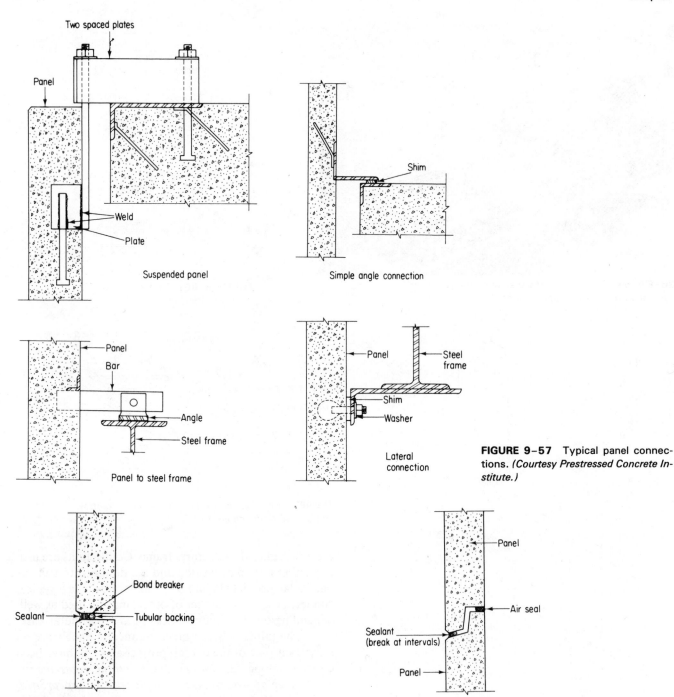

FIGURE 9-57 Typical panel connections. *(Courtesy Prestressed Concrete Institute.)*

FIGURE 9-58 One-stage joint. *(Courtesy Prestressed Concrete Institute.)*

FIGURE 9-59 Horizontal two-stage joint. *(Courtesy Prestressed Concrete Institute.)*

One of the major advantages of architectural precast concrete panels lies in the great choice of surface finishes which are available, with regard to both color and surface texture. Colors are varied by using *white cement, color pigments,* and *colored aggregates.* A variety of surface textures is produced by *casting on a smooth surface,* by using a *ribbed* or *textured form face,* by *casting on a sand-molded surface,* by *sand blasting,* by *exposing the aggregate* to various degrees on the panel surface, and by *grinding.* Color or surface texture variations on a single panel are often accentuated by the use of ribs and recesses. Figure 9-60 illustrates a panel with a highly ground and polished surface, while in Fig. 9-61, the opposite type of surface, an effective change in texture in a panel is brought about by changing the size of aggregate in one of the mixes and using a recess to outline the change.

FIGURE 9-60   Ground panel surface.

FIGURE 9-61   Changing panel surface texture.

The final step in the erection of an architectural precast concrete structure is the cleaning of the exposed surfaces. They should normally be thoroughly wet before cleaning to prevent cleaning solutions from being absorbed into the concrete. Then cleaning can usually be done with detergents, followed by a thorough rinsing. In some cases an alkali or acid cleaning agent must be used, but before this is done, tests should be made to ascertain the result of using such an agent.

For a full and comprehensive understanding of the subject of architectural precast concrete, reference should be made to the excellent literature on the subject published by the Prestressed Concrete Institute.

TILT-UP LOAD-BEARING WALL PANELS

Tilt-up construction is a technique for producing precast wall panels for one- or two-storey buildings at the job site, rather than in a precast plant. Originally developed for economic reasons, improvements in technique now make it possible to produce panels which are also architecturally pleasing.

Load-bearing panels may be produced by this sys-tem, which function in exactly the same way as architectural load-bearing members. There are three quite common types of panels used as load-bearing members, each intended to support a different type of roof system. Panels intended to carry prestressed roof slabs are cast with a continuous ledger near their top edge for this purpose (see Fig. 9-62). Panels which will support a steel beam roof frame are cast with weld plates in their top inner surface (see Fig. 9-63), while panels intended to support open web joists are cast with seated weld plates and shelf angles (see Fig. 9-64). If they are two-storey panels, they may have ties cast into their inner face at the first floor level in order to tie into a cast-in-place concrete floor (see Fig. 9-63).

At the base, panels may be attached to the foundation or slab by *weld plates,* by *dowels,* or with a *floor edge strip and dowels.*

The concrete floor slab for the building is often the most convenient place on which to cast tilt-up panels (see Fig. 9-65), but other casting beds may be set up near by if it is not convenient to use the floor.

When the floor slab is used, it must be level, smoothly troweled, and strong enough to carry the

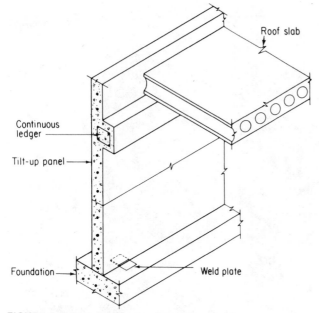

FIGURE 9-62 Load-bearing tilt-up panel with continuous ledger.

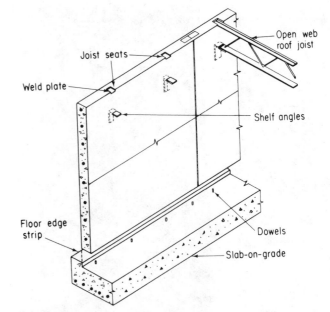

FIGURE 9-64 Load-bearing tilt-up panel with weld plates and shelf angles.

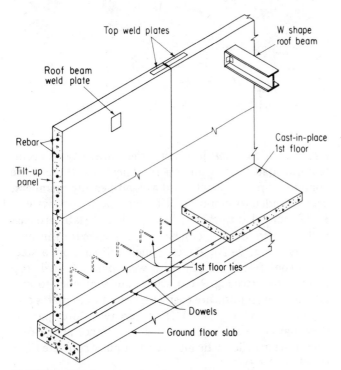

FIGURE 9-63 Load-bearing tilt-up panel with weld plates for roof beams.

FIGURE 9-65 Casting panels on floor slab. *(Courtesy Portland Cement Association.)*

materials trucks or mobile cranes which may be used during casting and erection. It must also be surfaced with a good bond-breaking agent in order to get a clean, trouble-free lift from the floor.

Panel edge forms must be bolted, pinned, weighted, or otherwise held down (see Fig. 9–66). If they have to have holes for rebars, they will usually be split to facilitate removel (see Fig. 9–67).

Window and door openings are formed by casting steel window or door frames into the panel or by setting a *form buck* into the panel form (see Fig. 9–68). In either case, such opening forms must be held down securely and braced internally to hold the square. Wood forms and exposed sash should be given a coat of bond-breaking compound.

Reinforcement may be in the form of bars or welded wire fabric. No. 4 bars are often used and a common re-

FIGURE 9-66 Hold-downs for panel forms. *(Courtesy Portland Cement Association.)*

FIGURE 9-68 Openings cast in tilt-up panels. *(Courtesy Portland Cement Association.)*

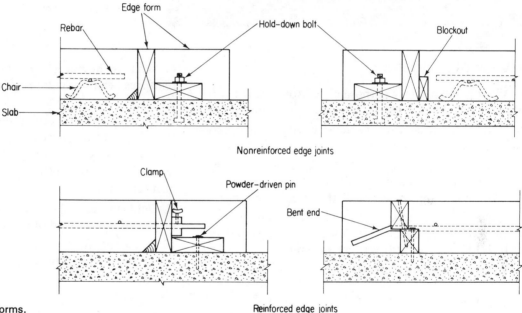

FIGURE 9-67 Panel edge forms.

quirement for steel area is 0.0015 to 0.0025 of the panel cross-sectional area. No. 5 bars are added around openings to strengthen edges and corners. If panels and columns are to be tied together and wire fabric is being used as reinforcement, *dowel bars* must be added for this purpose. Reinforcing mats are placed on *chairs* which will hold them near the center of the panel. This position, in most cases, will provide the greatest strength for the panel (see Fig. 9-67).

Concrete is placed and finished in the same way as for a floor slab. Finishing techniques will depend on the texture desired. Curing is begun as soon as possible and may be carried out by any of the accepted methods which are feasible in such a situation.

The method of tilting up will depend on whether the crane can operate on the floor slab or not and on whether the panel has been cast face up or face down (see Fig. 9-69). Lifting will be done by the use of inserts and bolts (see Fig. 9-55) or, in some cases, by the use of vacuum lifters.

Surface treatments and colors for tilt-up panels may be almost as varied as those found with architectural precast panels. Color variations may be achieved by using color pigments or white cement, and decorative designs are produced by using *wood strips,* form liners of foamed plastic or rubber matting, and by exposing aggregates. See Chapter 14 for details of surface treatments.

For further details on tilt-up construction, consult the literature on the subject published by the Portland Cement Association.

FIGURE 9-69 Lifting tilt-up panel. *(Courtesy Portland Cement Association.)*

REVIEW QUESTIONS

1. What is the primary purpose of reinforcement in **(a)** concrete girders and beams? **(b)** concrete columns?

2. What is the reason for including a window in a column form?

3. Twenty-four pieces of $2'' \times 6''$ are to be used as cribbing to make a circular wooden form. How much bevel should there be on the edges of the cribbing pieces?

4. What is the purpose of the following items used in beams and girder forming: **(a)** T-head shore, **(b)** ledger strip, **(c)** kicker, and **(d)** vertical stiffener?

5. What is a spandrel beam?

6. What is the purpose of the following items used to reinforce beams or columns: **(a)** stirrups, **(b)** chairs, and **(c)** lateral ties?

7. Differentiate between **(a)** *normal reinforcing* and *prestressing,* and **(b)** *pretensioning* and *post-tensioning*.

8. What factors are usually considered to determine whether a prestressed member will be pretensioned or post-tensioned?

9. Outline three major advantages of using precast structural members instead of a cast-in-place concrete frame.

10. Differentiate between *rigid* and *simply supported* connections when joining precast structural members.

Chapter 10

Structural Steel Frame

The use of structural steel is one of the three basic methods used in the construction of building frames for all kinds of permanent buildings but is perhaps of particular importance in the construction of high-rise buildings. Constant improvements in methods of production and fabrication have rendered its use virtually limitless, as may be observed from the array of steel buildings being erected everywhere (see Figs. 10–1 and 10–2).

The most common steel shapes used in building frame construction are S *(standard)* shapes, W *(wide flange)* shapes, *channels,* HSS *(hollow structural)* shapes, *structural* Ts, *angles,* HP *shapes,* and *plates.* These are all available in a great variety of sizes (in the case of S shapes and channels, the nominal size represents the depth of the member) (see Fig. 10–3), and each size is produced in several weights per lineal foot. Differences in weight are produced by altering the width of flanges and the thickness of webs in the case of S shapes, the width and thickness of flanges and the thickness of webs in the case of W shapes, and the thickness of channel webs, the thickness of the legs of angles, and the thickness of the shell of HSS. Consult a steel manual for complete details on the sizes and weights of various rolled shapes. Tables 10–1 and 10–2 illustrate the type of information available from steel manuals.

A great deal of detailing work is necessary before a steel frame can be erected. Each member is given an *erection mark* which it carries through the fabrication stage and which is used on the job to identify the member and its position in the frame. The erection mark is usually placed on the left end of horizontal members to eliminate the possibility of trying to place the member end for end or upside down (see Figure 10–27).

FIGURE 10–1 Steel frame high-rise office building. *(Courtesy Bethlehem Steel Co.)*

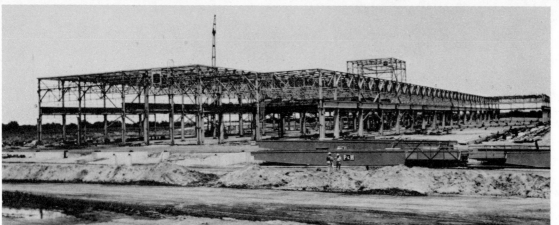

FIGURE 10–2 Steel framework for factory building. *(Courtesy Bethlehem Steel Co.)*

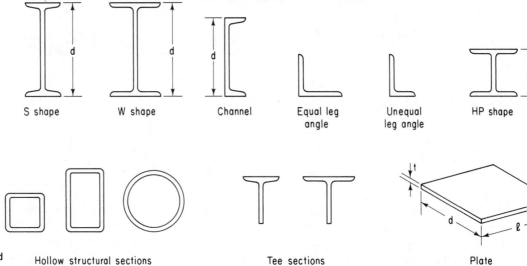

S shape   W shape   Channel   Equal leg angle   Unequal leg angle   HP shape

FIGURE 10–3 Standard rolled structural shapes.

Hollow structural sections

Tee sections cut from S and W shapes

Plate

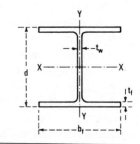

TABLE 10–1 Theoretical dimensions and properties for designing bearing piles[a]

Section number and nominal size (in.)	Weight per foot (lb)	Area of section A (in.²)	Depth of section d (in.)	Flange		Web thickness, t_w (in.)	Axis X-X		
				Width, b_f (in.)	Thickness, t_f (in.)		I (in.⁴)	S (in.³)	r (in.)
HP14 × 14 × 14½	117	34.4	14.23	14.885	0.805	0.805	1230	173	5.97
	102	30.0	14.03	14.784	0.704	0.704	1050	150	5.93
	89	26.2	13.86	14.696	0.616	0.616	910	131	5.89
	73	21.5	13.64	14.586	0.506	0.506	734	108	5.85
HP12 × 12 × 12½	74	21.8	12.12	12.217	0.607	0.607	566	93.4	5.10
	53	15.6	11.78	12.046	0.436	0.436	394	66.9	5.03
HP10 × 10 × 10	57	16.8	10.01	10.224	0.564	0.564	295	58.8	4.19
	42	12.4	9.72	10.078	0.418	0.418	211	43.4	4.13
HP8 × 8 × 8	36	10.6	8.03	8.158	0.446	0.446	120	29.9	3.36

[a]All shapes have parallel-faced flanges.
Source: Courtesy Bethlehem Steel Co.

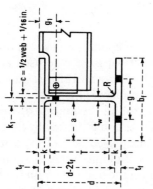

TABLE 10–2 Approximate dimensions for detailing of bearing piles[a]

Section number and nominal size (in.)[b]	Weight per foot (lb)	Depth of section, d (in.)	Flange Width, b_f (in.)	Flange Thickness, t_f (in.)	Web thickness, t_w (in.)	d − 2t_f (in.)	a (in.)	T (in.)	k (in.)	k¹ (in.)	g¹ (in.)	c (in.)	Usual flange gauge, g (in.)
HP14 × 14 × 14½, R = 0.60	117	14¼	14⅞	13/16	13/16	12⅝	7	11¼	1½	1 1/16	2¾	7/16	5½
	102	14	14¾	11/16	11/16	12⅝	7	11¼	1⅜	1	2¾	7/16	5½
	89	13⅞	14¾	5/8	5/8	12⅝	7	11¼	1 5/16	15/16	2½	3/8	5½
	73	13⅝	14⅝	1/2	1/2	12⅝	7	11¼	1 3/16	7/8	2½	5/16	5½
HP12 × 12 × 12, R = 0.60	74	12⅛	12¼	5/8	5/8	10⅞	5¾	9½	1 5/16	15/16	2½	3/8	5½
	53	11¾	12	7/16	7/16	10⅞	5¾	9½	1⅛	7/8	2½	¼	5½
HP10 × 10 × 10, R = 0.50	57	10	10¼	9/16	9/16	8⅞	4⅞	7¾	1⅛	13/16	2½	3/8	5½
	42	9¾	10⅛	7/16	7/16	8⅞	4⅞	7¾	1	¾	2¼	¼	5½
HP8 × 8 × 8, R = 0.40	36	8	8⅛	7/16	7/16	7⅛	3⅞	6⅛	15/16	5/8	2¼	5/16	5½

[a]All shapes have parallel-faced flanges.
[b]R = radius of fillet.
Source: Courtesy Bethlehem Steel Co.

FIGURE 10–4 Mobile crane erecting steel frame.
(Courtesy Bethlehem Steel Co.)

Studies must be made during the detailing stage to make sure that it is possible to place each member with the types of connections that have been designated for each end. Girders or beams are sometimes swung into place from the side and are sometimes lowered from

above. The type of connection used at one end of a beam must not conflict with the type used at the other.

Two methods are used for raising the members of a steel frame into place. One is the use of *mobile cranes* which operate around the outside of the building, as in Fig. 10–4. But erection is limited by the height and reach of the boom, and *guyed derricks* and *stiff-leg derricks,* which rest on the building frame and *tower cranes,* which may rest either on the building or on the ground, depending on type, are used to erect tall buildings.

DERRICKS AND CRANES

A guyed derrick (see Fig. 10–5) consists basically of a vertical mast and swinging boom attached to a rotating *bull wheel,* with the whole assembly being fastened to the building frame. *Guy lines,* running from the *guy cap collar* at the top of the mast, are attached to *hairpin anchors* (see Fig. 10–18) cast into the concrete foundation at designated points, to hold the mast in its vertical position. Figure 10–6 indicates the basic parts of a guyed derrick.

A stiff-leg derrick (see Fig. 10–7) has its *mast* held erect by a *stiff-leg,* attached to the mast at one end and to a horizontal *sill* at the other, thus forming a fixed triangle. A *boom* is attached to the bottom of the mast, and the whole assembly is mounted on a *bull-wheel,* which rotates in a circle (see Fig. 10–8).

A tower crane (see Fig. 10–9) is made up of a vertical *tower* with a horizontal *jib,* which can rotate through 360°, pivoted at the top. Figure 10–10 shows the basic parts of a tower crane, and the dimension lines shown indicate standard dimensions given for various sizes of

FIGURE 10–5 Guyed derricks. *(Courtesy Bethlehem Steel Co.)*

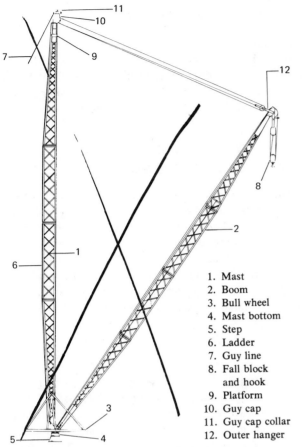

1. Mast
2. Boom
3. Bull wheel
4. Mast bottom
5. Step
6. Ladder
7. Guy line
8. Fall block and hook
9. Platform
10. Guy cap
11. Guy cap collar
12. Outer hanger

FIGURE 10-6 Diagram of guyed derrick. *(Courtesy Clyde Iron Works Inc.)*

FIGURE 10-7 Stiff-leg derrick at work on steel frame. *(Courtesy Bethlehem Steel Co.)*

cranes. The tower crane has displaced the derrick to a great extent in the erection of structural steel in high-rise buildings because it is much easier to set up, move from one floor to the next, and is more versatile than the derrick. For extremely heavy lifts, the derrick still has its place.

FIGURE 10-8 Diagram of a stiff-leg derrick. *(Courtesy Clyde Iron Works Inc.)*

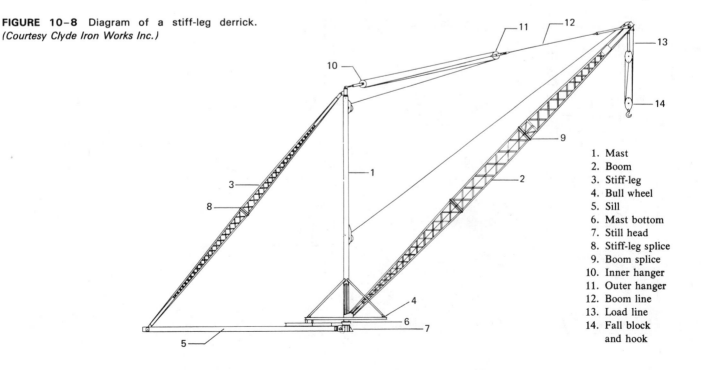

1. Mast
2. Boom
3. Stiff-leg
4. Bull wheel
5. Sill
6. Mast bottom
7. Still head
8. Stiff-leg splice
9. Boom splice
10. Inner hanger
11. Outer hanger
12. Boom line
13. Load line
14. Fall block and hook

Mobile cranes are normally used to erect the first lift of columns and beams and to raise the derrick or crane to its initial position. Special beams, known as *dunnage beams,* are built into the frame at the location of the crane or derrick to support the additional loads. Two floors are usually erected from one setting, and when this is completed the derrick or crane is raised to the new position.

STRUCTURAL STEEL CONNECTIONS

The two most common methods used for connecting steel members in structural steel frames are *bolting* and *welding.* Experience has shown that welded connections are usually less expensive when done in the shop under controlled conditions and bolted connections are usually better suited to field conditions. In most structural steel buildings, welds and bolts are used in combinations to produce the most practical and economical connections possible.

Bolts

Structural bolts may be classified according to *type of thread, type of steel, steel strength, shape of head and nut,* and *type of shank,* but for structural steel purposes, bolts are usually specified by their ASTM designation. Three types of bolts are usually specified: A307, A325,

FIGURE 10-9 Tower crane erecting steel frame. *(Courtesy Bethlehem Steel Co.)*

FIGURE 10-10 Diagram of tower crane. *(Courtesy Heede International Inc.)*

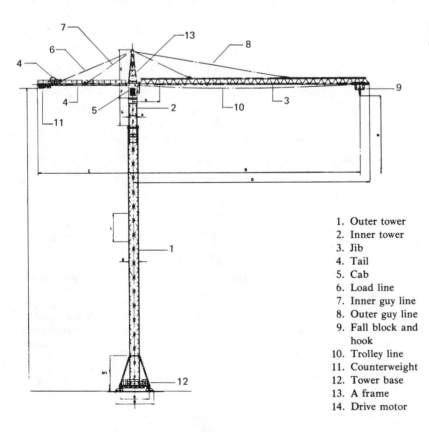

1. Outer tower
2. Inner tower
3. Jib
4. Tail
5. Cab
6. Load line
7. Inner guy line
8. Outer guy line
9. Fall block and hook
10. Trolley line
11. Counterweight
12. Tower base
13. A frame
14. Drive motor

and A490. They may have either square or hexagonal heads and nuts—hexagonal heads and nuts are preferred for the A325 and the A490 bolts as they are easier to grip with a wrench when being torqued.

The A307 bolt is a standard unfinished or *machine bolt*. It is of relatively low strength and is used in temporary and secondary-type connections only (clip angles, bridging, etc.). The A325 bolt is a *high-strength carbon steel bolt* and is the most common bolt used in structural steel connections. The A490 bolt is manufactured from *alloy steel* and is of a higher strength than the A325 bolt. The A490 bolt was developed to complement the new high-strength steels that are now being used in the construction industry. Figure 10–11 illustrates a bolt of this type and Table 10–3 gives the basic dimensions for high-strength structural bolts.

Two types of bolted connections are used in structural steel connections: bearing-type and friction-type. In the *bearing-type* connection the bolted parts bear on the shank of the bolt, thus using the bolt as a direct means of transferring load from one part to the other. Bearing-type connections are used in locations where the applied load acts primarily in one direction without much variation in magnitude. Because the connection depends on the connected parts to bear on the bolt, and the bolt hole being larger than the bolt, changes in direction of the load could produce movement in the connection. This movement could eventually cause the bolted connection to work loose producing an undesirable situation.

Friction-type connections transfer load from one member to the other by means of the friction developed between the connected surfaces. In this type of connection, the bolt acts as a clamp only, and is not used in the transfer of load from one member to the other as in the bearing-type connection. Since the friction between the connected parts determines the capacity of the connection, no movement in the connection is expected. This type of connection can be used where the load direction varies without producing any undesirable effects on the connection.

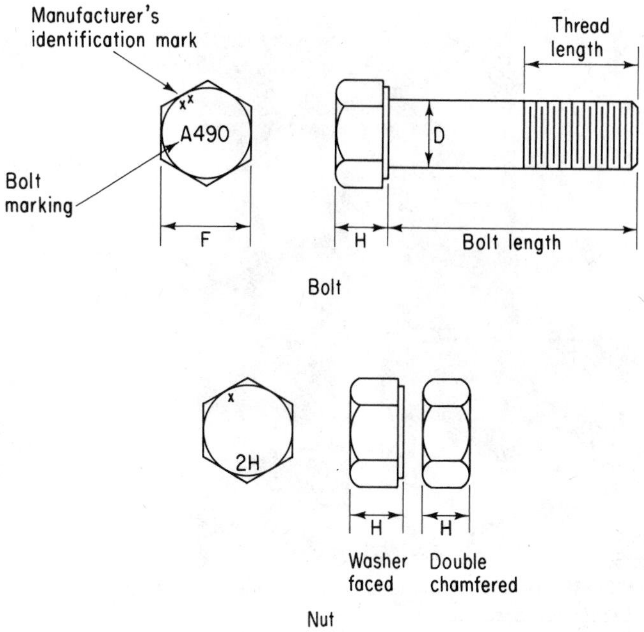

FIGURE 10–11 High-strength bolt details.

The bolts in all bolted connections must be properly tightened to produce good results. In a friction-type connection, especially, proper bolt tension is of utmost importance as the friction developed between the connected parts depends directly on the amount of bolt tension provided. That bolt tension is achieved by applying the necessary force to the nuts as they are turned down. The required force may be applied by using a *pneumatic*

TABLE 10–3 Basic dimensions—high-strength bolts—all types

Nominal bolt diameter	Heavy hex structural bolts: Thread length	Width across flats heavy	Height	Heavy hex nut[a]: Width across flats	Height	To determine required bolt length add to grip[b] (in.): No washers	1 washer	2 washers
1/2	1	7/8	5/16	7/8	31/64	11/64	27/32	1
5/8	1 1/4	1 1/16	25/64	1 1/16	39/64	7/8	1 1/32	1 3/16
3/4	1 3/8	1 1/4	15/32	1 1/4	47/64	1	1 5/32	1 5/16
7/8	1 1/2	1 7/16	35/64	1 7/16	55/64	1 1/8	1 9/32	1 7/16
1	1 3/4	1 5/8	39/64	1 5/8	63/64	1 1/4	1 13/32	1 9/16
1 1/8	2	1 13/16	11/16	1 13/16	1 7/64	1 1/2	1 21/32	1 13/16
1 1/4	2	2	25/32	2	1 7/32	1 5/8	1 25/32	1 15/16
1 3/8	2 1/4	2 3/16	27/32	2 3/16	1 11/32	1 3/4	1 29/32	2 1/16
1 1/2	2 1/4	2 3/8	15/16	2 3/8	1 15/32	1 7/8	2 1/32	2 3/16

[a]May be either washer-faced or double-chamfered.

[b]To get bolt length required if beveled washers are used, add to grip length the amount in "no washer" column plus 5/16 in. for each beveled washer. The length determined by the use of this table should be adjusted to the next longer 1/4 in.

Source: Courtesy Steel Company of Canada, Ltd.

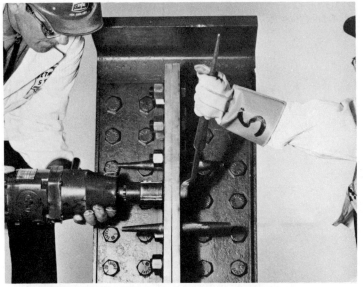

FIGURE 10-12 Pneumatic impact wrench. *(Courtesy Bethlehem Steel Co.)*

impact wrench, with a clutch set to apply the required tension (see Fig. 10-12). Nuts may also be tightened by means of a *torque wrench,* such as that used in Fig. 10-13, and the same type of wrench may be used to test the tension on nuts tightened with an impact wrench.

Another method of checking the tension on bolts is by the use of *load indicator washers.* These are hardened steel washers with a series of protrusions on one face, which bear against the underside of the bolt head when assembled (see Fig. 10-14). When the nut is tightened down, the protrusions are flattened and the gap between the bolt head and the washer reduced. At a specified average gap, measured by feeler gage, the bolt tension will not be less than that specified by various standards. Washers are produced for both ASTM A325 and A490 bolts.

Bolt holes for bolted connections are punched or drilled 1/16 in. larger in diameter than the diameter of the bolts to be used, in set patterns normally determined by the size of the connectors, and the type of member being

FIGURE 10-13 Tightening high strength bolts with torque wrench. *(Courtesy Bethlehem Steel Co.)*

FIGURE 10-14 Load indicator washer. *(Courtesy Bethlehem Steel Co.)*

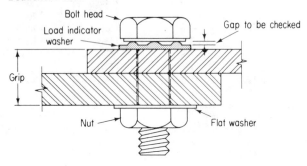

FIGURE 10-15 Bolt hole gauge and pitch.

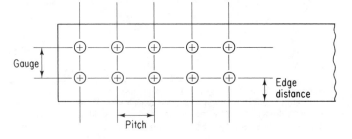

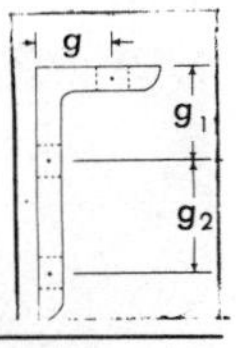

TABLE 10-4   Usual gauges for angles (in.)

						Leg (in.)								
	8	7	6	5	4	3½	3	2½	2	1¾	1½	1⅜	1¼	1
g	4½	4	3½	3	2½	2¼	1¾	1⅜	1⅛	1	⅞	⅞	¾	⅝
g¹	3	2½	2¼	2										
g²	3	3	2½	1¾										

connected. The distance between rows of bolts is usually known as the *gauge,* while the center-to-center distance between holes in a row is called the *pitch* (Fig. 10–15). Steel manuals generally provide standard gauge and pitch distances for various structural shapes. An example of the type of information available is indicated in Table 10–4, in which the usual gauges for angles are given.

Welds

With the development of good welding practices and procedures, the process of welding has given designers and fabricators an alternative method of connecting structural steel sections.

Two types of welds are most often used in welded connections: the *fillet weld* and the *groove or butt weld.* Figure 10–16 illustrates some of the more common groove and fillet welds that are used. Fillet welds are used wherever possible, as there is no edge preparation required on the material to be welded. When the full strength of the connected parts is to be developed, a groove weld is usually used. Welds that are used in critical areas must be inspected thoroughly to ensure that no defects are hidden below the surface. *X-rays* or *ultrasonic*

FIGURE 10-16   Typical welded joints.

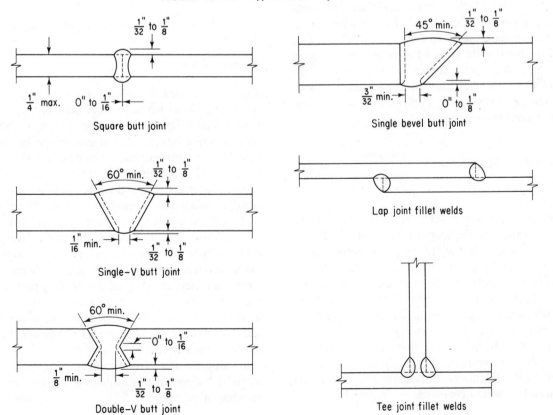

FIGURE 10–17 Field-welded connections. *(Courtesy Bethlehem Steel Co.)*

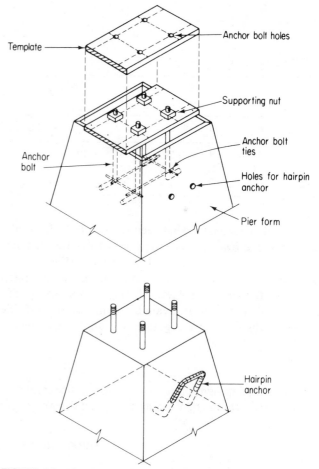

FIGURE 10–18 Anchor bolts and hairpin anchor.

testing are two methods that can indicate whether the weld is sound or defective.

As much welding as possible is done in the fabricator's shop, where better quality control can be achieved for less cost than when the welding is done on site. High-speed automatic and semiautomatic welding equipment used in the shop is much faster than the manual equipment used in the field. However, in certain instances, extensive welding is done in the field when necessary. Figure 10–17 shows beam-to-column moment connections that were field welded. In this application the connections were required to develop the full moment capacity of the beams. By using a full-penetration groove weld to connect the beam flanges to the columns, this requirement was achieved without adding additional gussets or stiffeners to the beam. The final result is a neat and efficient connection.

ERECTION OF STEEL FRAME

Anchor Bolts

The successful erection of a steel frame will depend to a considerable extent on the proper positioning of the *anchor bolts* which will tie the frame to the foundation. They are cast into the concrete foundation to hold the first members of a steel frame to be placed—the *column bearing plates*. These anchor bolts must be positioned very carefully, according to the plans, so that the bearing plates can be lined up accurately. To locate them properly, they must be held in position by a template (see Fig. 10–18) and the template accurately located on the foundation form. Hairpin anchors must also be located in the foundation where required.

Bearing Plates

The column bearing plates are steel plates of various thicknesses in which holes have been drilled to receive the anchor bolts. The holes are slightly larger than the bolts, so that some lateral adjustment of the bearing plate is possible. The angle connections by which the columns will be attached to the bearing plates are bolted or welded on them. Plates are marked on the axis lines of their four edges to allow positioning by a survey instrument, used to line them up accurately (see Fig. 10–19).

Shim packs may be set under the four corners of each bearing plate as it is placed. These are 3- to 4-in. metal squares of thicknesses ranging from $\frac{1}{16}$ to $\frac{3}{4}$ in. which are used to bring all the bearing plates to the correct grade level and to level each on its own base. The bearing plates are first leveled individually by adjusting the thickness of the shim packs. They are then brought to the proper grade by raising or lowering each by the same amount at all four corners. Finally, they are lined up in both directions by adjusting each on its anchor bolts as necessary.

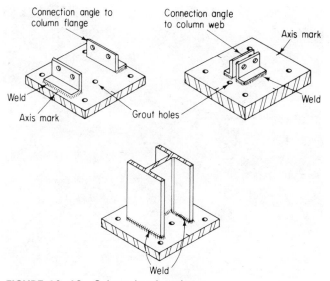

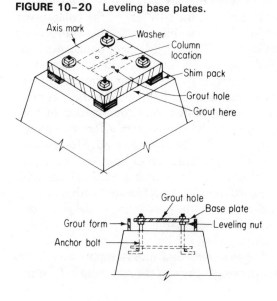

FIGURE 10-19 Column bearing plates.

An alternative to shim packs as a means of leveling base plates is the use of leveling nuts (see Fig. 10–20). They are placed on the anchor bolts under the base plate and can be adjusted to level the plate. After it is level, the plate is adjusted horizontally and the top nuts turned down to hold it in place.

After all bearing plates have been set and aligned, they must be *grouted*. This is done by completely filling the space between the bearing plate and the top of the concrete with mortar, preferably a nonshrinking type made with metallic aggregate.

A simple wood form is placed around the bearing plate about 2 in. away from its edges. Grout is poured or pumped into the space until it is completely filled. Large bearing plates have grout holes drilled near the cen-

FIGURE 10-20 Leveling base plates.

ter to make sure that no air becomes trapped under them. When grout can be seen in these holes, it indicates that the space has been filled.

Columns

First-tier columns are the next group of members to be erected. They are often two stories long, or in any case, the same size for at least two stories. Column lengths are such that joints—splices—will come 1½ to 2 ft above the floor levels. This is done to prevent splice connections from interfering with girder or beam-to-column connections. Column ends are milled to exactly the right length and to make sure that column loads will be evenly distributed over the entire bearing area.

HP shapes, as nearly square in cross section as possible (see Fig. 10–3), are often chosen for columns. Round or square HSS sections may also be used, though the round column may present some connecting problems. Columns may also be *built-up* by welding or bolting a number of other rolled shapes into a single unit (see Fig. 10–21).

Column Splices

Column sections are joined together by *splice plates* that are bolted or welded to the column flanges and, in special cases, to the webs. If the column flanges match reasonably well, it is common to butt the column ends

FIGURE 10-21 Built-up column sections.

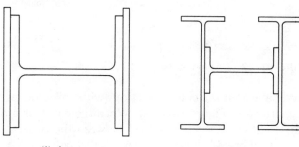

W shape
with covers

3 W shapes

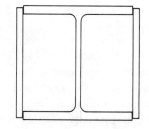

Box shape

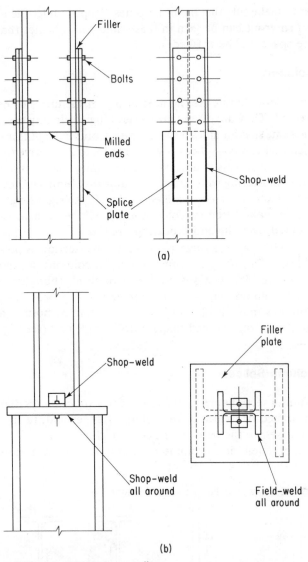

(a)

(b)

FIGURE 10-22 Column splices.

FIGURE 10-23 Typical structural steel frame.

directly and join the column sections with splice plates, as illustrated in Figure 10–22(a). When the columns are of different sizes, a plate is used between the two column ends to provide bearing for the smaller column. The smaller column is bolted in place with positioning bolts and then welded to the bearing plate to complete the splice [see Fig. 10–22(b)].

Girders

Girders—the primary horizontal members of a frame—span from column to column (see Fig. 10–23) and support the intermediate floor beams. They carry wall and partition loads as well as the point loads transmitted to them by the beams. Rolled W shapes or three-plate welded I shapes are normally used as girders. Connections between girders and columns can be one of two types: *pin-type* connections or *moment-resisting* connections.

Pin-type connections are designed to resist vertical loads only. These connections consist of angles or plates bolted or welded to the girder web and usually bolted to the supporting column flange or web (see Fig. 10–24). The *seated connection* in Fig. 10–24(c) has the angle seat welded to the column and the girder may be bolted or welded to the seat. A clip angle, usually 4 × 4 × ¼, is welded to the column and the girder top flange to provide lateral support for the girder. Because of their flexibility, pin-type connections cannot be relied upon to resist lateral loads and, therefore, provide no lateral stability to the structural frame. *Cross-bracing, shear walls,* or a *concrete core* must be used to absorb the effects of lateral forces and to provide the necessary stability.

Building frames that are designed as *rigid frames*—frames that depend on the stiffness of the column sections to provide lateral stability—use *moment-resisting* connections (see Fig. 10–25). Moment connections between the girders and the columns provide lateral stability in the structural frame by preventing rotation to occur between the girders and the columns. *Moment connections* may be welded, bolted, or a combination of bolts and welds. Figure 10–25 shows three different arrangements of bolts and welds that may be used. Moment connections that use a combination of bolts and welds; the bolts are designed to carry the shear or vertical load, while the welds are used to resist the effects of bending. In Fig. 10–25(c), the seat aids in the positioning of the girder during erection as well as resisting the vertical load due to the beam reaction.

Seated connections that must support very heavy loads are reinforced to prevent the seat from deforming.

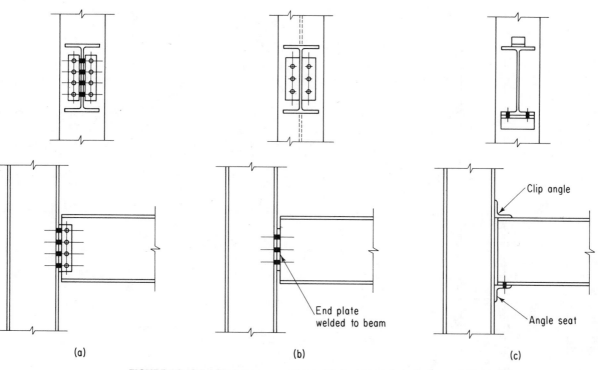

FIGURE 10–24 Pin-type connections: (a) double-angle bolted connections; (b) end-plate connection; (c) seated connection.

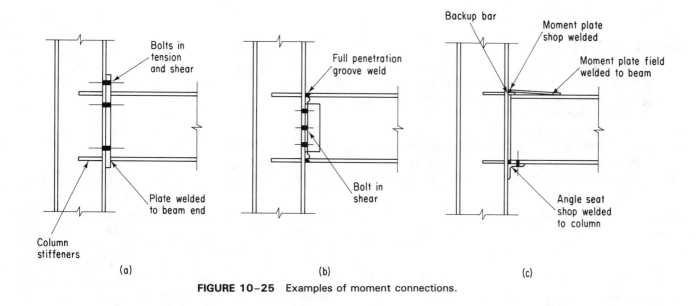

FIGURE 10–25 Examples of moment connections.

Figure 10–26 illustrates two common types of reinforced seats: an angle seat with a stiffener added and a tee section cut from a rolled W shape.

Many connections have been standardized to allow for easier detailing and fabrication. Most structural steel design manuals have tables to aid in the selection and design of good structural connections.

Beams

Beams, generally smaller than girders, may be connected either to a column or to a girder. Beam connections at a column are similar to the girder-to-column connections. In beam-to-girder connections, the main consideration is the shear connection, since the beams are considered

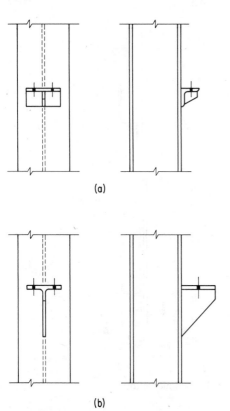

Blocked beam

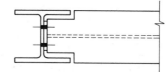

Blocked beam

Coped beam

FIGURE 10–26   Reinforced seated connections: (a) angle seat with stiffener; (b) tee seat.

FIGURE 10–28   Coped and blocked beam ends.

to be carrying the floor loads and transferring them to girders as vertical loads.

Since beams are usually not as deep as girders, there are several alternative methods of framing one into the other. The simplest one is to frame the beam between the top and bottom flanges of the girder, as shown in Fig. 10–27. If it is required that the top or bottom flanges of girders and beams be flush, it becomes necessary to cut away a portion of the upper or lower beam flange, as illustrated in Fig. 10–28. If the girder is an S shape, the

end of the beam is *coped*, while if it is a W shape, the end is *blocked*. In many cases, the shear connection angles are bolted or welded to the beam ends in the shop, and the member comes to the job ready for connection to the girder web.

CHANNELS AND ANGLES

Channels and angles, because of their shape, light weight, and wide range of sizes have many different uses in the structural frame. Channels are usually used as secondary framing members when loads and spans are not too great. They are used as *wall girts* and *roof purlins*—girts being the horizontal members attached to the columns to support siding (see Fig. 10–29) and purlins being the roof framing members spanning between the roof beams and supporting the roof deck. Other uses for channels are as framing for doors and windows, as stair stringers, and as web and chord members in trusses; as built-up sections they are used as eave struts and spandrel beams.

Angles are used as the connecting pieces for beams, as chord and web members in light trusses and joists, and as bracing (see Fig. 10–30). Angles are also used as reinforcing around openings and as supports for mechanical equipment.

FIGURE 10–27   Floor beams framed between girder flanges.

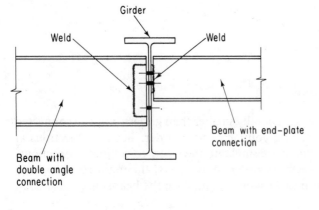

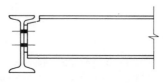

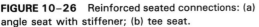

Girder

Weld

Weld

Beam with end-plate connection

Beam with double angle connection

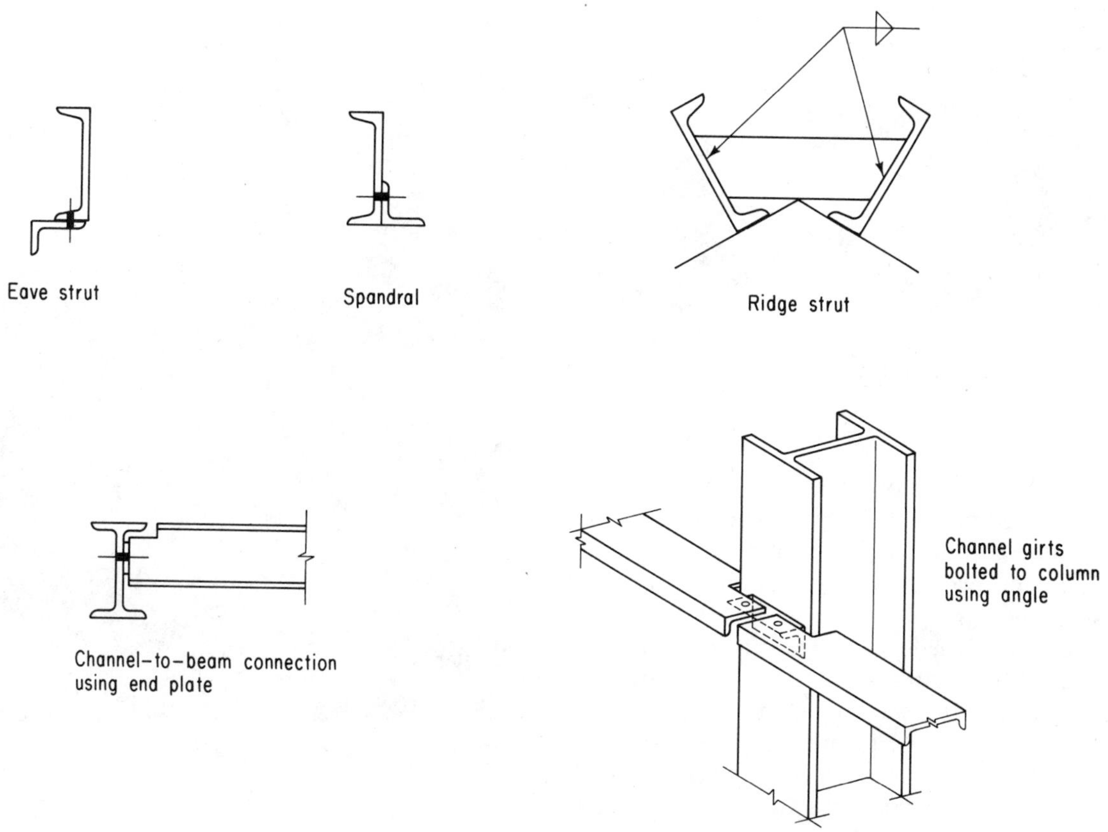

Eave strut

Spandral

Ridge strut

Channel-to-beam connection
using end plate

Channel girts
bolted to column
using angle

FIGURE 10-29 Channel uses and connections.

FIGURE 10-30 Double-angle cross-bracing.

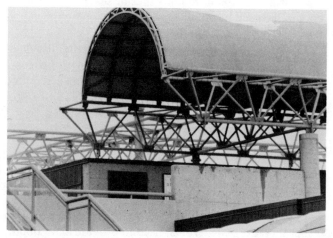

FIGURE 10-31 Space frame.

HOLLOW STRUCTURAL SECTIONS

Many buildings are designed using the structural frame as an architectural feature. The use of various colors on different parts of the frame can help create an air of spaciousness within the building. Because of their clean appearance, high strength, and good stability, hollow structural sections are popular with designers for use as column sections, trusses, and space frames (see Fig. 10-31).

OPEN-WEB STEEL JOISTS

Open-web steel joists are lightweight trusses that are used for supporting roofs and floors. They are normally fabri-

FIGURE 10–32 Open-web steel joist with double-angle top chord.

cated in depths that range from 10 to 48 in. and in lengths that range from 10 to 100 ft. Many manufacturers use an assembly-line approach for the fabrication of their joists to ensure uniformity, quality, and economy in the final product.

The makeup of the joist can vary from one manufacturer to another, but in general, the *webs* are usually made of bent rod for joists in the range 10 to 24 in. and a lightweight hollow tube is substituted for the rod in the deeper joists. The *chords* may be small angles (Fig. 10–32) or some special shape designed specifically for use as a chord section. Open-web steel joists are normally designed to support gravity loads only. Joists that must resist horizontal loads require special seats or cantilever ends, which can be obtained from the manufacturer upon request.

Joists used in a steel frame are normally field-welded to the supporting beams to maintain proper spacings. Continuous horizontal or cross-bridging is required at regular intervals along the span of the joists to provide vertical stability.

SPECIAL GIRDERS

In particular situations where regular rolled shapes are not suitable because they are either not deep or wide enough, special girders may be fabricated. There are two general types: *plate girders,* used to provide extra depth, and *built-up girders,* used to provide extra width.

Plate girders are made by using a plate of appropriate width and thickness for the web, and angles, riveted or welded at the top and bottom edges for flanges. Deep plate girders often require vertical stiffeners, in the form of angles welded to the web. In Fig. 10–33, a plate girder made from plate and channels and with vertical stiffeners is being hoisted into place as a spandrel beam.

Built-up girders are made by joining two or more regular rolled shapes side by side by means of plates or pipes known as *separators.*

FIGURE 10–33 Built-up girder being hoisted into position.

CONSTRUCTION PROCEDURES AND PRACTICES

There are two distinct stages in the construction of a steel frame. The first stage is performed in the shop, where the individual pieces are *cut to length, coped* or *blocked,* and *punched* or *drilled* for bolts. Detail material such as *gusset plates, clip angles,* and *end plates* for beams are attached so that the connections in the field can be made quickly.

The second stage begins when the fabricated steel arrives on the site. The field crew, with the aid of a crane (Fig. 10–34), assembles the individual pieces according to the erection drawings. The connections are aligned with *drift pins* and temporary bolts are used to hold the sections together. The building is plumbed using a building transit and held in alignment with temporary cable bracing equipped with *turnbuckles* (to allow for adjustments) until permanent bracing has been put in place and final bolting and welding have been completed. Figure 10–35 shows the final stages of completing a bolted connection—removal of a drift pin which was used to align the bolt holes—so that the final structural bolt can be put in place.

The conventional methods of steel erection—beginning at the bottom and working up do not always apply. Unusual designs sometimes require unorthodox erection schemes to be used, as illustrated in Fig. 10–36. Two con-

FIGURE 10–34 Steelworkers erecting steel frame. *(Courtesy Canadian Institute of Steel Construction.)*

FIGURE 10–36 Structural steel floor frames hang from straps. *(Courtesy Bethlehem Steel Co.)*

FIGURE 10–35 Final stage of completing bolted connection. *(Courtesy Bethlehem Steel Co.)*

crete support towers have steel tension hangers draped over reinforced concrete saddles from which the steel frame floors were suspended. The cantilever steel frames were preassembled on the ground, raised to the proper elevation, and suspended from the steel hangers. Nine of the 13 floors were erected in this manner, providing large column-free areas on each floor.

Under certain conditions, abnormally high structures require innovative construction methods to be applied. Figure 10-37 shows a Sikorsky sky-crane placing the steel sections of an antenna on top of a communications tower. The erection of steel requires skill, thoughtfulness, and constant attention to rules of safety. Some general safety rules to observe are as follows:

1. Do not attempt steel erection on rainy days. It is dangerous to work on wet steel.

FIGURE 10-37 Sikorsky sky-crane placing steel antenna sections atop a tower. *(Courtesy Canadian Institute of Steel Construction.)*

2. Wear shoes with sewn leather or rubber soles.
3. Do not allow anyone to stand beneath a loaded boom.
4. Do not allow anyone to ride a load.
5. Wear heavy gloves which do not have loose cuffs.
6. Do not wear loose clothing which could easily catch on projections or swinging objects.
7. Use shackles instead of hooks.
8. Use cable slings instead of chains for lifting steel.
9. Make sure that the brakes on hoisting drums are inspected daily.
10. Keep a constant lookout for frayed or broken strands in cables and repair or replace them if necessary.

REVIEW QUESTIONS

1. Explain the purpose of **(a)** *erection marks* on structural steel members, **(b)** the precise *location* of the erection mark on a beam, **(c)** a *hairpin anchor,* and **(d)** *dunnage beams.*

2. Outline two alternative methods of leveling bearing plates.

3. Explain why **(a)** *column splices* are made above floor level, **(b)** column ends are *milled* in the fabricating shop, **(c)** it is sometimes necessary to use a plate between the ends of two columns being connected, and **(d)** HP shapes are often chosen for columns.

4. With the aid of diagrams, illustrate how you would try to overcome the connecting problems involved if you were required to connect a W shape beam to a round column.

5. Explain the difference between **(a)** *coping* and *blocking,* as applied to steel beam ends; **(b)** *plate girders* and *built-up girders;* **(c)** *shear* connections and *moment* connections.

6 Using a scale drawing, illustrate how a bolted shear connection will be made between an S shape beam and an HP shape column. Use specific sizes for both.

Floor Systems and Industrial Flooring

The erection of the structural frame marks the completion of a phase in the construction of a building. The next step involves the installation of floors and is usually done in two operations: first, the building of a base, or *subfloor;* and second, the application of *finish flooring.*

Subfloors may be made of wood, concrete, or a combination of concrete and steel. The type used in any particular situation depends to a great extent on the type of structural frame involved. When a reinforced concrete frame is involved, the base slab is usually also concrete and the installation of floors is not really a separate phase since both frame and floor are placed monolithically. If timber or steel frames are used, the structural supports are erected prior to the start of all floor construction.

WOOD FLOORS

Wood floors are almost always used with timber frames (see Fig. 11–1) but may be used with steel frames.

Wood subfloors may be constructed in two ways. One method involves the use of 2″ × 4″s or 2″ × 6″s laminated on edge and at right angles to the floor beams. The pieces are spiked together side by side and no attempt is made to make the end joints meet over a support (although some of them should and, of course, the starting course end joints must do so).

When the floor frame is composed of wood beams, the laminations may be toenailed to the beams. In the case of a steel floor frame, a wooden nailer is first fastened to the top flange of the beams (Fig. 11–2) and the subfloor is then toenailed to the pad.

The other type of wood floor consists of heavy planks, thick enough to carry their load without excessive deflection, laid across the beams and spiked to them. Planks are usually tongue-and-grooved or *splined* for greater rigidity. End joints in these planks will come over a beam, as in Fig. 11–2.

CONCRETE FLOORS

There are two distinct types of concrete floors: slabs-on-grade and structural slabs. The *slab-on-grade* is a floor poured directly on a prepared subgrade to provide a clean and durable surface for the building occupants. The durability of the slab depends on the concrete strength and finish, but the strength of the slab depends on the strength of the supporting subgrade. The *structural slab,* which must span between supporting walls, columns, or beams, must be able to support its own dead weight as well as the superimposed load that it was designed to carry. Thus its strength depends solely on the concrete strength and the amount of reinforcing steel that has been used.

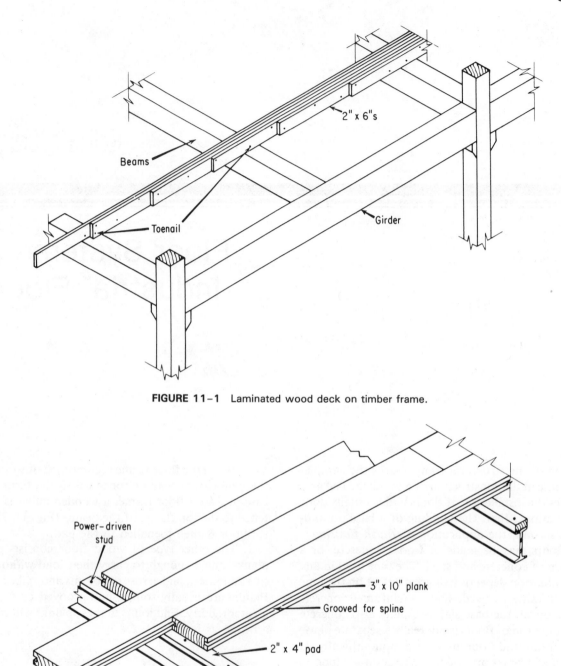

FIGURE 11–1 Laminated wood deck on timber frame.

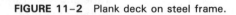

FIGURE 11–2 Plank deck on steel frame.

Concrete Slab-on-Grade

Concrete slabs-on-grade may be placed before any other portion of the building has been constructed (Fig. 11–3) on a prepared subgrade to provide a level surface on which the structural frame is erected. In the case where the frame is supported by footings and grade beams, the floor may be poured after the building has been closed in. In the first instance, side forms of wood or metal are placed, leveled, and staked around the perimeter of the building and screed strips are placed at convenient intervals to provide guides for the leveling of the concrete. In Fig. 11–3, a vibrating straightedge and drag are used to level the slab. A slab placed on grade after the walls and roof have been erected is common practice in commercial and industrial buildings, where large floor

FIGURE 11–3 Slab-on-grade concrete floor.

(a)

areas poured in sections require longer periods of protection from the weather.

Slabs-on-grade often provide less than adequate results when put into service due to excessive cracking, dusting, and spalling. To ensure that the slab will provide a long-lasting and serviceable surface, the Portland Cement Association puts forward the following recommendations:

1. The subgrade must be uniform and have adequate bearing capacity.

2. The concrete must be of uniform quality.

3. The slab thickness should reflect the anticipated loads.

4. Ensure proper jointing of the slab.

5. Provide good workmanship.

6. Provide a proper surface finish for the anticipated loads.

7. Use proper repair procedures if and when they are necessary.

A standard procedure to produce a good concrete slab-on-grade may be as follows:

1. Backfill all ditches and trenches under the floor with material similar to the surrounding subgrade, ensuring similar density and moisture content. Uniform compaction with some type of mechanical equipment (Fig. 11–4) is most important.

2. Isolate all columns from the floor slab by boxing them with square wood or metal forms, or with round fiberboard forms. Forms should be set to the level of the top of the slab (Fig. 11–5).

3. Isolate the walls from the slab by fastening strips of asphalt-impregnated fiberboard or other joint material

(b)

FIGURE 11–4 (a) Vibratory compactor; (b) gas-powered compactor.

not more than ½ in. thick around the walls, level with top of the slab.

4. Prepare any changes in slab thickness, as at doorways, to be as gradual as possible and at slopes of not more than 1 in 10.

5. Use a template with legs equal to the slab thickness to check the grade (see Fig. 11–5).

FIGURE 11-5 Leveling subgrade.

6. Cover the grade with a polyethylene moisture barrier, allowing generous lap between strips.

7. Place the reinforcement—mesh or rod—as specified in plans.

8. Set screed strips at the same elevation at convenient intervals throughout the area to be concreted (see Fig. 11-6).

9. Oil the screed strips.

10. Place the concrete as close to its final position as possible. Consolidate with an internal vibrator, especially at corners, walls, and bulkheads.

11. Straightedge the concrete to the level of the screed strips.

12. Smooth the surface with a long handled float or a darby (Fig. 11-7) to remove the high and low spots. Cover with damp burlap until ready for the next operation.

FIGURE 11-6 Typical screeds and bases.

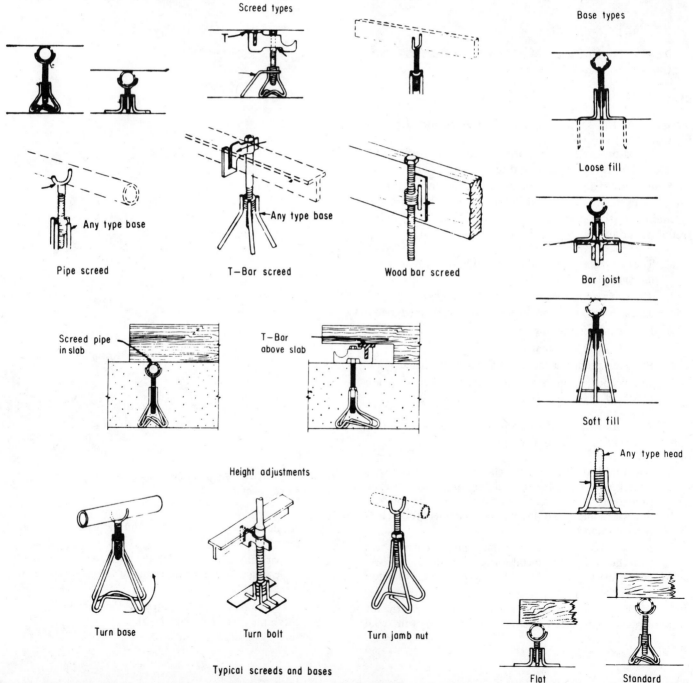

13. Float the surface with hand or mechanical floats as soon as the concrete supports the weight of a man.

14. If specified, apply metallic aggregate hardener (see directions for applying hardener, p. 286).

15. Trowel to a hard, dense surface with hand or power trowels (Fig. 11–8).

16. Cure by covering with (a) waterproof curing paper, (b) two coats of curing compound, (c) burlap kept moist at all times, or (d) a layer of damp sand (see Fig. 11–9).

FIGURE 11-8 Power trowel in action. *(Courtesy Portland Cement Association.)*

FIGURE 11-7 Darbying a concrete slab. *(Courtesy Portland Cement Association.)*

(a)

FIGURE 11-9 Concrete curing preparations: (a) plastic film covering; (b) wax compound covering. *(Courtesy Portland Cement Association.)*

(b)

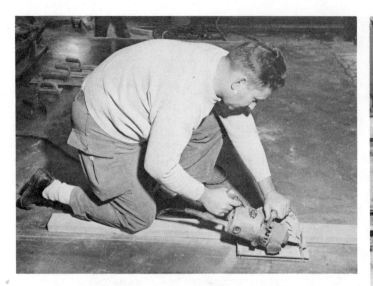

(b)

(c)

11-10 (a) and (b) Cutting control joints; (c) abrasive wheel for cutting concrete. *(Courtesy Portland Cement Association.)*

17. Remove the forms around columns and attach joint material to the vertical faces of the slab and the base of the columns. Fill with concrete, edge, and finish.

18. Cut control joints to a depth of at least one-fifth the slab thickness with a power saw every 20 to 25 ft in both directions (Fig. 11-10).

19. Caulk the joints with mastic joint filler.

20. Cure for at least 7 days before allowing regular traffic on the floor.

Structural Concrete Floor Slabs

A concrete floor supported by a structural frame may be one of a number of floor types now available. The one to be used for any particular building will depend on a number of factors, including (1) the intended use of the building, (2) the type and magnitude of the loads to which the floor will be subjected, (3) the length of the span be- tween supports, (4) the type of building frame, and (5) the number of storeys in the building. Concrete structural floor slab types include:

1. *One-way solid slabs:* One-way solid slabs (Figs. 11-11 and 11-12) are structural slabs that have their main reinforc- ing steel running in one direction only (parallel to the span). One-way slabs are used in reinforced concrete frames, structural steel frames (Fig. 11-13), or can be sup- ported by masonry. They are used as single spans or in multispan construction. One-way slabs are easy to form and the reinforcing steel is not complicated to place. Over short spans they can support reasonably high loads rela- tive to their thickness but become uneconomical on spans over 20 ft because of their weight.

2. *Two-way solid flat slabs:* Two-way flat slabs have load- supporting reinforcing steel running in two directions. Beams or walls are not necessary in the support of this type of slab, as it can be placed directly on supporting con-

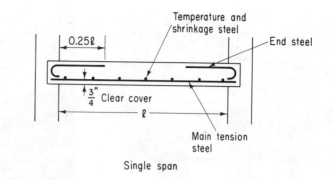

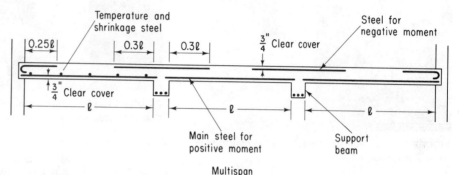

FIGURE 11–11 Reinforcing steel in solid one-way slabs.

Single span

Multispan

FIGURE 11–12 Beam-and-slab floor. (*Courtesy Portland Cement Association.*)

FIGURE 11–13 One-way slab supported by steel frame.

crete columns. Thickened sections over the columns known as *drop panels* (Fig. 11–14) ensure that sufficient shear capacity is available in the slab. In the past, *column capitals* (Fig. 11–15) have been used with or without drop panels. These have been eliminated to a great extent because of higher forming costs. The slab thickness depends on the span between columns—the minimum thickness between drop panels is 4 in. and the maximum is 12 in. The minimum size of drop panel used is one-third of the span and the total drop panel thickness is usually 1¼ times the slab thickness. The two-way slab system is used for heavier applications such as industrial storage buildings and parking garages. Maximum practical spans for the two-way slab are about 34 ft.

3. *Two-way flat plate slab:* The two-way flat plate slab is very similar to the two-way solid slab with the exception

FIGURE 11-14   Flat slab floor. *(Courtesy Portland Cement Association.)*

FIGURE 11-15   Column capital. *(Courtesy Portland Cement Association.)*

that it does not have a drop panel. Flat plates are very efficient from the point of forming, steel layout, and total thickness. The slab thickness varies from 5 to 10 in. and maximum practical spans reach about 32 ft. Because of its simplicity and minimum structural depth, the flat plate is popular in high-rise offices and apartments (Fig. 11-16).

4. *One-way joist slab:* This is a one-way structural system consisting of a series of *concrete ribs or joists* containing the reinforcing steel, cast monolithically with a relatively thin slab. The ribs, in turn, frame into supporting beams (see Figs. 11-17 and 11-18). The main purpose of using such a system is to reduce the dead load by concentrating the reinforcing steel in the ribs and eliminating the concrete between them. The slab thickness is usually $2\frac{1}{2}$ to $4\frac{1}{2}$ in. and the joist depth varies from 6 to 20 in. Standard pan forms produce between-rib dimensions of 20 and 30 in. Maximum joist spans are about 45 ft.

FIGURE 11-16   Flat plate floors. *(Courtesy Portland Cement Association.)*

FIGURE 11-17   Ribbed slab.

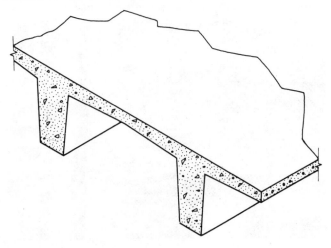

FIGURE 11–18 One-way joist floor. *(Courtesy Portland Cement Association.)*

5. *Waffle (two-way joist) slab:* In this type of floor, ribs run at right angles to one another, resulting in a two-way structural system. The forms, called domes, are available in two standard sizes, 19 × 19 and 30 × 30 in., but they can be custom-made for particular job requirements. The slab depth varies from 3 to 4½ in., and the rib depth varies from 8 to 20 in. The standard rib widths are 5 in. and 6 in. and the distance between rib centers is 24 in. or 36 in. This system is found to be economical for spans ranging from 25 to 50 ft (see Figs. 11–19 and 11–20).

Post-Tensioned Concrete Slabs

Adding *post-tensioning strands* to concrete slabs produces sections that can span over greater distances than unpre-

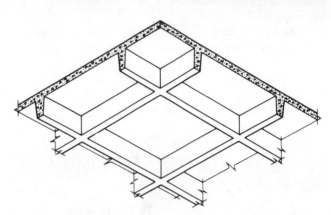

FIGURE 11–19 Two-way ribbed slab.

FIGURE 11–20 Under side of waffle slab. *(Courtesy Portland Cement Association.)*

stressed sections and yet are able to support greater loads with less deflection and virtually no cracking.

High-strength steel strands or tendons enclosed in protective plastic sheathing are distributed throughout the slab at predetermined intervals. Figure 11–21 illustrates a typical tendon with *slab anchor, wedge anchor,* and *protective sheathing.* The sheathing ensures that the concrete does not bond to the post-tensioning tendon. Once the concrete has cured sufficiently, the tendons are tensioned to a predetermined stress and anchored. Figure 11–22 illustrates strands being placed for a two-way post-tensioned flat plate slab. Note that the tendons are raised higher above the form in the vicinity of the supporting column than at the midspan of the slab. This arrangement ensures that the compressive forces induced in the concrete slab by the tendons counteract the tension stresses produced in the slab by anticipated loads. Figure 11–23 illustrates a completed arrangement of post-tensioning tendons used in conjunction with regular reinforcing steel.

Any concrete slab may be post-tensioned, including slabs-on-grade, where it has been found that the slab per-

FIGURE 11–23   Two-way slab ready for concrete.

formance improved even where subgrade conditions were less than ideal. The application of post-tensioning has become most popular in solid one-way and two-way slabs, one-way and two-way ribbed slabs, and in two-way flat plate slabs. Post-tensioning can be applied to advantage in all sizes of buildings, ranging from single-storey car parks to multistorey apartment blocks and office towers.

Floor Slab Forms

Forms for a beam-and-slab floor consist of slab forms and the shores which support them, beam forms and the T shores which support them, *kickers* between slab and beam forms (see Fig. 9–18), and supports for the edges of the slab form over the T shores.

The slab forms should be made first. The frames are of 2″ × 4″s in any required size, but one edge of the sheathing should be beveled at 45° and extended about 2¼ in. beyond the frame on panels which will meet the beam form (see Fig. 11–24).

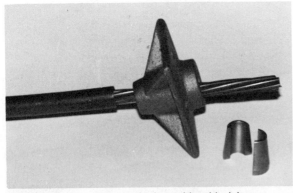

FIGURE 11–21   Post-tensioning cable with slab anchor, wedge anchor and sheathing.

FIGURE 11–22   Post-tensioning cables being positioned on chairs.

FIGURE 11–24   Frame for slab form panel.

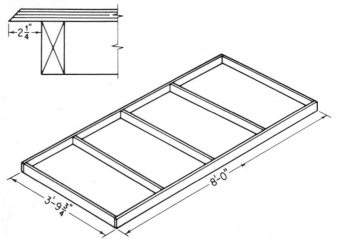

The column forms are next erected, plumbed, and braced. The T shores are set up between each pair of columns (see Fig. 9–23), using a chalk line from one column top to another as a guide. If a camber is required in the beam, raise the center shore the required amount, run the chalk line over it, and set the remainder of the shores to the line.

The beam form may now be built on the shore heads, and the slab form panels can be raised and supported. It is important to see that the *kickers* are in place between the beam sides and the edges of the slab forms. These kickers prevent the beam sides from being pushed out by the pressure of the concrete and allow easy removal of beam sides when the concrete has cured sufficiently.

The first step in the erection of formwork for a flat slab floor is to erect the shores and brace them temporarily. Stringers are then placed on the shores and joists laid across the stringers, leaving openings wherever there are to be drop panels. This framework is leveled by wedging or adjusting the length of the shores, and sheathing is finally laid over the joists. Where sheathing meets a drop panel, the edge is beveled and extended over the top of the drop panel in the same way that slab form sheathing extends over beam side forms (see Fig. 11–25).

Drop panels are built separately and are supported on separate shores or on sections of steel scaffolding set up to form a square supporting structure. Figure 11–26 illustrates one method of forming the drop panel. The sheathing must be cut to fit the column capital which will be part of the column form.

Ribbed slab floors are formed by the use of prefabricated metal or plastic forms—called *pans* or *tile*—held in position and supported in any one of several ways. Pans for one-way ribs are usually made from sheet metal, (see Fig. 11–30) but those used to form two-way ribs—a *waffle* floor—may be either sheet metal or plastic.

FIGURE 11–25 Typical flat slab formwork.

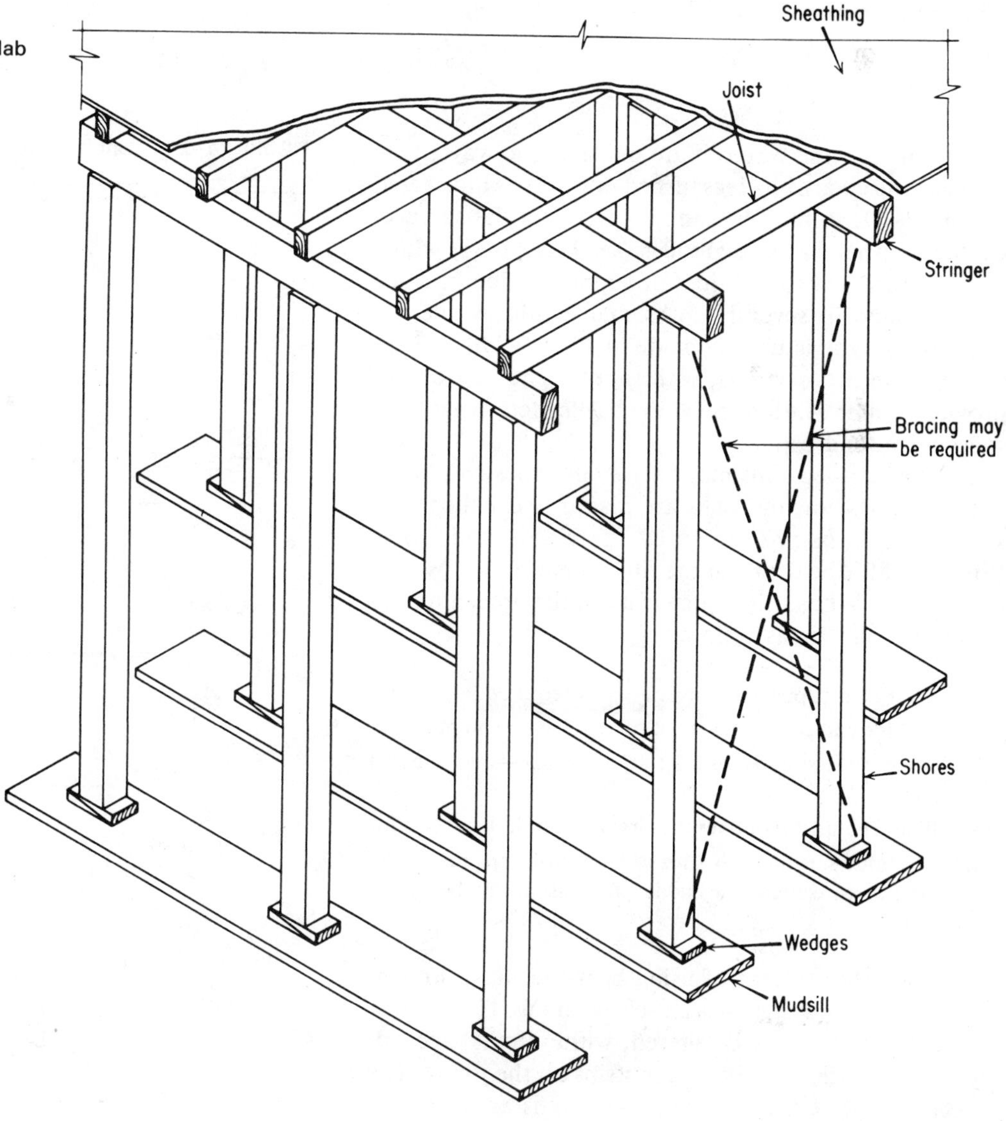

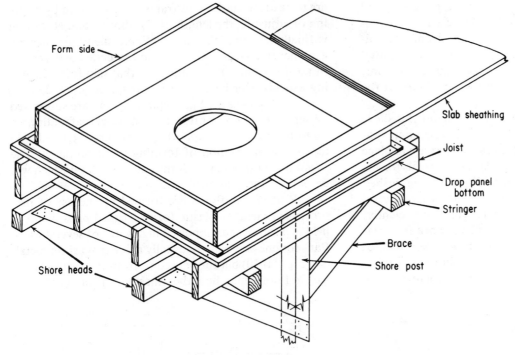

FIGURE 11-26 Drop panel form.

Pans for one-way ribs are made in two ways. One type has the ends of its legs turned out flat, as illustrated in Fig. 11-27, so that the pan may rest on a flat surface, while the other has straight legs (see Fig. 11-31) which must be fastened to the sides of supports. The *bent-leg* type is made in several widths and depths, while the straight-leg type is usually made in 20-in. widths. Both are made in 3-ft sections, with tapered end sections to provide greater width of rib at the beam. Both types have sheet metal end caps.

The bent-leg pans may be used in two ways. In the nail-down system, the pans are placed and nailed to a series of *soffit boards* supported on stringers and shores (Fig. 11-28). The depth of the ribs is determined by that of the pans. When this system is used, the procedure for setting up pans is as follows:

1. Erect and support soffit boards as required.
2. Nail the end caps to the soffit boards at each end of a bay.
3. Place the first pan—usually a tapered one—over the end cap.
4. Place the next pan over the first and lap at least two in.
5. Continue setting pans with at least 2-in. laps until they meet at the center. The middle pan must always be placed last.

The other system of using bent-leg pans—known as the adjustable system—is illustrated in Fig. 11-29. Here the depth of ribs may be altered, within limits, according to the position of the ledger strips on the joists. The procedure for setting up pans this way is as follows:

FIGURE 11-27 Bent-leg metal pan.

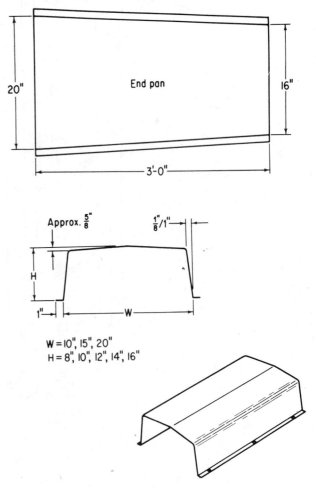

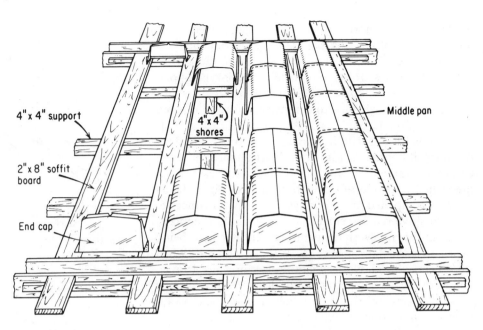

4" x 4" support

4"x 4" shores

2" x 8" soffit board

End cap

Middle pan

FIGURE 11-28 Metal pans in nail-down system.

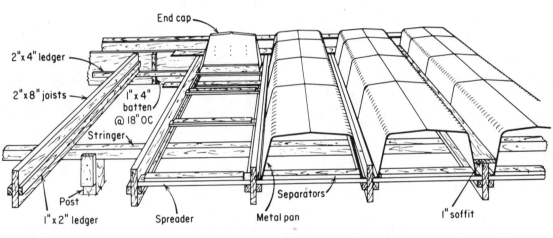

End cap

2" x 4" ledger

2" x 8" joists

1" x 4" batten @ 18" OC

Stringer

Post

1" x 2" ledger

Spreader

Separators

Metal pan

1" soffit

FIGURE 11-29 Metal pans in adjustable system.

FIGURE 11-30 Pans for one-way joist system in place.

1. Set up and level stringers are required.
2. Nail ledger strips to the joists at the specified level.
3. Set joists on the stringers on the required centers.
4. Cut spreaders to length (face to face of joists) and separators 2¼ in. shorter, with ends cut to the slope of the pans. Tack separators to spreaders.
5. Set the spreaders on the ledger strips, two per pan section.
6. Place end caps on spreaders at each end of bay.
7. Place metal pans on spreaders, using the same sequence as outlined above.
8. Cut soffit pieces from 1-in. material, wide enough to fit between pans and rest on the top of joists. The ends of these soffits must be widened to compensate for the taper on the end pans.

Straight-leg pans are supported by nailing them to the sides of the joists with double-headed nails (Fig.

11–31). Three or more sets of nail holes are drilled in the legs and the depth of ribs is altered by adjusting the position of the legs against the joists and using the appropriate set of holes. Joists to be used with this type of pan are usually built to form a standard rib width. The ends of the joists are widened to compensate for the tapered end pans (Fig. 11–32).

Pans for a waffle floor may be set on a solid deck or on soffit boards. In Fig. 11–33, soffit boards are used to support the *dome* pans, and a plywood deck form is used where a solid section of slab is required. Figure 11–34 shows a completed dome pan system ready for concrete.

Positive reinforcement in ribbed floors is concentrated in the ribs, with temperature reinforcement being placed over the pans. Figure 11–35 shows a typical ribbed floor being placed. Notice that the exterior walls, beams, and floor are being placed monolithically. Notice also the column dowels for second floor columns and the runways used to carry concrete buggies over the pans.

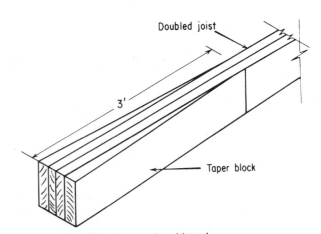

FIGURE 11–32 Joist ends widened.

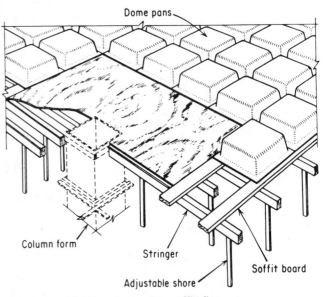

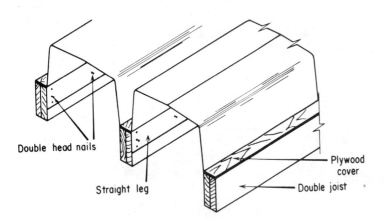

FIGURE 11–31 Straight-leg pans on double joists.

FIGURE 11–33 Dome pans for waffle floor.

FIGURE 11–34 Dome pans for waffle floor in place. *(Courtesy Portland Cement Association.)*

FIGURE 11-35 Ribbed slab floor forms ready for concrete.

FIGURE 11-36 Adjustable wood post.

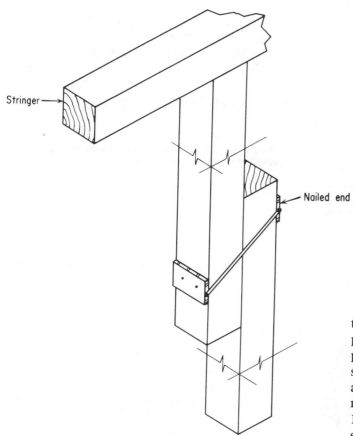

Stringer

Nailed end

FIGURE 11-37 Screw-type shore.

Forms for all these concrete floors must have some type of adjustable support. This is usually provided by posts or by sections of steel scaffolding. One-piece wood posts are adjusted by wedges under the bottom end, as shown in Fig. 11-25. Two-piece posts are made and used as illustrated in Fig. 11-36. Several kinds of adjustable metal posts are available, one of which is shown in Fig. 11-37. In Fig. 9-16, steel scaffolding is being used as form support.

I/D ("Integrated-Distribution") Floor System

The I/D floor (and roof) system has recently been developed by the Portland Cement Association and is specially designed for use in low-rise apartments, light commercial and industrial buildings, as well as residences.

The concept includes three basic components: (1) steel truss Ts, (2) forms to be left in place, and (3) cast-in-place concrete—and incorporates heating, plumbing, and electrical utilities in the system.

The truss Ts (see Fig. 11–38) do double duty as supports for the I/D forms and as the only reinforcement normally required for a simple-span system. They are supported on and anchored to load-bearing walls and intermediate shoring on 24-in. centers. Additional top reinforcement may be installed when required (see Fig. 11–43).

I/D forms (see Fig. 11–39) are lightweight units

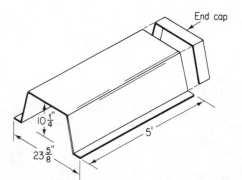

FIGURE 11–39   I/D form. *(Courtesy Portland Cement Association.)*

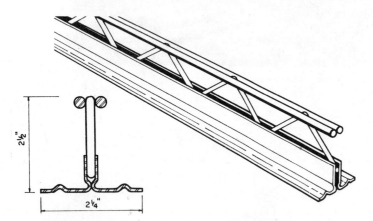

FIGURE 11–38   Steel truss T. *(Courtesy Portland Cement Association.)*

weighing from 10 to 12 lb/form, intended to remain in place when the floor is completed. They are easily cut by hand or saber saw for length adjustments or to form cross channels for utilities' connections.

Shoring should be spaced not more than 6 ft o.c. and should be left in place for a minimum of 7 days, depending on the temperature. Shores may be single jacks, properly spaced; stringers with scaffolding supports (see Fig. 11–40); or non-load-bearing interior partitions. Where the last are used, the truss Ts should not rest directly on the partition but should be supported by removable shims, to provide ¾-in. clearance (see Fig. 11–41).

Concrete, which should normally be designed for 4000 psi, may be placed by any of the commonly used methods (see Fig. 11–42 and consolidated with a spud vibrator or a vibrating screed. Care must be taken not to touch the sides of the forms with the vibrator.

The resulting floor (see Fig. 11–43) will weigh

FIGURE 11–40   Scaffold-supported shoring for I/D system. *(Courtesy Portland Cement Association.)*

FIGURE 11–41   Shims for truss Ts supported on non-load-bearing partition.

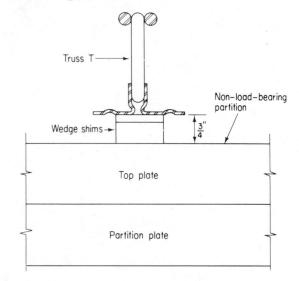

FIGURE 11–42 Placing concrete by pump for I/D floor system. *(Courtesy Portland Cement Association.)*

approximately 53 lb/sq ft with normal-weight concrete or 40 lb/sq ft with lightweight concrete. It provides significant acoustical control, is fire-resistant, is free from squeaks, and allows for the installation of mechanical services between ribs. For more complete details on the system, consult PCA literature on the subject.

FIGURE 11–43 Section through I/D floor. *(Courtesy Portland Cement Association.)*

Concrete Floors with Steel Frame

Concrete floors supported by a steel frame are often simple flat slabs. However, in some cases, concrete fireproofing for the beam may be incorporated with the slab, in which case the structure becomes a beam-and-slab floor. Though the forms shown in Fig. 11–44 are supported from below, the steel frame itself is frequently utilized to support the forms. Steel *hangers* are placed over the top web of the beams and carry the ends of wood joists

FIGURE 11–44 Forming for slab and fireproofing.

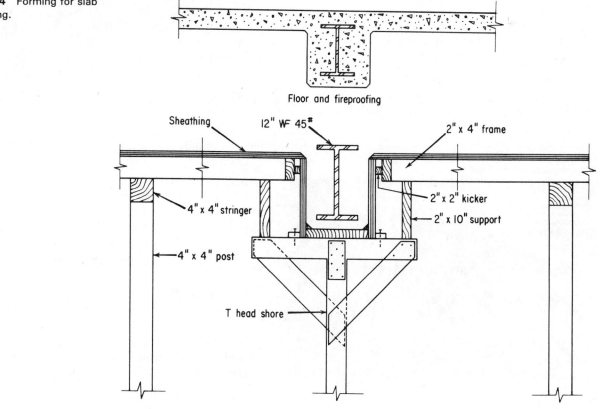

Floor and fireproofing

Sheathing

12" WF 45#

2" x 4" frame

4" x 4" stringer

2" x 2" kicker

2" x 10" support

4" x 4" post

T head shore

spanning from beam to beam. Plywood sheathing laid on the joists provides the base for the slab.

Hangers are available in a number of styles and lengths. Long *loop* hangers, such as those used in Fig. 11–45, allow the top flange and the upper portion of the beam webs to be encased by the slab. *Coil* hangers (Fig. 11–46 make it possible to keep the sheathing up tight under the top flanges, so that the entire slab is above the beams.

FIGURE 11–48 Cellular steel decking. (*Courtesy Bethlehem Steel Co.*)

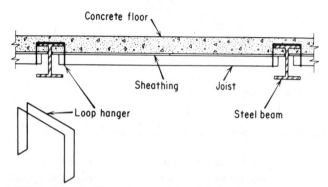

FIGURE 11–45 Floor form suspended by loop hangers.

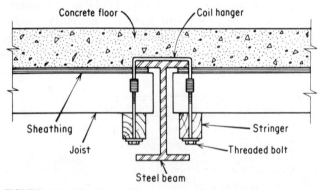

FIGURE 11–46 Floor form suspended by coil hangers.

FIGURE 11–47 Corrugated sheet steel decking. (*Courtesy Bethlehem Steel Co.*)

FIGURE 11–49 Spot-welding steel deck. (*Bethlehem Steel Co.*

Combination concrete and steel floors consist of a sheet steel base covered with a concrete slab. The base may be made of corrugated sheet metal laid over a steel or concrete floor frame (see Fig. 11–47) or it may be one of several styles of specially formed steel decking. Illustrated in Fig. 11–48 is a *cellular*-type deck, but other types consist only of the upper, folded sheet.

Decking of this kind is fastened to a steel frame by spot welding (Fig. 11–49), by plug welding, or by self-tapping screws—for which holes must be predrilled. Connections to a concrete frame may be made by welding to cast-in connection strips or by the use of power-driven pins. Notice in Fig. 11–47 that the studs which fasten the decking to the frame will also act as anchors to the concrete slab.

INDUSTRIAL FLOORING

The structural floors previously described generally require some type of *finish floor*. It may be *wood* (block or strip flooring), *concrete, terrazzo, terra cotta tile, mastic flooring,* or *resilient tile*.

Wood Floors

Strip flooring for industrial use is made from both hardwood and softwood in thicknesses of from 2 to 4 in. Widths of strips vary from 1½ in. for hardwood flooring to 12 in. for heavy softwood flooring. Strips may be single or double tongue-and-grooved or grooved on both edges for *splines*. Strip flooring may be laid over plank or laminated wood subfloors by direct nailing (see Fig. 11–50 or over a concrete subfloor by nailing to *sleepers* which are laid in asphalt or partially cast into concrete (see Fig. 11–51).

Concrete Floors

Concrete finish floors are widely used in industrial and commercial buildings and may consist of a single, finished slab or of a base and topping slab.

A two-course slab is sometimes specified in order to produce a harder, denser, more durable surface. If this is the case, the base slab should be left about 1 in. below

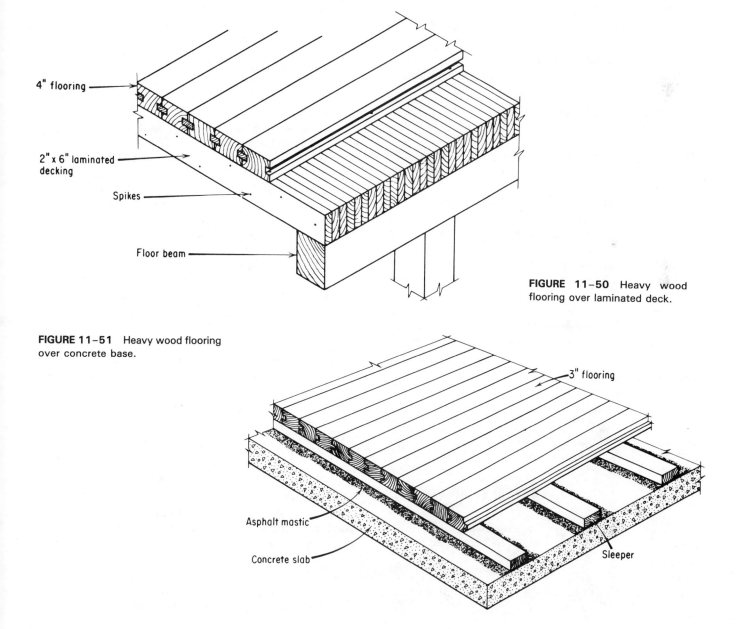

4" flooring

2" x 6" laminated decking

Spikes

Floor beam

FIGURE 11–50 Heavy wood flooring over laminated deck.

FIGURE 11–51 Heavy wood flooring over concrete base.

3" flooring

Asphalt mastic

Concrete slab

Sleeper

the finish grade. Just before the concrete sets, the surface should be roughened with a stiff broom to provide a better bond with the topping.

The topping mix should be made from the hardest, densest aggregates available, with a high design strength and a very low slump (not exceeding 1 in.). Just prior to the application of the topping, the base slab should be washed and a coat of cement paste applied with a stiff brush. The topping mix is spread over the surface, raked, leveled, and tamped (preferably with a power tamper). Notice in Fig. 11–52 that in spite of the dry mix, the machine is bringing enough moisture to the top to produce a smooth surface. Final floating and troweling is best accomplished when done by power machinery.

A very durable surface may be produced on a concrete floor slab by the introduction of metallic aggregate into the topping. Metallic aggregate is composed of iron particles which have been size-graded to within the range of #4–100 mesh sieves, specially processed for ductility, and mixed with a cement-dispersing agent, calcium lignosulfonite.

A series of operations for producing a floor of this type follows:

1. Provide expansion joints around columns and at least every 50 ft in both directions. Reinforce the top of the joints with angle iron (see Fig. 11–53).
2. Design the base concrete for at least 3500 psi.
3. Set the screeds for the base slab 1 in. below the finish grade, place the slab, and leave the surface with a raked finish.
4. Clean and saturate the base slab. It should be saturated from 4 to 6 hr before placing the topping.
5. Set the *screed level pats* for the topping a few hours before the topping slab is to be placed. These are small mounds of mortar set over the surface at frequent intervals with their tops leveled off to the finish grade.

FIGURE 11–52 Power tamper and float. (*Courtesy Portland Cement Association.*)

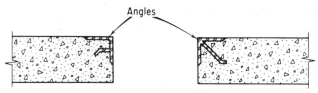

FIGURE 11–53 Reinforcing for expansion joints.

6. For old or hard base coats, a slush bond coat should be brushed into the surface just before placing the topping. This may be made up of 100 lb of expanding metallic aggregate to one bag of cement, mixed with water to form a slurry (about 16 lb of the metallic aggregate will cover 100 sq ft).
7. Design the topping mix. A good design consists of one sack of cement, enough water (in pounds per lb) to produce the specified strength of concrete, and a sufficient quantity of clean, well-graded F.A. and ¾-in. maximum C.A. in equal amounts to produce a mix with not more than 1 in. of slump. This will usually require about 1½ cu ft or 150 lb each of F.A. and C.A. However, there must be enough moisture available to apply the metallic aggregate.
8. Place the topping mix. Rake, screed, level, and tamp. A grill-type tamper is preferred (Fig. 11–54). Remove the level pats as placing progresses, and fill the holes.
9. Immediately after tamping, float the surface to bring moisture to the top.
10. First metallic aggregate shake. Mix one sack of portland cement and two sacks of metallic aggregate together dry and shake evenly over the surface at the following rates; for light duty floors—45 lb/100 sq ft; for moderate duty floors—75 lb/100 sq ft; for heavy duty floors—90 lb/100 sq ft. If the floor is to be finished with mechanical equipment, these amounts may be increased.
11. After the shake has absorbed the surface moisture, tamp the surface. Then follow with a wood float.
12. Apply a second metallic aggregate shake as above, tamp, and float.
13. Immediately after floating, trowel the floor with a steel trowel.
14. As soon as the surface becomes hard enough to ring under it, burnish with a steel trowel to a hard, dense finish.
15. Cure the finished floor as required (see Fig. 11–55).

When a nonslip concrete floor is specified, it may be produced by the addition of an aggregate which is composed of ceramically bonded *aluminum oxide abrasive* to the leveled surface. The following procedure is recommended:

1. Allow the leveled concrete surface to set until it will bear some weight.
2. Soak the abrasive aggregate in water for 10 min before applying.

FIGURE 11-54 Consolidating and leveling a concrete floor. (*Courtesy Portland Cement Association.*)

FIGURE 11-55 Well-finished metallic aggregate floor. (*Courtesy Master Builders Ltd.*)

3. Shake the aggregate evenly over the surface at the rate of ¼ lb/sq ft.
4. Float the aggregate into the surface but do not trowel smooth.

A colored surface may be produced by the addition of a prepared coloring material. It consists of a synthetic mineral oxide of a given color, mixed with fine silica sand or silicon carbide. The coloring procedure is as follows:

1. As soon as the slab is placed, float the surface with a wood float.
2. Shake the coloring material evenly over the surface at the rate of ¼ lb/sq ft.
3. Float the surface until the color is evenly distributed and worked into the topping.
4. Immediately apply another shake at the same rate.

5. Float again until the color has been worked into the surface to a depth of approximately ¼ in.
6. When the concrete has set sufficiently, trowel smooth as required.

Terra Cotta Tile

Tile flooring consists of a layer of some type of ceramic tile, from ¼ to 1 in. thick, laid over a concrete base slab. The tile is laid in a mortar bed spread evenly over the surface, and the spaces between tiles are filled with grout.

The procedure for laying terra cotta tile is as follows:

1. Wash and saturate the base slab. If the concrete is old or hard, apply a slush bond coat of cement and water with a stiff brush.
2. Stretch two lines across the floor at right angles to one another so that the area is divided into four equal parts.
3. Mix the bedding mortar in the proportion of 1 part cement to 3 parts sand, with enough water to make a plastic, workable mix.
4. Start at the intersection of the two lines, apply a layer of mortar, and lay the first row of tiles to the line. Tap each one firmly so that it is well bedded in mortar.
5. Lay a second row at right angles to the first, along the second line.
6. Lay the remainder of the tiles in that quarter, keeping them as level as possible and maintaining an even spacing between tiles.
7. Repeat in the opposite quarter and then lay the other two quarters.
8. When the tiles are set, prepare a grouting mix using the same proportions of cement and sand but make the mortar a little more fluid. Pour some over a section of tiled

surface and rub it into the spaces between tiles with a piece of heavy burlap.

9. Rub off the excess grout, and when it has set sufficiently so as not to pull out of the spaces, clean the surface of the tiles thoroughly.

Mastic Floors

Mastic flooring materials are applied to the base floor in a stiff plastic state by spreading, rolling, and troweling. Three types of material are commonly used: *magnesite, asphalt,* and *epoxy resin* compounds.

Magnesite flooring is composed of calcined *magnesium oxide* and *magnesium chloride,* mixed with water to form a stiff plaster. The material is applied in two coats, each about ¼ in. thick. It may be installed over either wood or concrete; when used over a wood subfloor, metal lath is laid first to provide a better bond.

The first coat of magnesite should contain some fibrous material such as asbestos or nylon to give it greater strength. The second coat—which may contain color, fine marble chips, or an abrasive aggregate—is placed as soon as the base coat has set. It is screeded, compacted, and troweled in much the same way as concrete topping.

Asphalt flooring consists of emulsified asphalt containing asbestos fiber mixed with portland cement and stone aggregates to form a stiff plastic material. It is spread over the base floor and rolled smooth and level. Asphalt flooring may be laid over either wood or concrete bases, though the application is somewhat different in each case. For a wood base, a layer of 15-lb asphalt-saturated felt is laid first with 4-in. laps. Stucco wire is laid over the paper and nailed down, followed by a coat of asphalt primer.

A fill mix is then applied to level the surface and cover the stucco wire. It consists of 1 part portland cement, 1 part emulsified fibrated asphalt, and 5 parts clean, coarse sand. The final coat is the same as that used over a concrete floor.

A concrete base slab should first be coated with asphalt primer thinned with equal parts of mineral spirits. When it is dry, a slurry coat consisting of 1 gal of portland cement, 1 gal of emulsified asphalt primer, and 15 gal of water is painted over the surface. The flooring mix consists of 1 part portland cement, 2 parts emulsified fibrated asphalt, 2½ parts clean coarse sand, 4½ parts ⅛- to ⅜-in. pea gravel, and not more than 1 part water. This mixture is spread, leveled, compacted, and rolled (to a depth of from ⅝ to ¾ in.) to a smooth finish.

Epoxy resin is a synthetic now used for several purposes in construction. It may be used as a basic ingredient of floor topping. It is made of two components—a liquid resin and a curing agent. These must be mixed in prescribed proportions when the topping mixture is made.

Two types of floor topping can be made with this plastic. Kiln-dry, salt-free sand and portland cement may be mixed with it to produce a smooth floor topping. The mix is made up of 182½ lb of liquid resin, 17½ lb of curing agent, 160 lb of portland cement, and 640 lb of 30-50 mesh sand. The topping may be colored by adding pigments to the liquid portion. The surface is first primed with an epoxy prime coat, and the mix is then spread over the floor, floated, and troweled smooth to a depth of from ⅛ to ¼ in. Mixes should be limited to 100-lb batches because of the limited pot life of the mixed topping.

A terrazzo topping may be prepared by mixing marble chips with liquid plastic. Coloring pigment may also be added. These ingredients are mixed in the following ratios: 160 lb of liquid plastic, 87 lb of calcium silicate, 35 lb of rutile titanium dioxide paste, and 18 lb of hardener per 700 lb of marble chips. The surface must again be primed, and the mix must be applied and troweled to a depth of from ³⁄₁₆ to ¼ in. After 24 hr, the surface is ready for grinding and polishing.

Resilient Tile

Resilient tile flooring includes a number of products such as asphalt, linoleum, cork, vinyl or vinyl-asbestos, and rubber. All require a smooth surface—either wood or concrete—and a special mastic. Asphalt and vinyl-asbestos tiles are particularly suited for concrete slabs on grade, since both are laid in moistureproof asphalt mastic. Laying should in all cases begin at a line stretched across the center of the floor.

TERRAZZO FLOORS

Terrazzo is defined as a composition material, cast-in-place or precast, which is used for floor topping and wall treatments. It consists of marble chips, seeded or unseeded, with a binder or *matrix* which may be cementitious, noncementitious, or a combination of both. The terrazzo is cast, cured and then ground and polished or otherwise finished. *Rustic terrazo* is a variation in which quartz, quartzite, onyx, or granite chips may be substituted for the marble chips, and, in place of grinding and polishing, the surface is washed with water or otherwise treated to expose the stone chips.

There are a number of types of terrazzo installations, including *sand cushion, bonded, monolithic,* and *structural,* in which portland cement paste is used as the binder for the topping (see Fig. 11–57). In addition, there are a number in which emulsions of plastic resins such as *epoxy, polyester,* and *polyacrylate* or combinations of these with portland cement paste are used as the binder.

In any of these types of installations, *divider strips*

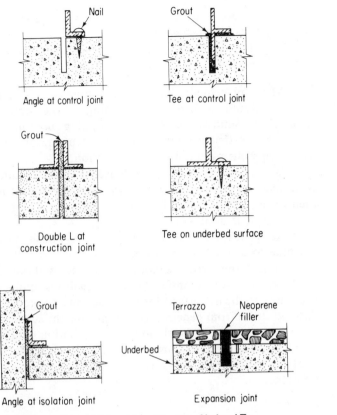

FIGURE 11-56 Divider strips. (*Courtesy National Terrazzo and Mosaic Association Inc.*)

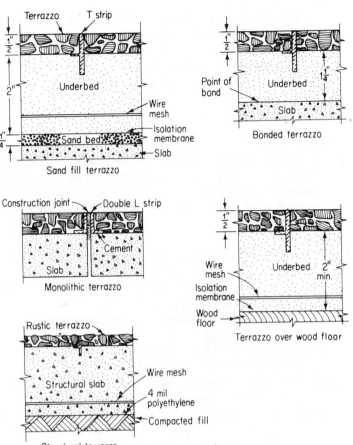

FIGURE 11-57 Terrazzo installation types. (*Courtesy National Terrazzo and Mosaic Association Inc.*)

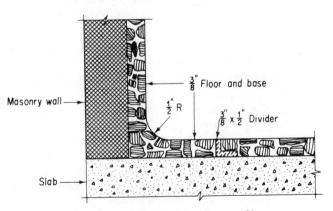

FIGURE 11-58 Polyacrylate terrazzo. (*Courtesy National Terrazzo and Mosaic Association Inc.*)

are used to control cracking of the terrazzo surface due to expansion and contraction of the base floor. Divider strips are of two basic types: those that are grouted into openings cut in the base slab and those that are cemented or nailed to the surface (tees and angles) (see Fig. 11–56). They are available in half hard brass, white alloy zinc, and plastic. In addition, there is an expansion joint strip available, which is recommended for use over all expansion joints in the base slab.

Divider strips should be installed over all *control joints, construction joints,* and *isolation joints* in the base slab and, in addition, may be set on the surface where there are no joints or cuts, in order to divide the floor area into squares or rectangles containing 200 to 300 sq ft. Rectangles should be not more than 50% longer than wide and squares not more than 16 by 16 ft.

Terrazzo Installation

Terrazzo installation procedures will vary somewhat, depending on the type of installation (see Fig. 11–57). For example, monolithic terrazzo, as the name implies, is placed at the same time as the base slab and bonded directly to it (see Fig. 11–58). As a result, any structural movement occurring in the slab is likely to cause dam-

age to the finished floor. To minimize any such damage, the following steps should be taken when placing the slab:

1. For on-grade slabs (structural terrazzo) the fill should be thoroughly compacted.
2. For on-grade slabs particularly, a power screed should be used to bring some cement paste to the surface to seal off surface pores.

3. All plumbing and electrical conduits should be placed neatly and rise at right angles through the slab.

4. When the slab has set enough to walk on, it should be broomed to roughen it to ensure a good bond.

5. In large areas, the concrete should be placed in alternating slabs, to extend more control over expansion and contraction.

For any type of terrazzo installation, great care must be taken in placing the subfloor. In most cases, specifications will limit the variation in level to not more than ¼ in. in 10 ft, though in the case of monolithic terrazzo, for example, the variation will be limited to ⅛ in. in 10 ft.

The installation of sand cushion terrazzo is a typical operation. It is normally 2½ to 3 in. thick, with ½-in. terrazzo topping, 2- to 2¼-in. underbed, and ¼-in. sand cushion (see Fig. 11–57).

To install sand cushion terrazzo, proceed as follows:

1. Cover the entire slab surface with clean, dry sand to a maximum depth of ¼ in. and level uniformly.

2. Install isolation membrane over sand with 3-in. overlap at ends and edges.

3. Install welded wire reinforcement, overlapping two squares at ends and edges and stopping the mesh 1 in. short of expansion joints.

4. Place concrete underbed. This is a mixture made from 1 part portland cement to 4½ parts sand, with just enough water to provide workability at as low a slump as possible.

5. Screed the under bed to ½ in. below the finished floor level.

6. Install control joint strips precisely over all joints in the slab and divider strips as shown in the drawings and trowel firmly along the edges to assure positive anchorage.

7. Slush the underbed with neat cement paste, colored as specified for the topping, and broom the paste into the underbed surface.

8. Place the terrazzo mixture in the panels formed by divider strips. The topping is mixed at the rate of 1 lb of cement to 2 lb of chips, with approximately 0.6 lb of water and color pigment as specified to produce the color required, to a uniform, workable consistency.

9. Trowel the mixture to the level of the top of divider strips.

10. Seed the troweled surface with additional chips in the same proportions as contained in the terrazzo mix and trowel.

11. Roll the seeded surface with heavy rollers until all excess water has been extracted.

12. Trowel to a uniform surface, disclosing the lines of the divider strips.

After placement has been completed, curing material should be installed according to the manufacturer's recommendations and the topping cured until it develops sufficient strength to prevent lifting of terrazzo chips during the grinding operation.

When the terrazzo has hardened, it is wet-ground with machine grinders to a smooth, even surface. It is then washed, and a grout of cement and water, colored if necessary, is applied to fill any voids. After about 3 days the grout is removed and the surface is polished by machine to a satisfactory finish.

Finally, the surface is thoroughly washed and rinsed and allowed to dry completely, and a sealer is applied according to the manufacturer's instructions, to prevent dirt and stains from penetrating and marring the surface.

For complete information and installation details for all types of terrazzo consult *Terrazzo Technical Data*, published by the National Terrazzo and Mosaic Association, Inc.

REVIEW QUESTIONS

1. Briefly describe two major types of wood subfloors.

2. Explain the difference between **(a)** a *flat slab* and a *flat plate* concrete floor, **(b)** a *one-way joist* slab and a *waffle* slab floor, and **(c)** structural floor slab and a slab-on-grade.

3. Outline the reasons for **(a)** isolating columns from a concrete floor around them, **(b)** providing a *key* between two adjacent concrete floor slab sections, **(c)** covering the subgrade with a moisture barrier before casting a concrete slab-on-grade, and **(d)** providing chairs for reinforcing bars in a concrete floor slab.

4. Explain the purpose of **(a)** *metallic aggregate* in a concrete floor topping, **(b)** using *block flooring* in a wood floor specification, **(c)** a *column capital* in a flat slab floor design, and **(d)** using a *ribbed slab* floor design.

5. Define each of the following terms: **(a)** *broken bond* in terrazzo flooring, **(b)** *mastic* flooring, **(c)** and *sand cushion* in terrazzo flooring.

Roof Systems and Industrial Roofing

The primary purpose of a roof is to protect a building's interior, but it may also be used to contribute to a building's exterior appearance.

The completed roof consists of several components, including the *roof frame, roof deck, vapor barrier, insulation, waterproof roofing material, flashing* and *drains, construction* and *control joints, walks, parapets, gravel stops,* etc.

A great many factors enter into the design of a roof, such as weather, appearance, height, area, and style of the frame. As a result, many roofing systems have been designed, some used very frequently, others less so.

FLAT ROOF SYSTEMS

The simplest roof system is the frame for a flat roof, in *wood, steel,* or *reinforced concrete*. The frame supports a roof deck, which, in turn, is covered with a waterproof roofing material.

Wood Roof Frame

In heavy timber frame construction, the roof members for a flat roof are framed to columns as outlined in Chapter 8. A positive roof anchoring system must be employed to protect against displacement by wind. Figures 8–9 to 8–11 illustrate methods of attaching roof beams to masonry exterior bearing walls. Roof anchors are again

used. Note the *anchors* on the bearing plates (Fig. 8–10) to hold the roof beams in place. The framing of roof members in steel or reinforced concrete frame structures is essentially the same as that used for floors.

In place of regular roof beams, *joists* may be used to frame a flat roof (see Fig. 12–1). Wood joists, 12 or 16 in. o.c., span from girder to girder or from one bearing wall to another. Spans will generally be limited to about 24 ft.

A third type of wood roof frame is made up of prefabricated units like those illustrated in Fig. 12–2. The units are 16 in. wide; vary in depth from 6 to 16 in., depending on the span; and have a maximum span of about 30 ft. They may be handled singly or made up into panels 32 or 48 in. wide. Panel ends may rest on bearing walls or be carried on supporting girders. Concrete and unit masonry bearing walls and concrete or steel supporting girders must be capped with a wood pad, to which the panels may be secured.

Open Web Joist Roof Frame

Open web joists are light structural members composed of top and bottom chords, connected by rod or tubular steel webs. There are two main types in general use: open web *steel* joists (see Fig. 12–3) and open web *combination* joists, in which the chords are of wood and the webs of tubular steel (see Fig. 12–4).

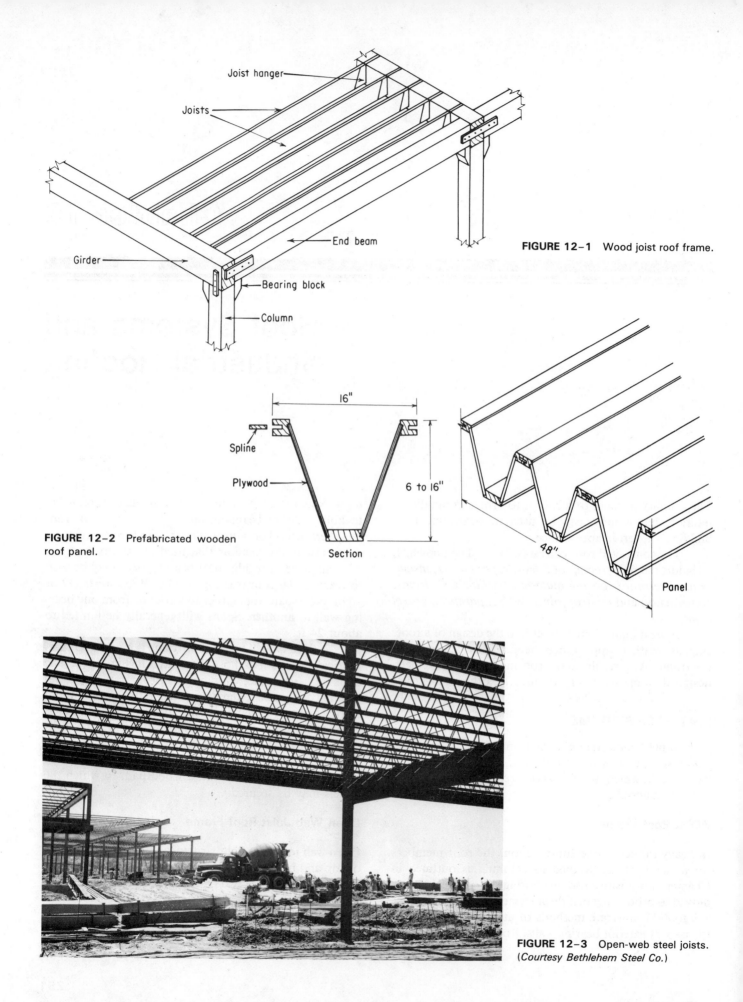

FIGURE 12-1 Wood joist roof frame.

Joist hanger

Joists

End beam

Girder

Bearing block

Column

FIGURE 12-2 Prefabricated wooden roof panel.

16"

Spline

Plywood

6 to 16"

Section

48"

Panel

FIGURE 12-3 Open-web steel joists. (*Courtesy Bethlehem Steel Co.*)

FIGURE 12-4 Combination wood and steel open-web joists. (*Courtesy Trujoist Western Ltd.*)

Open web steel joists up to 96 ft in length and 4 ft in depth, spaced 12 to 30 in. o.c., span from girder to girder (as shown in Fig. 12–3) or from one bearing wall to another. Joist ends may be welded to a steel girder or to a connection plate cast into the top of a concrete girder or wall, bolted to a wood or concrete wall, or built into pockets—*chases*—in a brick wall.

Combination joists span up to 50 ft with single top and bottom chords on the flat, as in Fig. 12–4, and up to 100 ft with double top and bottom chords on edge. A wooden pad on the bearing surface of the supports simplifies the anchoring of this type of joist.

Roof overhangs may be incorporated into the system by using web joists with ends which project (can-

tilever) beyond the outside bearing member, as in Fig. 12–5.

Structural Steel Roof Frame

In a structural steel frame building, the roof frame will very likely be fabricated with the same type of member as used in the rest of the building but in a smaller size or lighter weight. In the building frame illustrated in Fig. 12–6, the roof is framed with steel girders and light intermediate beams. Projecting eaves may be formed in conjunction with a steel frame, as indicated in Fig. 12–7.

Instead of using regular rolled steel shapes, a steel roof frame may be built up with light steel framing

FIGURE 12-5 Sloping soffit formed with overhanging open-web joists. (*Courtesy Trujoist Western Ltd.*)

FIGURE 12-6 Structural steel roof frame. (*Courtesy Bethlehem Steel Co.*)

FIGURE 12-7 Projecting eaves with steel frame. (*Courtesy Bethlehem Steel Co.*)

members, like those illustrated in Fig. 12-8. The main joists are usually tack-welded in place, while the cross members ride in joist hangers and are held in place with screws.

Monolithic Frame and Roof Slab Systems

Flat slabs, flat plates, ribbed slabs, and waffle slabs have all been described in the discussions on floor systems in Chapter 11, and the same techniques may be used to produce roof slabs (see Fig. 12-9).

FIGURE 12-8 Light steel joists.

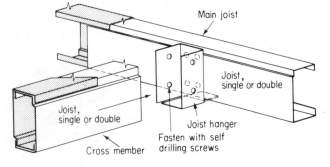

Main joist

Joist, single or double

Joist, single or double

Joist hanger

Fasten with self drilling screws

Cross member

FIGURE 12-9 Flat plate roof.

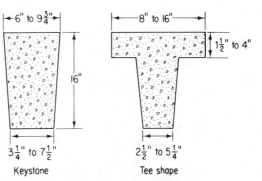

FIGURE 12-10 Prestressed concrete joists. (*Courtesy Portland Cement Association.*)

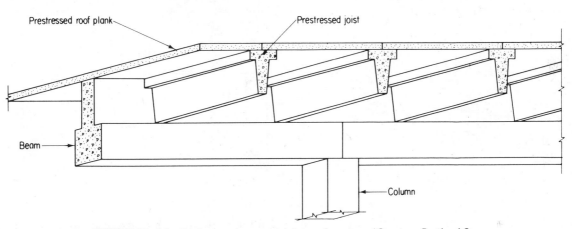

FIGURE 12-11 Prestressed concretejoist roof system. (*Courtesy Portland Cement Association.*)

Flat slabs are well suited for heavy roof loads and may be particularly useful for accommodating rooftop parking. Flat plates provide a continuous ceiling and offer complete flexibility for partition arrangement.

Precast Concrete Roof Systems

1. *Prestressed Concrete Joists.* For projects involving relatively light loads and short spans, *prestressed concrete joists* may be used as a roof frame and, in combination with prestressed concrete planks or other structural decking systems, produce an economical and functional roof. The joists are available in two basic shapes: *keystone* and *tee shape* (see. Fig. 12-10).

The keystone joist is available in depths of from 6 to 18 in., with spans up to 36 ft, while the tee joist comes in depths of from 8 to 20 in., with spans from 20 to 60 ft. Figure 12-11 illustrates typical roof construction with tee joists and prestressed planks.

Joists may be supported on bearing walls or on prestressed interior and exterior beams cast as part of the

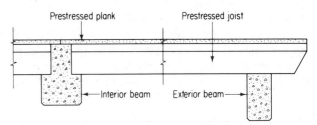

FIGURE 12-12 Support system for prestressed concrete joists.

system (see Fig. 12-12). Table 12-1 gives typical prestressed beam sizes for beams used in a prestressed concrete joist roof system.

2. *Single-Tee Roof Slab.* Single tees are among the most common products of the precast concrete industry and widely used as a roof system. They may be set edge to edge, as in Fig. 12-13, or spaced, with the spaces between filled with cast-in-place concrete or concrete planks.

TABLE 12-1 Typical prestressed beam sizes (in.) for concrete joist system[a]

| Span of beam (ft) | 16 in. keystone joist, 4 ft o.c., 28-ft span | | 16 in. tee joist, 4 ft o.c., 36-ft span | | 20 in. tee joist, 4 ft o.c., 50-ft span | |
	Interior beam, $b \times h$	Exterior beam, $b \times h$	Interior beam, $b \times h$	Exterior beam, $b \times h$	Interior beam, $b \times h$	Exterior beam, $b \times h$
20	12 × 20[b] **12 × 18**[b]	12 × 16	12 × 24	12 × 18 **12 × 16**	12 × 28	12 × 20
28	12 × 28 **12 × 26**	12 × 22 **12 × 20**	12 × 32 **12 × 30**	12 × 24 **12 × 22**	12 × 40 **12 × 38**	12 × 28
36	12 × 40 **12 × 36**	12 × 28 **12 × 36**	12 × 42 **12 × 40**	12 × 32 **12 × 30**	12 × 52 **12 × 48**	12 × 38 **12 × 36**

[a]Sizes in boldface have deflected strands.
[b]Joist must be notched for this beam if tops are flush.
Source: Courtesy Portland Cement Association.

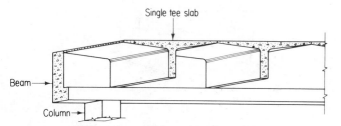

FIGURE 12-13 Single-tee roof slab system. (*Courtesy Portland Cement Association.*)

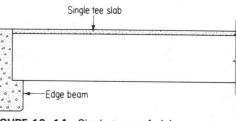

FIGURE 12-14 Single-tee roof slab support.

Flange widths range from 4 to 10 ft although the 8-ft section is most commonly used. Depths range from 12 to 48 in. or more, with the 36-in. depth being a popular one. Spans are normally from 30 to 100 ft, though longer spans are possible under certain circumstances.

The ends of single tee roof slabs are usually supported on edge beams, as illustrated in Fig. 12-14, though they may be carried on bearing walls as well. Prestressed edge beams are cast in typical sizes, indicated in Table 12-2.

3. *Double-Tee Roof Slab.* Double tees are probably the most widely used prestressed concrete products

in the medium span range of buildings (see Fig. 12-15). They are available in various widths and depths, depending on the manufacturer. Widths are generally 4, 5, 6, 8, and 10 ft, with the 4- and 8-ft sections probably the most popular. Depths range from 6 to 36 in., but in most cases, the maximum depth available will be the most economical. The 4-ft double tees are commonly used for spans up to 60 ft, while the 8-ft units, which usually have a deeper web, span 70 ft or more.

Supports for double tees consist of bearing walls and interior beams or edge beams and interior beams (see Fig. 12-16). Typical prestressed beam sizes are indicated in Table 12-3.

TABLE 12-2 Typical prestressed beam sizes (in.) for single-tee slab system

Span of beam (ft)	12-in. single-tee, 30-ft span	16-in. single tee, 45-ft span	20-in. single-tee, 60-ft span	28-in. single-tee, 75-ft span
20	12 × 16[a]	12 × 16[a]	12 × 24[a]	
24	12 × 20	12 × 24	12 × 28	12 × 32[a]
32	12 × 28	12 × 32	12 × 40	12 × 44
40	12 × 36	12 × 40	12 × 48	12 × 56

[a]The stem of the tee should be notched for this beam.
Source: Courtesy Portland Cement Association.

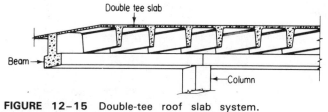

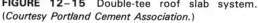

FIGURE 12–15 Double-tee roof slab system. (*Courtesy Portland Cement Association.*)

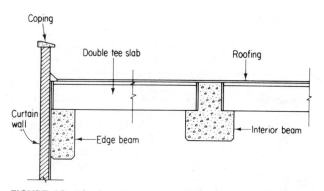

FIGURE 12–16 Double-tee roof slab supports.

TABLE 12–3 Typical prestressed beam sizes (in.) for double-tee slab system

Span of beam (ft)	10-in. double tee, 30-ft span		12-in. double tee, 40-ft span		14-in. double tee, 50-ft span	
	Interior beam	Edge beam	Interior beam	Edge beam	Interior beam	Edge beam
20	12 × 20	12 × 16	12 × 22	12 × 18	12 × 26	12 × 20
25	12 × 24	12 × 20	12 × 28	12 × 22	12 × 30	12 × 26
30	12 × 28	12 × 24	12 × 32	12 × 28	12 × 36	12 × 32
35	12 × 32	12 × 28	12 × 38	12 × 34	12 × 44	12 × 38
40	12 × 38	12 × 34	12 × 44	12 × 38	12 × 50	12 × 42

Source: Courtesy Portland Cement Association.

4. *Cored Slabs.* The production of *hollow core slabs* is highly mechanized, and as a result their cost is relatively low. Depths range from 4 to 12 in., but 6- and 8-in. depths are most commonly used. Widths range from 16 to 48 in., in limited increments, depending on the manufacturer. Spans may reach to 50 ft, but the general range is below 35 ft.

Cored slabs are quick and easy to install and provide a finished, flush ceiling (see Fig. 12–17). They may be used with almost any type of framing system, including prestressed edge and interior beams (see Fig. 12–18).

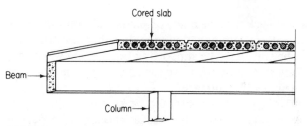

FIGURE 12–17 Cored-slab roof system. (*Courtesy Portland Cement Association.*)

FIGURE 12–18 Cored-slab support system. (*Courtesy Portland Cement Association.*)

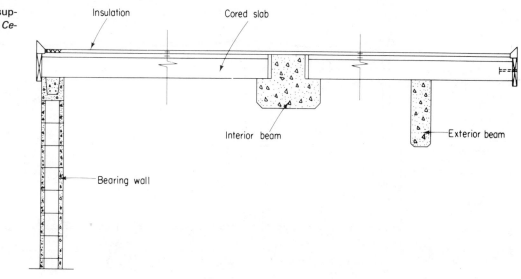

TABLE 12–4 Typical prestressed beam sizes (in.) for cored slab roof

Span of beam (ft)	6-in. slab, 28-ft span, DL + LL = 80 psf		8-in. slab, 34-ft span, DL + LL = 90 psf		10-in. slab, 40-ft span, DL + LL = 100 psf		12-in. slab, 46-ft span, DL + LL = 110 psf	
	Interior	Exterior	Interior	Exterior	Interior[a]	Exterior	Interior[a]	Exterior
20	12 × 20	12 × 18	12 × 24	12 × 20	12 × 24	18 × 18	12 × 28	18 × 20
24	12 × 24	12 × 20	12 × 28	12 × 24	12 × 28	18 × 22	12 × 32	18 × 24
28	12 × 28	12 × 24	12 × 32	12 × 28	12 × 34	18 × 26	12 × 38	18 × 28
32	12 × 32	12 × 26	12 × 38	12 × 32	12 × 38	18 × 28	12 × 44	18 × 32

[a]Beam has heavy prestressing.

Source: Courtesy Portland Cement Association.

Typical precast beam sizes to suit various depths of slab and slab load are listed in Table 12–4.

ROOF DECKS

The roof frame supports a roof deck of *wood, concrete, steel, gypsum,* or lightweight *cellular concrete,* anchored to the frame and providing a solid surface for the application of roofing material.

Wood Deck

A wood deck may be made of planks laid on the flat or 2-in. material laminated on edge. In either case, the deck must be spiked to the supporting beams, and when they are other than wood, a wooden pad must be provided on their bearing surface.

Solid wood decks should be kept back ½ in. from parapet walls or from the fascia where a flush deck is used with a cant strip to cover the gap. In the case of a flush deck, the cant strip will also act as a gravel stop.

Concrete Roof Deck

A concrete roof deck may be placed over *removable wood forms,* over *steel decking* laid on the roof frame, or over *paper-backed wire mesh* stretched over roof framing members spaced 16 in. o.c.

The removable wood forms are used in conjunction with a steel frame and are usually supported as illustrated in Figs. 11–44 and 11–45. Steel decking will be similar to that shown in Figs. 11–46 and 11–48. Paper-backed wire mesh consists of 1- to 1½-sq in. mesh to which is attached a layer of strong, heavy, impregnated paper. One end of the mesh is fastened to the frame at the wall, stretched across the framing members, and stapled to each one. Concrete is then placed over the surface, with the mesh acting as reinforcement (see Fig. 12–19).

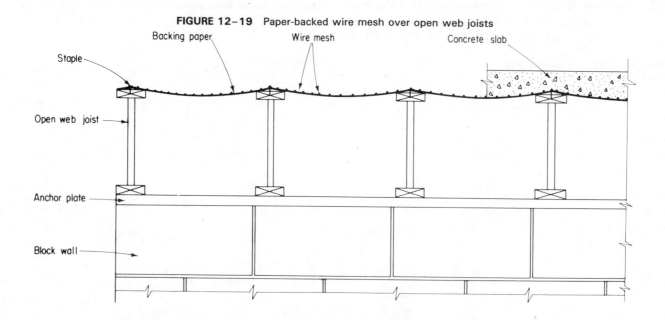

FIGURE 12–19 Paper-backed wire mesh over open web joists

Steel Roof Deck

A steel deck is most frequently used in conjunction with a steel frame and is laid and covered as shown in Fig. 12–20. The decking may be secured by *spot welding,* by *self-tapping screws,* by *powder-driven pins,* or by *nails* if the frame is wood.

Gypsum Roof Deck

Gypsum decks are made from poured-in-place gypsum or precast gypsum planks. To use poured-in-place gypsum, the roof frame is made of structural Ts spaced 33 in. o.c.; ½-in. gypsum board 32 in. wide is then laid over the flanges of the steel members (Fig. 12–21). Welded wire mesh is draped over the Ts, and freshly mixed gypsum is poured and screeded to the proper level.

Gypsum planks are *plain* or *metal edged.* Both are supported on steel roof framing members, but in the case of metal edged planks it is not necessary for the ends to rest on a beam.

Cellular Concrete Roof Deck

Cellular concrete is used for roof decks as precast slabs 18 in, wide, 4 in. thick, and up to 10 ft in length. Slabs may be used over any type of roof frame and are fastened with nails, bolts and washers, galvanized metal

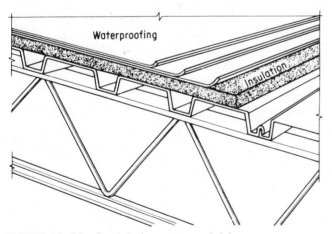

FIGURE 12–20   Steel deck on open web joists.

FIGURE 12–21   Form for gypsum deck.

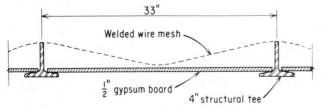

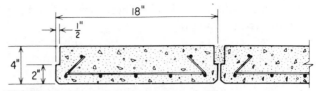

FIGURE 12–22   Cellular concrete roof slab.

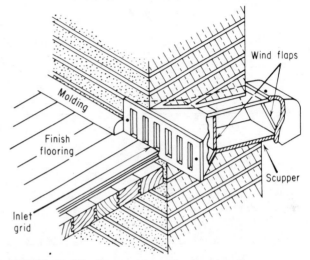

FIGURE 12–23   Scupper through parapet wall.

clips, or powder-driven steel pins. Grooves between slabs are later filled with grout (see Fig. 12–22).

Flat Roof Drainage

Provisions must be made for draining a flat roof. This may be done by installing roof drains at various places and sloping the roof surface toward them. The roof may also be left perfectly flat with a parapet all around so that it retains water. Roof drains must then project several inches above the surface so that excess water may drain. The roof is sometimes sloped in one direction so that water may flow into a collecting gutter. Another method is to provide drainage outlets through parapet walls. These outlets are called *scuppers,* and their installation is illustrated in Fig. 12–23.

TRUSS ROOF

Another method of framing the roof of a building is by the use of *trusses.* These are structural frames designed to carry the roof loads and transmit them to the bearing walls on which they rest. They are generally made of straight members arranged and fastened together in triangular form, so that the stresses in the members caused by loads at the *panel points* are either compressive or tensile. A number of truss styles have been designed and Fig. 12-24 illustrates some of them.

Trusses consist basically of upper and lower *chords,* tension and compression *webs,* and the connectors used to fasten the members together (see Fig. 12-25).

Two methods are used to connect the members of a wooden truss. In one case members are butted together, and in the other they are lapped. If members are butted to one another, a wood or steel *gusset plate* or *nail plate* is fastened over the joint (see Fig. 12-26). A wood plate is fastened with nails and glue, a steel plate with bolts. A nail plate is a flat plate studded with spikes on one side. It is laid over the joint, and the spikes are driven or pressed into the wood. If the members overlap one another, *timber connectors* and bolts are used to connect them. The connectors are set in circular grooves cut in

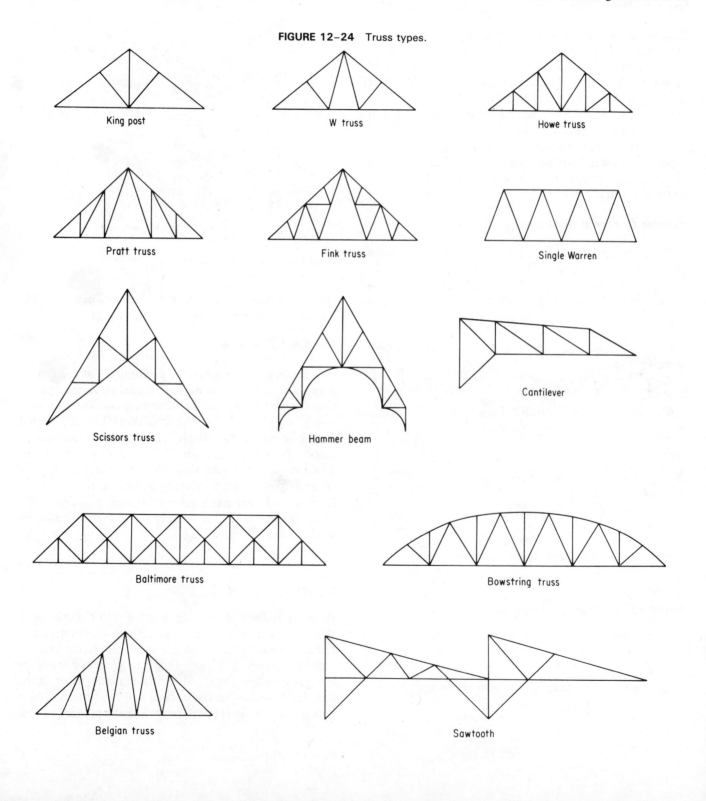

FIGURE 12-24 Truss types.

King post

W truss

Howe truss

Pratt truss

Fink truss

Single Warren

Scissors truss

Hammer beam

Cantilever

Baltimore truss

Bowstring truss

Belgian truss

Sawtooth

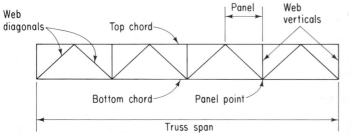

FIGURE 12-25 Parts of a truss.

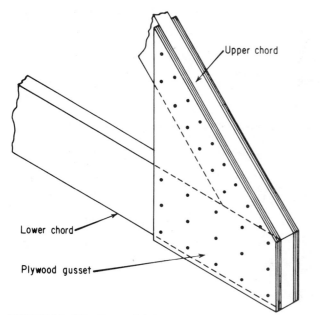

FIGURE 12-26 Gusset joint.

the meeting faces (Fig. 12–27), and a bolt through the assembly holds the two members tightly together. Grooves are cut to a predetermined depth by a special tool which drills the bolt hole at the same time. (See Fig. 8–4).

FIGURE 12-28 Bowstring trusses with joists.

FIGURE 12-27 Cutting grooves for split rings.

Roof sheathing may be applied directly to trusses, but is more often supported by *purlins* which span from truss to truss at panel points or by joists resting on the upper chord (Fig. 12–28).

FIGURE 12–29 Stressed skin roof panel. (*Courtesy Trujoist Western Ltd.*)

Rather than using sheathing supported by purlins or joists, large trusses may be covered by stressed-skin panels, such as that shown in Fig. 12–29, which will span from truss to truss and fasten directly to the upper chords.

FOLDED PLATE ROOF

A folded plate roof is one in which the roof slab has been formed into a thin, self-supporting structure, usually made of either wood or concrete. Two of the most common shapes are *W shape* and *V shape* (see Fig. 12–30).

This style of roof is being used more and more to provide large areas of column-free space. Such roofs are capable of long spans, and their cantilever projections can be used to advantage in exterior design, as well as to counterbalance the span.

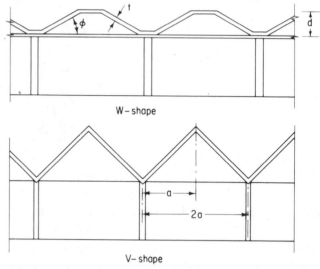

FIGURE 12–30 Folded plate roof shapes.

FIGURE 12–31 Form for concrete folded plate.

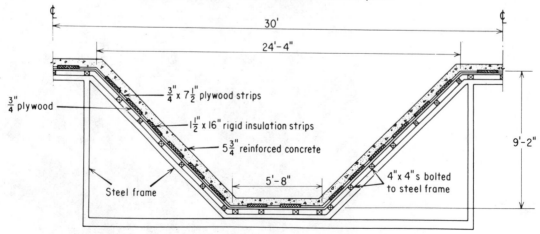

For reinforced concrete folded plates, the slope of plates is from 25 to 45° maximum, with a shell thickness of from 3 to 6 in. The span-to-depth ratio varies from 1:10 to 1:15.

A concrete roof of this type can be made with precast panels or may be cast in place. Figure 12–31 shows details of typical formwork for a cast-in-place concrete folded plate roof.

Span data for W-shaped and V-shaped folded plates, based on dimensions shown in Fig. 12–30, are given in Tables 12–5 and 12–6.

A wood folded plate roof is made up of panels bolted together to form a complete roof structure. The top and bottom chords of each panel are usually laminated members, and the ribs are $2'' \times 4''$s, $2'' \times 6''$s, or $2'' \times 8''$s, as the case may be, 16 in. o.c., with a plywood skin on both sides of the panel. Figure 12–32 shows details in the construction of a typical panel.

TABLE 12–5 Span data for W-shaped folded plate

Span (ft)	Φ (deg)		d (ft)		$2a^a$ (ft)	t (in.) (average)	Reinforcement (lb/sq ft of projected area)
	Min.	Max.	Min.	Max.			
40	30	45	2½	5	20	3	1.5–2.0
60	—	45	—	6	25	3	2.0–3.0
	30	—	4	—		3½	
75	—	45	—	7½	30	3	2.5–4.0
	30	—	5	—		4	
100	—	45	—	10	40	4	4.0–6.0
	30	—	6½	—		5	

[a]Values shown may vary with architectural design.
Source: Courtesy Portland Cement Association.

TABLE 12–6 Span data for V-shaped folded plate

Span (ft)	Φ (deg)		d (ft)		$2a^a$ (ft)	t (in.) (average)	Reinforcement (lb/sq ft of projected area)
	Min.	Max.	Min.	Max.			
40	25	45	2¾	4	20	3	1.2–1.6
60	—	45	—	6	20	4	1.9–2.7
	25	—	4	—		6	
75	—	45	—	7½	25	4	2.6–3.7
	25	—	5	—		6	
100	—	45	—	10	30	5	4.0–5.2
	25	—	6¾	—		6	

[a]Values shown may vary with architectural design.
Source: Courtesy Portland Cement Association.

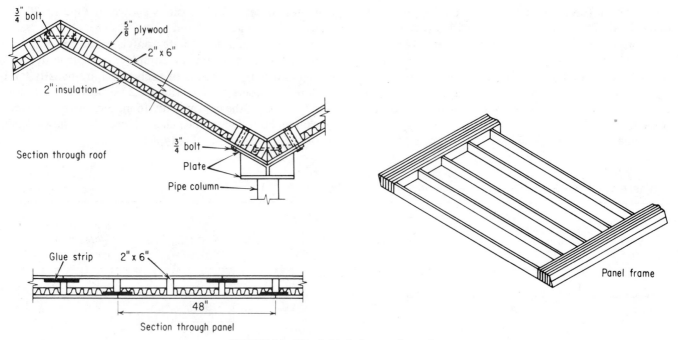

Section through roof

Glue strip 2" x 6"

48"

Section through panel

FIGURE 12-32 Folded plate roof panel.

LONG-BARREL SHELL ROOF

A long-barrel shell is a curved, thin-shell roof, made from either concrete or wood. There is a variety of shapes, ranging from a low-rise arc to a high-rise cylindrical arch (Fig. 12–33). These roofs provide long, uncluttered spans beyond the normal range of other structural roof systems, as well as providing unique architectural features.

Forms for concrete shells consist of a series of light bowstring trusses of the proper span and height, supported on shores or metal scaffolding and sheathed over the top with plywood. Shells are normally reinforced, but post-tensioning may be introduced into the shell and transverse beams to allow a reduction in shell thickness.

Chord widths range to about 60 ft, while lengths vary from 40 to 160 ft. The usual span-to-depth ratio varies from 1:10 to 1:15. Shell thickness is normally about 3 in. Long-barrel shell span data for concrete shells are given in Table 12–7, based on dimensions shown in Fig. 12–33.

Wooden long-barrel shells are made up of a frame shaped to fit the curve (see Fig. 12–34), covered inside and out with a plywood skin. Each barrel rests on *valley beams,* like those shown in Fig. 12–35.

FIGURE 12-33 Multiple long-barrel shells. (*Courtesy Portland Cement Association.*)

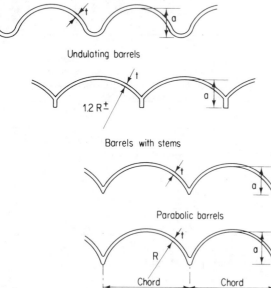

Undulating barrels

Barrels with stems

Parabolic barrels

Cylindrical barrels

TABLE 12-7 Span data for concrete long-barrel roof shells

Span (ft)	Chord width (ft)	a (ft)	R (ft)	t (in)	Reinforcement (lb/sq ft of projected area)
80	30	8	25	3	3.5
100	30	10	30	3	4.0
120	35	12	30	3	4.5
140	40	14	35	3	5.0
160	45	16	35	3½	6.5

Source: Courtesy Portland Cement Association.

FIGURE 12-34 Frame for long-barrel shell. (*Courtesy Plywood Mfg. Association of B. C.*)

FIGURE 12-35 Valley beams under long-barrel shells. (*Courtesy Plywood Mfg. Association of B. C.*)

UMBRELLA-STYLE SHELL ROOFS

Umbrella-style roofs, unlike folded plates and long-barrel shells, are nearly always approximately square or circular in plan. They include such shapes as *hyperbolic paraboloid* (H/P) *inverted umbrella shells. H/P umbrella shells, H/P saddle-shaped shells,* and *dome shells.* All are *thin-shell* structures that, because of their shape and the manner in which they are constructed, are largely self-supporting units. Depending on the type, units require one or more supporting *columns* or *abutments.* Most of them are particularly useful for one-storey buildings in which large spans are required for both length and width. Such shells are normally produced from either wood or concrete.

H/P Inverted Umbrellas

This is a popular umbrella-style roof, shaped as shown in Fig. 12–36. This popularity is due to the fact that such roofs are economical in material, require only one column per unit, have a simplicity of structural action, and are pleasing to the eye. They may be square or rectangular and lend themselves easily to repitition, thus making

FIGURE 12-36 Inverted umbrella roof.

form reuse feasible. Ribs and edge beams extend above the shell.

Shell thickness is from 3 to 3½ in., with the usual spans from 30 to 50 ft.

H/P Umbrellas

Hyperbolic paraboloid umbrella shell roofs can be economically designed for a variety of building purposes (see Fig. 12-37). The low stresses in a shell of this type require only a minimum of concrete, and in fact, the shell thickness often depends on the concrete cover needed for the reinforcement.

Shell thickness is from 3 to 3½ in. and spans from 30 to 50 ft. The dead load of the shell is approximately 50 lb/sq ft of projected area. Umbrella shell span data for concrete umbrella shells are given in Table 12-8.

H/P Saddle-Shaped Shells

A hyperbolic paraboloid saddle-shaped shell (see Fig. 12-38) is a three-dimensional slab in which strength and rigidity are attained by curving it in space, not by increasing the slab thickness. The doubly curved surface allows the transfer of loads to supports entirely by direct forces, so that all parts of the shell are uniformly stressed.

Shell thickness is from 2¾ to 4 in. and the shell *projection* (see Fig. 12-39) from 50 to 220 ft, with spans ranging from 50 to 160 ft. A *single-saddle* roof requires only

FIGURE 12-37 Umbrella roof.

TABLE 12-8 Span data for concrete umbrella shell roof

Span (ft)	Rise, h (ft)	Thickness, t (in.)	Average wt. of steel, including edge beams (lb/sq ft)	Approx. col. size (in.)	
				Round	Square
30 × 30	4-6	2½	2-2½	18	16
40 × 40	5-8	2½	2-2½	24	22
50 × 50	6-10	3	2½-3	30	28
50 × 60	7-11	3	3-3½	—	—
50 × 70	8-12	3½	3-3½	—	—

Source: Courtesy Portland Cement Association.

FIGURE 12–38 H/P saddle-shaped roof.

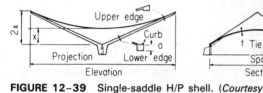

FIGURE 12–39 Single-saddle H/P shell. (*Courtesy Portland Cement Association.*)

two abutments (see Fig. 12–39) for support. However, the shape may be varied to suit particular circumstances. In Fig. 12–41, for example, there are three supports, while the roof in Fig. 12–40 requires four.

Span data for concrete hyperbolic paraboloid saddle-shaped shell roofs are given in Table 12–9, based on dimensions shown in Fig. 12–39.

FIGURE 12–40 Four-point support for H/P saddle roof.

TABLE 12–9 Span data for concrete hyperbolic paraboloid shell roofs

Span (ft)	Projection (ft)	Min. x^a (ft)	Average a (ft)	Average t (in.)	Reinforcement (lb/sq ft of projected area)
50	50–70	3–5	1	2¾	2–3
60	60–85	4–6	1	2¾	2–3
75	75–105	6–9	1½	3	3–4
100	100–140	9–13	2	3¼	3–4
125	125–175	13–20	2½	3½	4–5
150	150–210	17–25	3	4	5–7

^{a}Maximum feasible limit = S/5.
Source: Courtesy Portland Cement Association.

FIGURE 12–41   H/P roof frame with cables.

One of the problems involved in building hyperbolic paraboloid roofs is the forming of the roof deck. It is not in one or two planes, as is the case with more conventional roof systems, but presents a doubly curved surface, which might be likened to a warped parallelogram.

One of the solutions lies in the use of strands of cable to outline the surface. These are stretched from one perimeter member to another (see Fig. 12–41) and put under light tension. The surface is then produced by covering the cables with a layer of rigid material, usually foamed plastic insulation, over which is placed a thin layer of concrete (see Fig. 12–42).

Dome Shells

Dome shells may be *freestanding* structures or may be designed in *clusters*. In the latter case, adjustable segmental forms are used to create the shape, while in the case of a freestanding dome, it may be formed on a compacted mound of earth and lifted into final position by a series

FIGURE 12–42   Foamed plastic and concrete roof surface.

FIGURE 12-43 Dome shell with tension ring.

FIGURE 12-44 Dome shell uniformly supported.

of jacks. A tension ring is usually required at the circumference of the shell (see Fig. 12-43). Domes may be uniformly supported, as in Fig. 12-44, or may touch the earth at as few as three points.

Thickness of the shell ranges from 3 to 4½ in. and the spans from 50 to 500 ft or more. The shell thickness is usually increased from 50 to 75% near the periphery of the dome. Dome span data are given in Table 12-10, based on dimensions shown in Fig. 12-45.

FIGURE 12-45 Dome shell diagram. (*Courtesy Portland Cement Association.*)

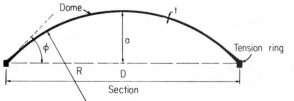
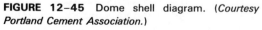

TABLE 12-10 Span data for dome shell roofs

D (ft)	t (in.)	φ (deg.)	a (ft)	R (ft)
100	3	30	13.4	100
		45	20.7	70.7
125	3	30	16.8	125
		45	25.9	88.4
150	3½	30	20.1	150
	(3)	45	31.0	106
175	4	30	23.5	175
	(3½)	45	36.2	123.7
200	4½	30	26.8	200
	(4)	45	41.4	141.4

Source: Courtesy Portland Cement Association.

FIGURE 12-46   Raising dome with hydraulic jacks.

Figure 12-46 illustrates an insulated concrete shell being raised into position by hydraulic jacks onto its supporting columns. The dome shape was produced by placing concrete over insulation and reinforcing steel supported by a compacted earth form.

CABLE SUPPORTED ROOFS

Large open-area buildings such as stadiums, arenas, and airplane hangers normally require complex structural components such as trusses, space frames, or domes to support their roof membranes. This creates a large volume of space that is of minimum use but still requires heating and ventilating.

A cable-supported roof derives its structural support from the tension capacity of the steel cables. The cables are usually anchored to a compression ring built into the exterior walls of the building. This arrangement produces a roof support system that has minimum depth, and as the cables are flexible, the roof shape can be adapted to any building shape that may be desired. Figure 12-47 illustrates a novel shape for an arena using a cable-supported roof.

In the illustrated example, the cables were placed in two directions, producing a supporting grid. Over this grid, lightweight precast concrete panels, complete with insulation, were placed, and all the voids between the panels were then filled with grout. The grouted panels were then covered with a waterproof membrane. Along with the shallow depth of the roof framing, the roof shape allows the building height to be kept to a minimum,

FIGURE 12-47   Cable-supported roof.

reducing the volume of air to be heated and ventilated, and yet provides an unobstructed view of the playing surface for every seat in the building.

STRESSED-SKIN PANEL ROOF

Stressed-skin panels are structures fabricated from lumber and plywood, in which the plywood skins act integrally with the lumber frame. The longitudinal framing members and the skins must be continuous in the longitudinal dimension, or adequately spliced. Skins are attached to the framing members with nails and glue so as to resist the developed shear stress (see Fig. 12-48).

Such panels can be produced in a great variety of shapes and are often used in building a shell-type plywood

FIGURE 12–48 Applying skin to curved stressed-skin panel. (*Courtesy Plywood Mfg. Association of B. C.*)

FIGURE 12–49 Stressed-skin panels with plywood web beams. (*Courtesy Plywood Mfg. Association of B. C.*)

roof. Plywood web beams may be introduced into such panels to provide extra stiffness in cases where there is to be only one support under each section (see Fig. 12–49).

INDUSTRIAL ROOFING

As indicated earlier in the chapter, the final operation in the construction of a roof is the application of the *roofing*. It is probably the most important element in a building, for it must protect the structure and its oc-cupants while exposed to a wide variety of climatic conditions.

There are many types of roofing from which to choose—the type selected for a particular building will depend on a number of factors. These factors include (1) the basic design and suitability of materials for that design; (2) the location and type of vapor barrier to be used; (3) the characteristics of the particular insulation to be used; (4) the probable expansion, contraction, and deflection of the roof deck; (5) the durability of roofing to be selected; (6) the fire resistance of the roofing; (7) the value of the roofing as a decorative feature.

Roofing materials include *shingles and shakes, terra cotta tile, sheet metal, asbestos-cement boards, built-up roofing, monoform roofing,* and *liquid envelope roofing*. The first three materials are used primarily in residential construction and will not be discussed here.

SHEET METAL ROOFING

Various methods are used to make roofing sheets, but two basic types of sheet are generally made: corrugated (see Fig. 12–50) and flat. Galvanized steel, aluminum, and Galbestos are all used to make corrugated roofing sheets of varying width, depth and pattern of corrugation, and allowable span, depending on the gauge and material used.

Corrugated sheet metal roofing sheets are normally supported on wood or steel purlins, properly spaced according to the gauge of the metal and the roof load involved. Table 12–11 indicates maximum purlin spacings for galvanized corrugated steel roofing sheets of various gauges.

There are two common laying orders for roofing sheets, as illustrated in Fig. 12–51. When laying according to order (b), be sure to tuck the corner of the #3 sheet under #2, that of #5 and #4, etc. Laying should start at the leeward end of the building so that side laps will have better protection from wind-driven rain. The top edges of eave sheets should extend at least 1½ in. beyond the back of steel purlins and 3 in. beyond the center line of timber purlins. At side laps (where the edge corrugation of adjacent sheets is opposite in direction), the underlapping side should finish with an upturned edge and the overlapping side with a downturned edge.

Sheets should extend at least one corrugation over the gable, and there should be 3 in. of overhang at the eaves. End laps between sheets should generally be 6 in. and side laps 1½ corrugations, but these may be increased to 9 in. and 2 corrugations for extreme conditions. For low-pitched roofs, it is good practice to seal the laps with a suitable caulking compound.

Special nails with a ring or screw-type shank should be used for fastening corrugated sheets to wood purlins. They should be zinc-coated, with a lead underhead, and long enough to provide adequate holding power. Nails should be driven at the top of corrugations, but care must be taken not to drive them so far as to flatten the corrugations—thus preventing the next sheet from fitting

FIGURE 12–50 Corrugated sheet metal siding and roofing. (*Courtesy Dominion Foundry & Steel Co.*)

TABLE 12-11 Maximum permissible purlin spacing for galvanized corrugated sheet metal[a]

	Superimposed load (psf)									
	10		15		20		25		30	
	Pitch of roof (in./ft)									
Gauge	3	5	3	5	3	5	3	5	3	5
14	11'3"	15'9"	10'0"	14'3"	9'3"	13'3"	8'9"	14'0"	8'3"	(11'9")
16	10'6"	15'0"	9'6"	13'6"	8'9"	12'6"	8'3"	11'0"	7'9"	(10'6")
18	9'9"	14'0"	8'9"	(12'3")	8'0"	(10'9")	7'6"	(10'0")	(7'3")	(9'3")
20	9'3"	(12'9")	(8'3")	(10'9")	(7'3")	(9'6")	(6'6")	(8'6")	(6'0")	(8'0")
22	9'0"	(11'3")	(7'6")	(9'9")	(6'6")	(8'6")	(6'0")	(7'9")	(5'6")	(7'0")
24	8'0"	(10'6")	(6'9")	(8'9")	(5'9")	(7'6")	(5'3")	(6'9")	(4'9")	(6'3")
26	(7'3")	(9'6")	(6'0")	(8'0")	(5'3")	(6'0")	(4'9")	(6'3")	(4'3")	(5'9")

[a]Figures enclosed in parentheses are governed by stress values; the remainder are governed by deflection values.

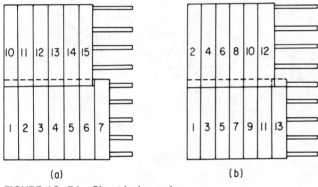

FIGURE 12-51 Sheet laying order.

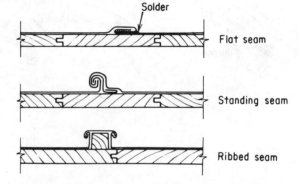

FIGURE 12-53 Seams for sheet metal roofing.

properly. Sheets are fastened to steel purlins with stainless steel self-tapping screws and aluminum washers (Fig. 12-52).

Flat sheets of metal—terne plate, galvanized iron, copper, lead, zinc, aluminum, Monel metal, or stainless steel—are applied over solid backing on either a flat deck or a pitched roof. Sheets are applied in strips which run up the slope of pitched roofs and are locked together by one of three types of joints (called *seams*) in common use. These seams allow the metal to expand without buckling. Figure 12-53 illustates these three types of seams.

ASBESTOS-CEMENT ROOFING

Roofing sheets of asbestos-cement are made in several designs, four of which are shown in Fig. 12-54. Corrugated board and transitile are used on sloping roofs over

FIGURE 12-54 Asbestos-cement roofing sheets.

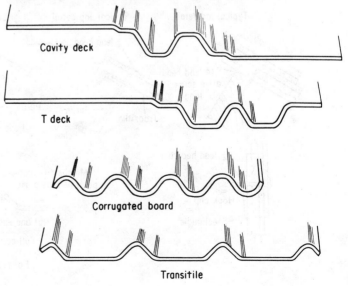

FIGURE 12-52 Corrugated roofing over steel purlins.

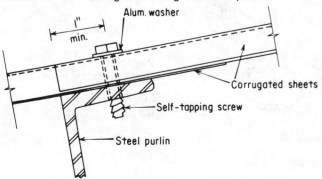

wood or steel purlins, and are laid with a 6-in. end lap and one corrugation side lap. Various types of fasteners are available for attaching these corrugated boards to the frame, and some are illustrated in Fig. 12–55.

There should be expansion joints in transitile roofing—wherever they occur in the structure or every 100 ft, if possible. Figure 12–56 illustrates two types of expansion joint battens in common use, but the curved batten is preferable.

Cavity decks and T decks are attached to the frame and to one another with self-tapping screws, as in Fig. 12–57.

FIGURE 12–55 Fasteners for asbestos-cement roofing.

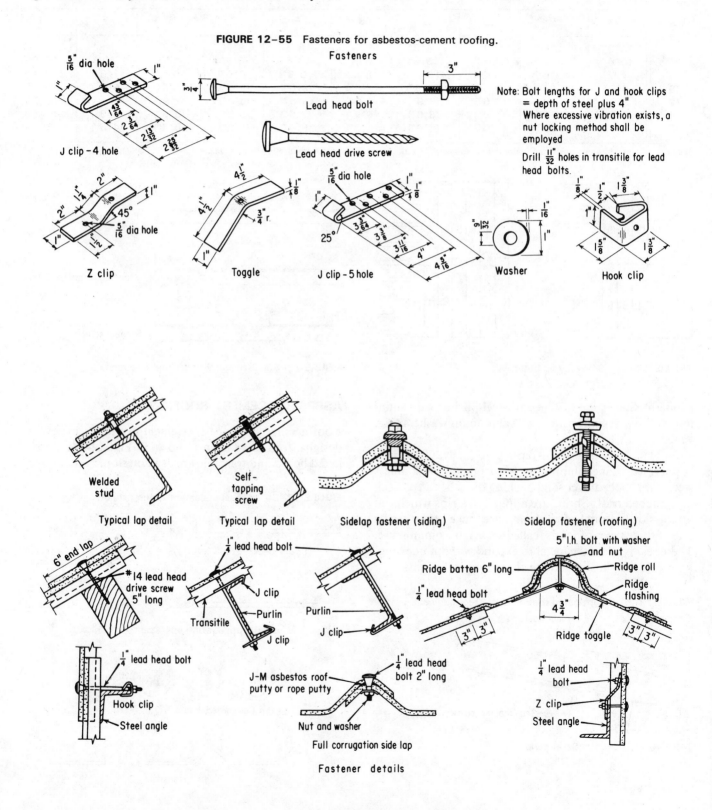

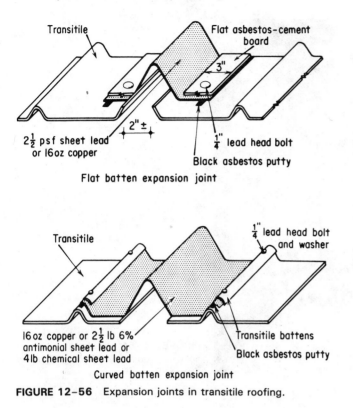

FIGURE 12–56 Expansion joints in transitile roofing.

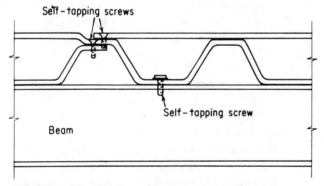

FIGURE 12–57 Attaching asbestos-cement decking to roof frame.

BUILT-UP ROOFING

Built-up roofing is probably the most widely used material for flat or nearly flat roofs. It consists of layers of roofing felt bonded with tar or asphalt. It is laid down to conform and bond to the roof deck and to seal all angles formed by projecting surfaces, and thus constitutes a single, flexible, waterproof membrane. The membrane is usually turned up to make a *base flashing* on vertical surfaces in order to form a large, watertight tray.

Insulation and vapor barriers are a part of most built-up roofing installations. The use of a vapor barrier

is essential wherever insulation is part of the roof structure. The proper design and installation of a vapor barrier is particularly important in northern regions. Without it, moisture vapor from within the building will penetrate the insulation, condense, destroy the insulating value, and eventually result in blistering or other damage to the roof covering.

The vapor barrier may be located at various points in the roof assembly, depending on the type of construction. It may be over a suspended ceiling, under the roof deck, or, most commonly, over the roof deck just under the insulation (Fig. 12–58). The insulation is sometimes completely enveloped in vapor barrier wrapping, in which case the joints between insulation boards must be adequately sealed.

Insulation is applied over the vapor barrier or over the deck by cementing it with asphalt. Roof deck insulation is of the rigid board type and may be *cellular glass* (see Fig. 12–58), *rigid urethane, glass fiberboard, corkboard, expanded polystyrene, wood fiberboard,* or *cellular concrete.*

A five-ply membrane for wood decks is one of the earliest standard specifications to be developed for built-up roofing. It consists of a layer of sheathing paper and two plies of dry felt nailed to the deck to secure the roof covering and prevent bitumen from dripping through the wood joints. Additional plies of felt are applied alternately with layers of bitumen, with a heavy top coating of bitumen embedded with gravel or slag.

Sheathing paper and dry-applied felts are not required when this type of roof covering is applied to concrete roof decks. One layer of felt is eliminated, but an extra layer of bitumen is necessary to secure the covering to the deck. This constitutes the standard four-ply built-up roofing.

Other specifications are modifications of the four- and five-ply standards, designed to meet varying conditions or to arrive at minimum specifications for roofing with reduced life expectancy. The factors to be considered in developing a specification include (1) slope of the roof, (2) type of roof deck—single or assembled units, (3) nailability of the roof deck, (4) presence of insulation and vapor barriers, (5) life expectancy of the membrane, (6) type of bitumen to be used, and (7) type of paper to be used.

Five basic types of membrane are recognized, depending on the types of paper and bitumen used:

1. Asphalt felts, asphalt and gravel.
2. Tarred felts, pitch and gravel.
3. Asbestos felts, asphalt and asphalt smooth flood coat.
4. Seventeen or 19-in. selvage roofing.
5. Cold process roofing.

FIGURE 12–58 Foamglas rigid insulation. (*Courtesy Pittsburgh-Corning Corp.*)

The basic specifications for each of these membranes, based on a 100-sq ft wood deck without insulation, are as follows:

1. *Incline:* ½ to 2 in./ft. (see Fig. 12–59)
   (a) *Dry sheathing:* 5 lb/100 sq ft.
   (b) *Felts:* 2 layers, 15# asphalt impregnated, dry applied; 2 layers, 15# asphalt impregnated, mopped on.
   (c) *Asphalt:* 3 moppings, approx. 20 lb/mopping; topping, approx. 65 lb, poured; for ⅛ - to 1-in. slope—140° asphalt; for 1- to 2-in. slope—170° asphalt; if over 2-in. slope, use 210° asphalt.
   (d) *Nails and caps:* 1 lb of 1¼-in. barbed nails; ½ lb of flat caps.
   (e) *Surfacing:* 400 lb of gravel *or* 300 lb of slag, ¼- to ⅜-in. size, dry, well rounded, free of dust.

2. *Incline:* 0 to 1 in./ft. (see Fig. 12–60)
   (a) *Dry sheathing:* 5# dry sheathing paper.
   (b) *Felts:* 2 layers, 15# tarred, dry applied; 3 layers, 15# tarred, mopped on.
   (c) *Pitch:* standard roofing pitch, not heated over 400°F or mopped at less than 350°F, 3 moppings, approx. 25 lb/mopping; topping, 80 lb, poured.
   (d) *Nails and caps:* 1 lb of 1¼-in. barbed nails; ½ lb of flat caps.
   (e) *Surfacing:* 400 lb of gravel *or* 300 lb of slag.

3. *Incline:* ½ to 8 in./ft.
   (a) *Dry sheathing:* 9# waxed kraft.
   (b) *Base sheet:* 25# asphalt felt.

   (c) *Felts:* 2 layers, 15# perforated asphalt felt, mopped on; 2 layers, 15# perforated asbestos felt, mopped on.
   (d) *Asphalt:* 4 moppings, approx. 20 lb/mopping; flood coat, 25 lb/flooding; for ⅛- to 1-in. slope—140° asphalt; for 1- to 2-in. slope—170° asphalt; over 2-in. slope—210° asphalt.
   (e) *Nails and caps:* ¾ lb of 1-in. barbed nails; ½ lb of flat caps.
   (f) *Surfacing:* flood coat as above.

4. *Incline:* 2 to 8 in./ft.
   (a) *Dry sheathing:* 5# dry sheathing paper.
   (b) *Felts:* 1 layer, 15# asphalt felt, dry applied; 2 layers, 15# asphalt felt, mopped on; 2 layers, 120#, slate surfaced, 19 in. or 17 in.; selvage, mopped on.
   (c) *Asphalt:* 4 moppings. approx. 20 lb/mopping, 210° asphalt.
   (d) *Nails and caps:* 1½ lb of 1¼-in. barbed nails; ¾ lb of flat caps.

5. *Incline:* ⅛ to 8 in./ft.
   (a) *Felts:* 3 layers, cold process, 53# felt, adhesive applied (backnailed and lap adhesion only).
   (b) *Adhesive:* Approx. 2½ gal of asphalt adhesive between laps.
   (c) *Nails and caps:* 2 lb of 1-in. flathead galvanized nails; 1 lb of flat caps.
   (d) *Surfacing:* 4 gal of Static Asphalt Fibrated Emulsion.

A double pour top coating is desirable on low-pitched roofs. It is accomplished by embedding approx-

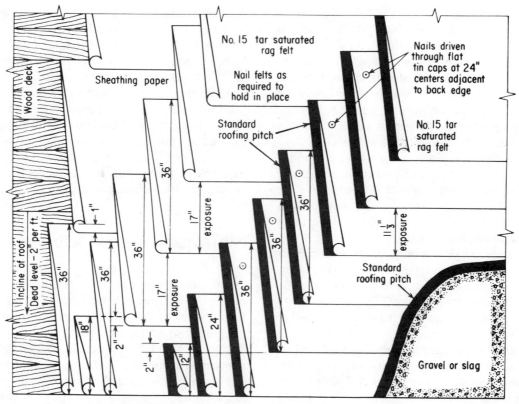

FIGURE 12–59   Five-ply felt and gravel roof of wood deck.

FIGURE 12–60   Four-ply felt and gravel roof on concrete deck.

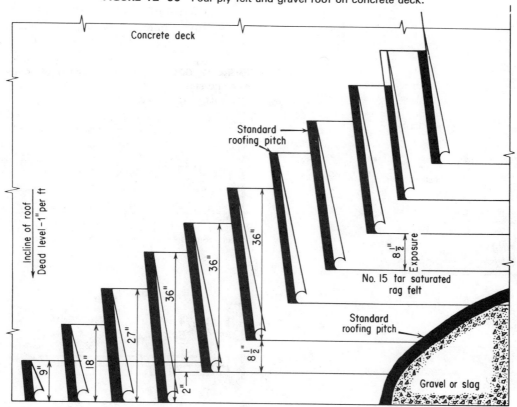

imately 200 lb of gravel in a top pour of 50 lb of bitumen, and later repeating the operation with 300 lb of gravel in 75 lb of bitumen on a second pour. This gives added protection against standing water and melting snow or ice. The second pour can be placed after all other phases of roof construction have been completed and provides an opportunity to check for possible damage.

Overheating bitumens during application must be avoided. The temperature range is 325 to 400 °F for coal-tar pitch and 400 to 450 °F for hot asphalt. If applied below these temperatures, the workability and cementing action is seriously reduced, while if poured above the maximum temperature, the material loses valuable oils by distillation and forms only a thin film which cannot provide adequate cementing action.

Careful attention should be given to interruptions in the roof surface at parapets, penthouses, vents, pipes, chimneys, and drains. Bends in felt used to make base flashing must be supported with cant strips. Base flashing should extend a minimum of 8 in. above the level of the deck and should be protected at the top edge with metal counterflashing. Since there is danger of rupture caused by movements between horizontal and vertical surfaces, base flashing should not be fastened directly to vertical surfaces such as parapet walls. An 8-in. board should be placed behind the cant strip against the wall, but not fastened to it (Fig. 12–61). The base flashing can then be nailed and cemented to the board and the top edge of the board and the base flashing covered by the counterflashing attached to the wall. Through-wall flashing must be provided in masonry walls extending above the roof to act in conjunction with counterflashing.

The need for expansion joints in the roof will depend on the size and design of the building. Expansion joints in the building itself must be extended to the roof. An expansion joint should be provided at each junction of the main portion of the building with a wing, and on the main roof if any dimension exceeds 150 ft. It is also desirable to provide an expansion joint at each change of joist or roof direction.

PROTECTED-MEMBRANE ROOFS

In a conventional built-up roof the waterproof membrane, usually composed of felts and bitumen, is placed on the outer surface of the insulation. It is then covered with a layer of gravel or slag for protection. With the membrane so near the outer surface of the roof system, it must be able to resist all the stresses exerted on it by temperature changes, weather changes, and traffic-related loads. In a protected-membrane roof the waterproof membrane and the vapor barrier are combined and placed below the insulation, giving the membrane considerably more cover than in the conventional roof (see Fig. 12–62).

Using this arrangement of materials it has been found that the membrane has a better chance of beginning its service life undamaged and be less affected by temperature changes. Because large fluctuations in

FIGURE 12–61   Flashing at parapet wall.

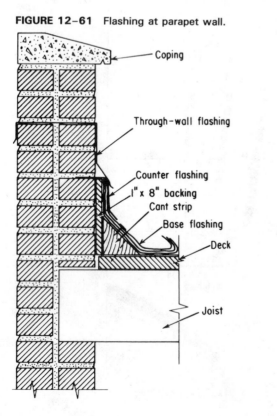

FIGURE 12–62   Protected membrane roof on concrete and steel roof decking.

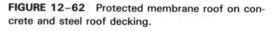
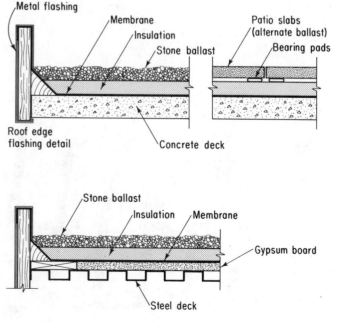

temperature are eliminated, control joints are normally not required in the roof.

With the advantages there are also disadvantages. The insulation, now on the outside of the waterproof membrane, must be able to withstand the effects of the weather and traffic-related loads. A number of different insulations have been used—rigid-type insulations composed of polystyrene foam, PVC foam, glass foam, urethane foam, glass fiber, wood fiber, perlite fiber, and cork. Also, fill-type insulations bonded with asphalt have been considered. It was found that all served well, but only the polystyrene foam could be used without some additional treatment to minimize the absorption of moisture. The one drawback of the polystyrene is that it must be protected from ultraviolet rays.

The top covering of the roof system must be of sufficient weight to prevent displacement during a heavy rain or high wind. For 2 in. of polystyrene foam, ¾- to 1¼-inch stone ballest with a weight of 12 lb/sq ft (1¾ in. depth) with an additional 5 lb/sq ft (½ in. depth) for each additional inch of thickness of insulation is recommended. In areas where wind may produce scouring, paving stones, or patio slabs should be used. Where additional traffic is anticipated, wood decking or patio slabs can be used to provide additional protection for the insulation (Fig. 12–63).

Inverted or protected roof systems must have good drainage to perform as intended. The covers must be applied so that they allow evaporation of the water that may seep down into the insulation. Roof slopes should be maintained to encourage fast drainage and roof drains must be placed so that ponding does not occur on the roof.

FIGURE 12–63 Patio blocks used as insulation protection.

MONOFORM ROOFING

Monoform roofing is a reinforced asphaltic membrane which is applied to a roof with a special type of spray applicator. The membrane consists of a specially formulated asphalt compound reinforced with glass fibers (see Fig. 12–64).

Base felts are applied as they are to any built-up roof. A special pressure gun is used to cut the glass fiber reinforcing into designated lengths and to spray out the fibers together with liquid asphalt. The result is a one-piece reinforced membrane of great strength and flexibility which is evenly spread over the entire surface. This method is particularly useful for contoured and steep-pitch roofs, to which a conventional built-up roof is sometimes difficult to apply.

FIGURE 12–64 Polysulfide polymer waterproofing membrane being applied by spray. (*Courtesy Thiokol Chemical Corp.*)

LIQUID ENVELOPE ROOFING

Liquid envelope roofing is a vinyl-based compound, available either clear or in a range of colors. It is sprayed on in enough applications to build up a 15- to 40-mil dry film which is impervious to moisture and humidity, flexible, and elastic through a wide range of temperatures. It may be applied on wood, concrete, steel, insulated surfaces, and asphalt or tar if the surface is first coated with a suitable primer. A continuous film can be carried up the face of a parapet or over the edges of a roof to reinforce flashing.

FABRIC ROOFS

Fabric roofs have been used in temporary structures for centuries. The early fabrics used were of natural materials and so had strength and durability limitations. Their use as roof material for permanent structures was never considered seriously when other more durable materials were available. Spurred on by the needs of the space program, new fabrics have been developed using new synthetic fibers of extreme toughness and durability. Because of

these properties and their light weight and flexibility, these space-age fabrics are being used as an alternative to the conventional materials used for roofing. Large-area structures such as stadiums (see Fig. 12–65) fieldhouses, arenas, and aquatic centers all over the world have incorporated fabric roofs with excellent results.

The fabric used is usually of a *polyester, nylon, or fiberglass* base coated with a *polymer, vinyl, or silicone* to produce an extremely strong, durable, and waterproof material. The use of additional liners produces insulation values that are comparable to conventional roofing systems. Many of the fabrics are translucent, allowing sunlight to pass through them, reducing the need for artificial lighting, yet reflecting much of the sun's radiated heat. Studies have shown that initial costs of these roof systems are much the same or lower than that of conventional roof systems.

Two basic methods have been developed to support fabric roofs: *cable-suspended roofs* and *air-supported roofs*. The cable-supported roof depends on a structural frame from which the roof cables are suspended over which the fabric is stretched. Figure 12–66 illustrates an aquatic center covered with a fabric roof. The central supporting rib is of reinforced concrete.

FIGURE 12–65 Inside the world's largest air-supported fabric roof stadium. (*Courtesy British Columbia Place.*)

FIGURE 12–66 Aquatic center covered with a fabric roof supported by a reinforced concrete arch.

In the case of an air-supported structure, fans provide sufficient positive pressure inside the building to support the roof structure. Steel cables are usually placed over the fabric in a checkerboard fashion to provide stability for the envelope during high winds. Figure 12–67 illustrates an air-supported roof on a stadium. The total weight of the roof, including the roof panels, cables, lighting, and sound system, is 280 tons. Sixteen fans rated at 100 horsepower each supply the necessary air pressure to keep the roof inflated.

FIGURE 12–67 Air-supported fabric roof on a stadium. Note the stabilizing cables anchored to the compression ring. (*Courtesy British Columbia Place.*)

REVIEW QUESTIONS

1. Differentiate between **(a)** *roof decking* and *roofing,* **(b)** *parapet* and *gravel stop,* **(c)** *prestressed concrete joist* and *prestressed roof slab,* and **(d)** *double T* roof slab and *cored* slab.

2. Explain why **(a)** long spans are made possible by the use of precast roof framing members, **(b)** the cost of hollow core roof slabs is relatively low, **(c)** a solid wood roof deck should be kept back ½ in. from parapet walls, and **(d)** *scuppers* are sometimes required in parapet walls surrounding flat roofs.

3. In truss work, explain what is meant by **(a)** a panel point, **(b)** a split ring, **(c)** a purlin, and **(d)** a gusset.

4. Outline the basic features of a folded plate roof.

5. Describe clearly **(a)** a practical method of forming for a long-barrel shell roof, and **(b)** the form of a hyperbolic paraboloid roof.

6. (a) Describe, by means of diagrams, three types of *seams* used to join sheet metal roofing sheets together. **(b)** Explain why these seams are used for joining, rather than *soldering* or *welding.*

7. Explain what is meant by **(a)** a *five-ply built-up* roof, **(b)** *cold process* roofing, **(c)** *monoform* roofing, and **(d)** *liquid envelope* roofing.

8. What is the basic difference between a conventional built-up roof and a protected-membrane roof?

9. What are the two methods used to support fabric roofs? For what purpose are the steel cables required in each type of system?

Masonry Construction

Building with masonry—stone, brick, etc.—involves two types of construction. In one type, the exterior walls are built first and the remainder of the building is framed into them in such a way that those walls transmit the loads of the building directly to the foundations. They are called *bearing walls*. In the other type, the load-bearing framework of a building is constructed of steel, concrete, or timber, and masonry may be utilized to close the building; the walls are simply *curtain walls*.

Historically, *brick* and *stone* were the foremost masonry materials. *Clay tile, architectural terra-cotta,* and *concrete blocks* of many kinds now also enjoy wide acceptance. All are unit masonry products and require mortar to bond them together. Whether in bearing walls or in curtain walls, the basic techniques involved in building with any type of masonry are similar, though the details of construction may depend on the type of unit used or the type of wall being built.

BUILDING WITH CONCRETE BLOCKS

Unit Block Construction

The basic techniques used in building with concrete blocks have been illustrated and described in many books and manuals and need not be repeated here. The allowable height and minimum thickness of block bearing walls are specified by building codes and depend on the type of wall and its eventual use.

Blocks are available in sizes that are some multiple of 4 in. A standard block is 8 in. high (actual height is 7⅝ in., allowing for a ⅜-in. mortar joint) and 16 in. long (actual length 15⅝ in.). Widths of block units range from 4 in. (3⅝ in. actual) to 12 in. (11⅝ in. actual). Pilaster blocks are usually 16¾ × 16¾ in. square or 15⅝ × 18¼ in. depending on the type. This allows the designer and the builder a good range of sizes with which to work. In addition to the standard sizes, many suppliers have a wide range of specialty units to deal with nonstandard situations [Fig. 13.1(a)].

Load-bearing walls may be single wythe (one block thickness) or multiwythe (two or more thicknesses). Multiwythe walls may be solid or cavity type [Fig. 13.1(b)]. In the cavity wall, an air space is maintained between the wythes, which in many instances is filled with insulation to increase the thermal efficiency of the wall. Exterior multiwythe walls normally incorporate some other type of material, such as brick or stone, to act as a finishing veneer rather than to support load.

Concrete block bearing walls must be provided with horizontal or vertical supports at right angles to the face of the wall. Vertical support may be provided by cross walls, pilasters, or buttresses; horizontal support is provided by floors or the roof. The greatest distance separating supports is usually 20 times the nominal wall thickness for walls of solid units, 18 times the nominal thickness for walls of hollow blocks, and 18 times the combined thicknesses of inner and outer wythes for cavity walls.

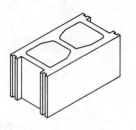

Standard

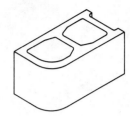

Corner sash

Bullnose corner

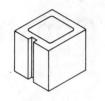

Half block

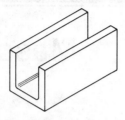

Lintel

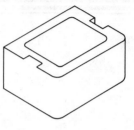

Pilaster

(a)

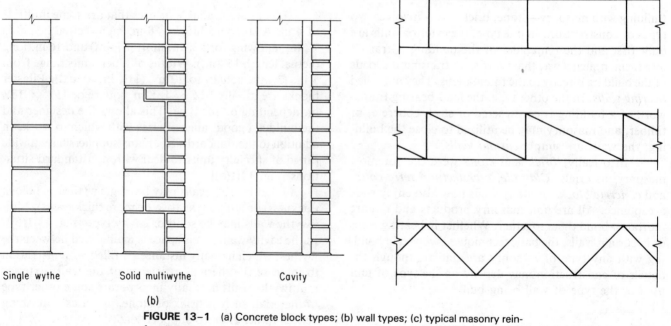

Single wythe   Solid multiwythe   Cavity

(b)

FIGURE 13–1   (a) Concrete block types; (b) wall types; (c) typical masonry reinforcement.

To provide maximum stability and to increase the
strength of block walls, the following practices are recommended:

1. Choose blocks of good dimensional stability.
2. Keep blocks dry at all times, particularly during site stor-
    age and laying operations.

3. Use horizontal joint reinforcement, especially around wall
    openings.
4. Provide control joints at appropriate locations.
5. Use bond beams where feasible.

Continuous horizontal joint reinforcement is provided by embedding specially fabricated wire reinforce-

ment [Fig. 13–1(c)] in the horizontal mortar joints. The reinforcement may be used continuously in every or every other joint throughout the building but should in any case be used for two consecutive joints above and below openings (unless there are control joints) and should extend at least 24 in. beyond the opening.

Control Joints

Control joints are vertical joints which separate walls into sections and allow freedom of movement. They should occur at intervals in long, straight walls, where abrupt changes in wall thickness take place; at openings (Fig. 13–2); at intersections of main walls and cross walls; and at locations of structural columns or pilasters in main walls.

To produce a control joint, place building paper on a coat of asphalt paint on the ends of the blocks on one side of the joint. Fill the core with concrete or mortar to provide lateral stability (Fig. 13–3). Single wire ties may also be used across the joint. Rake the mortar in the control joint to a depth of ¾ in. and caulk with a suitable compound. Caulking may be omitted on the interior face and the joint finished with a deep groove.

Bond Beam

The primary purpose of bond beams is to act as a continuous tie for exterior block walls where control joints are not required. They may also be used between control joints in larger buildings. Bond beams can also act as structural members, transmitting lateral loads to other structural members, and can provide bearings for beams or joists.

A bond beam is a continuous, cast-in-place, reinforced concrete beam running around the perimeter of a building or between control joints. It is formed by us-

FIGURE 13–3 Control joint in block wall. (*Courtesy Portland Cement Association.*)

ing standard lintel blocks, deeper bond beam blocks, or standard blocks with a large portion of the webs cut out. In Fig. 13–4, the bond beam is being utilized as a lintel over an opening. Notice the shoring required in the opening.

Lintels over openings in block walls may consist of precast concrete units, as in Fig. 13–5, or may be cast in place. In the latter case (illustrated in Fig. 13–6), lintel blocks are used again and are supported by shoring. When greater strength is required, two-course lintels may

FIGURE 13–4 Reinforced bond beam.

FIGURE 13–2 Control joints at window.

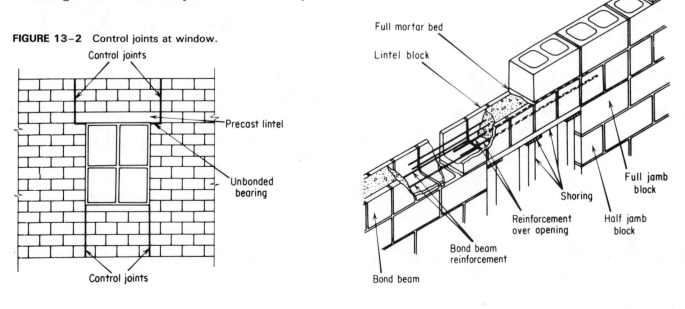

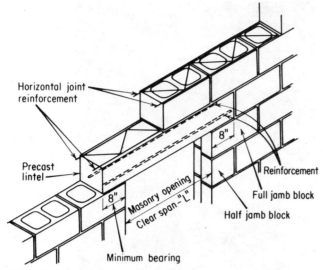

FIGURE 13-5 Precast concrete lintel.

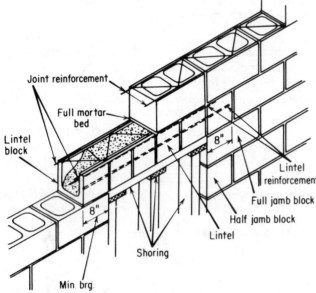

FIGURE 13-6 Cast-in-place concrete lintel.

FIGURE 13-7 Two-course lintel.

be employed. In Fig. 13-7, standard blocks with cutaway webs are being used to form the lintel.

The large number of shapes available makes a great variety of patterns possible for a block wall. Consult a concrete block manual for illustrations or patterns. In addition to the standard blocks, *screen* blocks are available to add to the variety or to build open walls which are to act as solar screens. Figure 13-8 illustrates three common solar screen shapes. Manufacturers' brochures are available which will provide a complete list of such units.

Concrete blocks are also used to build backup walls for brick, tile, or stone facing. Various types of metal ties are used to bind the facing to the backup, and the commonly used ones are illustrated in Fig. 13-9.

Customized Concrete Block Masonry

One of the more recent developments in the concrete block industry has been the production of block with a face which conforms to an architect's design for a particular building—walls of *customized concrete block masonry*. Figure 13-10 illustrates a few of the block faces in use, but a wide variety of designs is possible and a block producer should be consulted to ascertain what designs his plant is capable of producing. Figures 13-11 through 13-15 illustrate some of the customized block faces which have been used in concrete block walls.

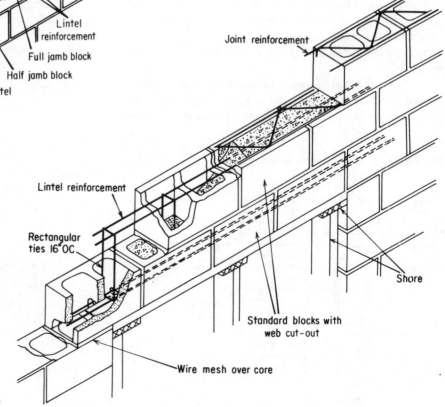

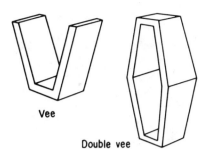

Vee

Double vee

Double X

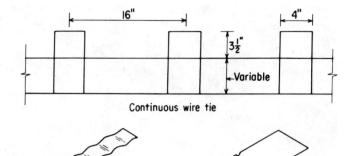

Continuous wire tie

Galvanized strip tie

Rectangular wire tie

FIGURE 13-8 Typical screen blocks.

FIGURE 13-9 Typical metal ties.

FIGURE 13-10 Customized concrete block units. (*Courtesy National Concrete Masonry Association.*)

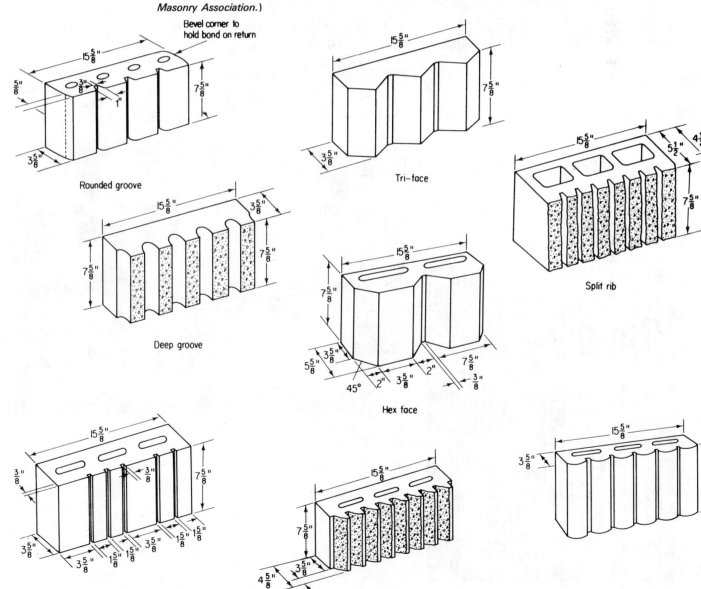

Rounded groove

Tri-face

Split rib

Deep groove

Hex face

Random scored

Split sawtooth

FIGURE 13-11 Screen block wall. (*Courtesy National Concrete Masonry Association.*)

FIGURE 13-12 Fluted face block wall. (*Courtesy National Concrete Masonry Association.*)

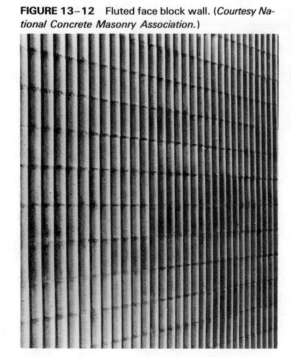

FIGURE 13-13 Scored-face block. (*Courtesy National Concrete Masonry Association.*)

FIGURE 13-14 Split block in stack bond. (*Courtesy National Concrete Masonry Association.*)

FIGURE 13-15 Split-rib block. (*Courtesy National Concrete Masonry Association.*)

Prefabricated Concrete Block Wall Panels

Recent innovations in concrete masonry construction include the utilization of *new mortar systems* and the development of *prefabricated concrete block wall panels.*

Mortar Systems

One of the new mortar systems involves the use of a very strong adhesive known as *organic mortar.* It is applied to masonry joints with a gun and replaces the usual portland cement mortar. The joint is only about $\frac{1}{16}$ in. thick and requires the use of blocks which have very precise dimensions. In many cases, blocks are precision-ground on their bearing surfaces to provide those close tolerances. A variation of this system is the use of an organic additive to portland cement mortar, to increase its flexural strength from 2 to 2½ times.

Another system, which is known as *surface bonding,* uses a mortar which is composed basically of portland cement reinforced by the addition of glass fibers to increase the tensile strength. Blocks are stacked without mortar in the joints, and the wall is plastered on both sides with the surface-bonding mortar.

Tongue-and-groove concrete blocks are made for use with both mortar systems, to facilitate alignment and provide interlocking action between the blocks. The usual procedure calls for a full 8 by 16-in. face dimension because of the very thin or completely eliminated mortar joints.

Panel Fabrication

Wall panels are fabricated by two different methods. In one, block layers (masons) position the units in the wall panels, while the other involves the use of a block-laying machine. The manual system can be employed either in a plant or at the job site, in all-weather enclosures on what are called *launch pads,* but the machine operates in a prefabricating plant or factory.

A block-laying machine lays block in either running or stack bond, inserts the joint reinforcement, makes the bed and head joints, and tools them. It thus assembles wall panels about 16 times as fast as they could be built in place, using the same manpower employed in the machine operation. It will handle blocks from 6 to 12 in. thick and is capable of producing panels up to 12 by 20 ft in face dimensions. Vertical reinforcement then has to be inserted in the block cores and the cores grouted. When they are ready, panels are transported to the job site by truck.

Under the manual method, blocks may be laid with conventional mortar, with organic mortar, or with no mortar. In the last case, surface bonding will be employed. Regardless of the mortar system used, panels will be reinforced with steel and grouted. Normally, panels up to 8 by 30 ft in size will be laid up by this method.

At the building site, a crane with lifting devices and special equipment to prevent breakage lifts the panels into position in the structure. Walls are set and braced, and if they are load-bearing, the floor panels are set directly on their top edges. *Wall-to-wall* and *wall-to-floor* connections are made through *access pockets* left in the panels and in the floor. Finally, after the connections are complete, the pockets are filled with grout.

Tying Systems

With conventional block walls, the tying systems between walls and floors or walls and columns have been relatively simple (see Fig. 13–16). The development of block panel construction has meant the need for strong connections between adjoining walls and between walls and floors. Horizontal edges (wall-to-floor edges) may be connected by *spliced reinforcement,* by *tension connections,* by *bolted-tension connections,* by *bolting,* or by *welding.* Vertical edges (wall-to-wall edges) may be connected by *steel plates,* by *lapped reinforcement,* or by *formed-lapped reinforcement,* among others.

Wall-to-Floor Ties

The spliced reinforcement system is much the same as that used to join precast concrete panels. The bottom ends of reinforcing bars have a coupling nut attached, and when the walls are in place, a short, straight bar, threaded

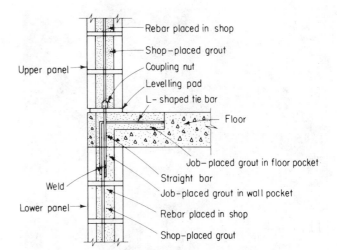

FIGURE 13–17 Spliced reinforcement connection. (*Courtesy National Concrete Masonry Association.*)

on one end, is attached to the upper bar by the coupling nut and to the bottom one by welding. An L-shaped bar, placed after the walls are up, is laid with one end in a pocket in the floor and the other down through the block core to tie the wall and floor together (see Fig. 13–17).

The tension connection provides a way of tying the panels together, through the block cores, from top to bottom of the building. Panels are brought to the job site with loose vertical bars in their cores. Each bar has a *threaded eyebolt* attached at the top and a steel *lifting plate and nut* at the bottom, to act as a lifting device. After the panel has been set in place, the nut on the bottom end of the top bar is removed and a *turnbuckle* is used to join the corresponding top and bottom bars together. The cores in the bottom panel are then grouted, but the ones in the upper panel are left empty until the next panel above is installed. Finally, the last panel in the tier must have a plate at the top end of the bars to complete the tension connection (see Fig. 13–18).

In the bolted-tension connection, vertical reinforcement bars are grouted into the panels in the plant, with threaded eyebolts attached at both top and bottom ends. When the bottom panel is in position in the wall, a connecting device consisting of a *turnbuckle with a yoke at each end* is attached to the end of each bar. Then, when the next panel is installed, the connection is attached to the top bars and the turnbuckle tightened to join the two panels. Finally, grout is placed in the cores and access pockets (see Fig. 13–19).

Bolted and welded connections are similar to those used with precast concrete panels. Anchor plates are set into the top and bottom of panels during fabrication, and these are welded or bolted to field-placed connections attached to the floor slabs. Access pockets are grouted after the connections are complete.

FIGURE 13–16 Anchors for conventional block curtain wall.

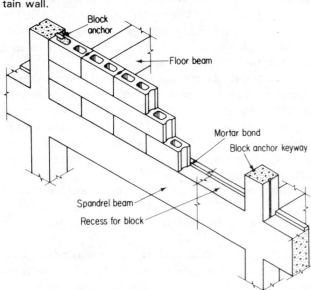

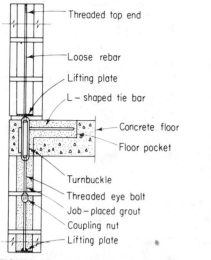

FIGURE 13–18   Tension connection. (*Courtesy National Concrete Masonry Association.*)

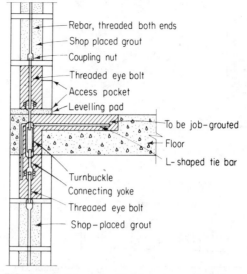

FIGURE 13–19   Bolted-tension connection. (*Courtesy National Concrete Masonry Association.*)

Wall-to-Wall Ties

Several types of vertical or wall-to-wall connections are used, including a steel bar connection, a *lapped reinforcement connection,* and a *lapped reinforcement with formed joint connection.*

In the first, a flat steel bar, with its ends bent at right angles, is preset in the top edge of one panel. A cutout in the top end block of the adjoining panel makes installation easier. After the second panel is in place, the end core is grouted to anchor the bar.

Lapped reinforcement consists of hairpin-shaped bars placed in alternate block courses, with their loops projecting 2 in. past the vertical edge of the panel. Modified end blocks are required to accommodate them, as shown in Fig. 13–20. After the two adjoining panels are aligned, a vertical bar is threaded through the loops and the core formed by the edges of the two panels is grouted.

The lapped reinforcement with formed joint is similar to the above except that regular end blocks are

FIGURE 13–20   Vertical panel edge splices. (*Courtesy National Concrete Masonry Association.*)

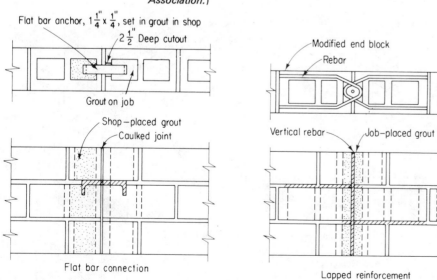

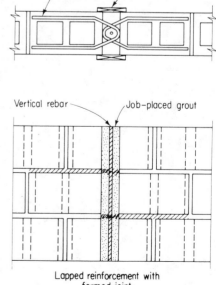

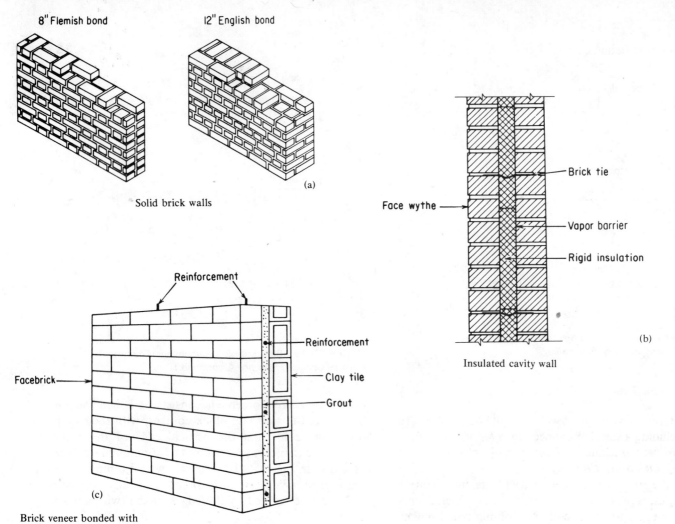

8" Flemish bond

12" English bond

(a)

Solid brick walls

Brick tie

Face wythe

Vapor barrier

Rigid insulation

(b)

Insulated cavity wall

Reinforcement

Reinforcement

Facebrick

Clay tile

Grout

(c)

Brick veneer bonded with
reinforced grout core

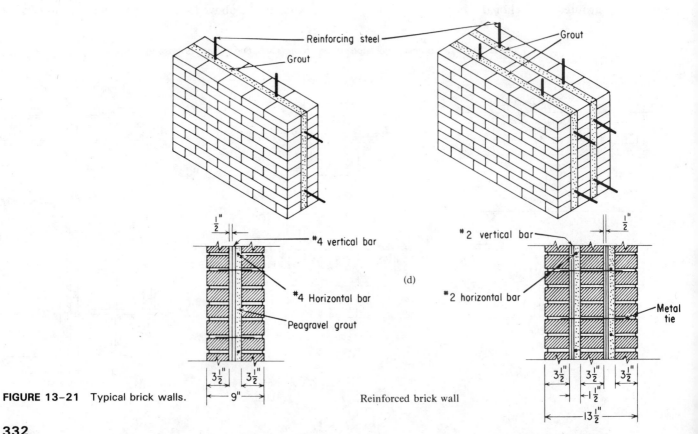

Reinforcing steel

Grout

Grout

(d)

$\frac{1}{2}''$

#4 vertical bar

#4 Horizontal bar

Peagravel grout

$3\frac{1}{2}''$ $3\frac{1}{2}''$

9"

#2 vertical bar

$\frac{1}{2}''$

#2 horizontal bar

Metal
tie

$3\frac{1}{2}''$ $3\frac{1}{2}''$ $3\frac{1}{2}''$

$1\frac{1}{2}''$

$13\frac{1}{2}''$

Reinforced brick wall

FIGURE 13-21 Typical brick walls.

used and the panels are spaced 2 to 3 in. apart when in position. Forms then have to be used to grout the space, after the vertical bar has been inserted.

BRICK CONSTRUCTION

Brickwork Fundamentals

Building with brick may involve the construction of *solid brick walls,* walls of *reinforced brick, brick cavity walls,* or *brick veneer* walls (see Fig. 13–21). The terminology used in connection with various positions of brick in a

wall (see Fig. 13–22) applies in all cases, and almost any of the numerous *pattern bonds* available (see Fig. 13–23) may be used with any one of these walls. The *structural bond* involved will depend on the type of wall. The third type of bond involved in brickwork, the *mortar bond,* is a function of the type of mortar used.

The same types of mortar are used for laying brick as are used for concrete block, namely; *sand* or *sand-and-lime mortar, organic mortar,* and *surface-bonding mortar.* The last two were described in the foregoing section on block construction.

Sand mortars vary according to the proportions of ingredients used in making them, and the kind used for

FIGURE 13–22 Common terms used in brickwork.

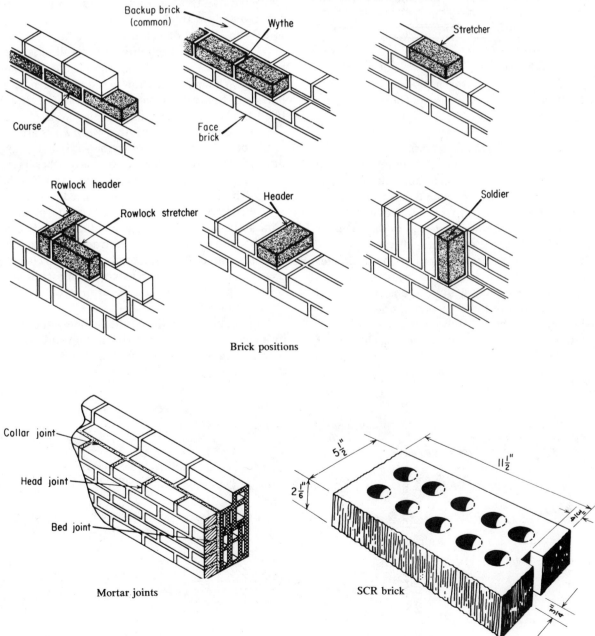

Brick positions

Mortar joints

SCR brick

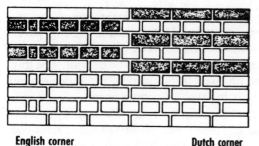

English corner   **Dutch corner**

Flemish bond

Dutch corner   **English corner**

English bond

FIGURE 13-23 Two common brick pattern bonds.

a specific job may be designated by type or by compressive strength. Table 13-1 outlines the composition and strength of various sand and sand-and-lime mortars. Types O or K may not be used where masonry is to be in direct contact with the soil, where an isolated pier is to be built, or where the wall is exposed to the elements on all sides, as in the case of a parapet wall.

When brick walls are required to have greater resistance to compressive, tensile, or shear forces than usual, they may be reinforced with steel rods, as indicated in Fig. 13-21(d). Two or three *wythes* of brick are laid, the starting wythe to be built up not more than 16 in. ahead of the others. Metal ties connect each pair of

wythes; reinforcing steel (½ in. for a two-wythe wall or ¼ in. for a three wythe wall) is placed in the cavities, vertically and horizontally, and the cavities filled with *pea gravel* grout. The depth of any one grout placement should not exceed about 4 ft.

Brick walls may be further strengthened by the use of bond beams (Fig. 13-24). In the case of a three-wythe wall, the two outside wythes act as the form. Side forms must be provided for a bond beam in a two-wythe wall. Vertical reinforcing should extend through the beam.

A brick cavity wall consists of two wythes of brick separated by a continuous air space not less than 2 in. wide. Metal ties are used to connect the two wythes. They should be spaced 36 in. o.c. horizontally with rows of ties spaced 16 in. o.c. vertically. The spacing should be staggered in alternate rows. The facing wythe is always a nominal 4 in. thick, while the interior one may be 4, 6, or 8 in. thick, depending on height, loads, etc.

If insulation is required in the wall, it is usually installed in the cavity, and 2-in. rigid insulation is normally used. Whether insulation is required or not, a vapor barrier in the form of an asphalt coating is usually applied to the cavity face of the inner wythe.

Care should be taken that *weep holes* are provided at the bottom of the outer wythe so that the cavity will drain. Oiled rods, sash cord, or rolled fiberglass insulation may be used in the vertical joints to form the drain holes. The sash cord or fiberglass need not be removed.

Brick veneer construction consists of applying a 4-in. wythe of brick as facing material over a backup wall. The backup wall may be of frame construction or a masonry wall of common brick, structural clay tile, concrete block, cellular concrete blocks, or cast-in-place concrete. In any case, the load-bearing properties of brick are not utilized.

Structural stability of the facing wythe is obtained by anchoring the veneer to the backup with metal ties or by grouting it to the wall with a reinforced grout core, usually 1 in. thick.

TABLE 13-1 Mortar Types

Type of mortar	Portland cement	Masonry cement	Lime	Aggregate	Compressive strength (psi)
M	1	1	—	Not less than 2¼ or more than 3 times the sum of volumes of cement and lime used.	2500
	1	—	¼		
S	½	1	—		1800
	1	—	¼–½		
N	—	1	—		750
	1	—	½–1¼		
O	—	1	—		350
	1	—	1¼–2½		
K	1	—	2½–4		75
	—	—	1		

FIGURE 13-24 Reinforced bond beams.

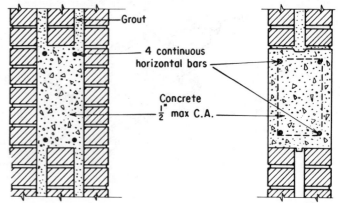

Grout — 4 continuous horizontal bars — Concrete ½ max C.A.

Lintels

The support of the masonry over openings in brick walls is an important consideration. That support will probably be provided by a *lintel,* which may be a *reinforced concrete unit,* a *reinforced brick structure,* or a built-up structural steel member (see Fig. 13–25).

Reinforced brick lintels are similar to a bond beam. A support must be made on which to build the lintel, and the brick is laid in running bond with full bed and head joints. Horizontal reinforcing steel may be laid in the mortar joints as the units are laid, or the lintel may be built as illustrated in Fig. 13–25. In the latter case, the grout is placed to the level of only one brick course at a time.

A number of built-up structural steel shapes may be used as lintels, depending on the thickness of walls, the length of spans, the loads involved, etc.

Flashing

One item of vital importance in the construction of brick walls is the installation of *flashing.* The purpose of flashing is to exclude moisture or to direct any moisture which may penetrate the wall back to the exterior. To perform satisfactorily, flashing must be a permanent material and must be properly installed. Most flashing is made of sheet metal (copper, lead, aluminum, or galvanized iron), fabric saturated with asphalt, or pliable synthetic material. Copper and bituminous material are sometimes combined to form flashing material.

Two types of flashing are in common use, *external* and *internal.* External flashing prevents the penetration of water at points where walls intersect flat surfaces, such as roofs. Internal flashing is built into and usually concealed in the wall to control the spread of moisture and to direct it to the outside. This is sometimes called *through* flashing.

The points at which flashing should be installed are (1) above grade in exterior walls; (2) under and behind window sills; (3) over lintels; (4) over spandrel beams; (5) at projections or recesses from the face of a wall; (6) under parapet copings; (7) at intersections of wall and roof; (8) at projections from the roof, such as ventilators, penthouses, etc.; and (9) around chimneys and dormers.

External flashing for roof and wall intersections was described in Chapter 12 and illustrated in Fig. 12–61. Flashing at various vulnerable points in brick structures is shown in Fig. 13–26.

FIGURE 13–25 Brick lintels.

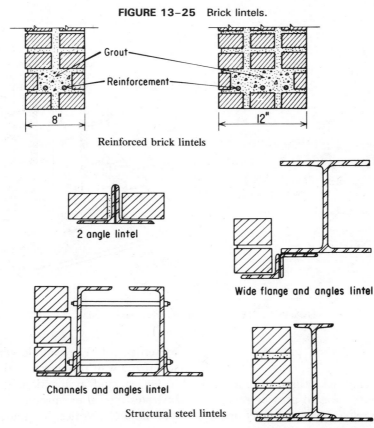

Reinforced brick lintels

2 angle lintel

Channels and angles lintel

Structural steel lintels

Wide flange and angles lintel

I beam and plate lintel

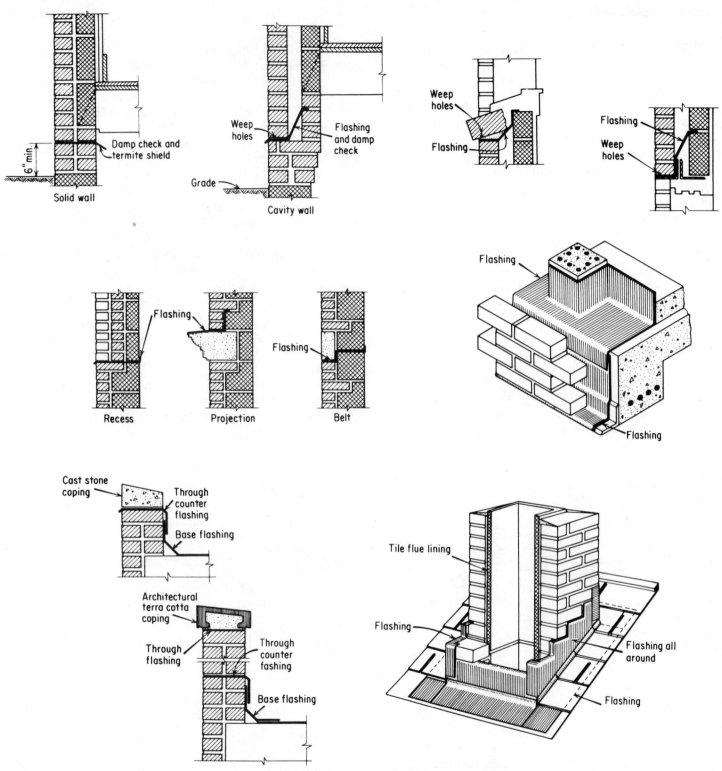

FIGURE 13-26 Flashings in brick construction.

Expansion Joints

Expansion joints are needed in brick construction under certain conditions. Their purpose is to provide separations in a structure so as to relieve the stresses set up by changes in temperature and moisture conditions or caused by settlement. They should generally be located at offsets if the wall running into the offset is 50 ft or more in length, and at junctions in walls. Figure 13-27 illustrates typical expansion joints in combination brick and tile walls. Notice the molded copper water stop used in the vertical joint on the outer face. Caulking is later

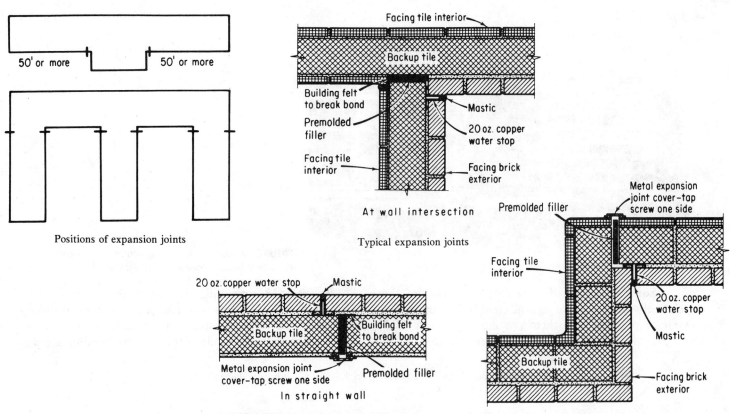

FIGURE 13-27 Expansion joints in brick construction.

Positions of expansion joints

Typical expansion joints

At wall intersection

In straight wall

applied to seal the joint. A premolded expansion joint material is used to separate the sections of backup wall, and the vertical joint on the face is covered with a metal plate which is fastened to one section only.

LOAD-BEARING BRICK WALLS

Until recent years, the design of brick buildings was based on the empirical requirements of building codes with regard to the minimum allowable wall thickness and maximum height. These regulations often placed economic restrictions on *load-bearing brick walls* for buildings over four or five stories in height.

New developments are now taking place in which the design of load-bearing brick walls is based on a realistic structural analysis of the proposed building. As a result, relatively thin, load-bearing brick walls are being extended to as much as 18 stories (see Fig. 13-28). Designs are including the interior as well as the exterior walls for load-bearing purposes, and consequently the thickness of all walls can be reduced. In addition, brick construction allows a considerable degree of flexibility in design, and designers take advantage of this to distribute the floor loads over as many bearing members as possible. For example, in the buildings shown in Fig. 13-29, the load-bearing masonry walls radiate outward from a central utility core like the spokes of a wheel.

FIGURE 13-28 Typical load-bearing brick wall structure. *(Courtesy International Masonry Institute.)*

337

FIGURE 13-29 Brick building with interior bearing walls. *(Courtesy International Masonry Institute.)*

Some examples of the structural details of these relatively thin load-bearing walls and the variety of floor systems possible with them are shown in Figs. 13-30, 13-31, and 13-32. The building represented in Fig. 13-30(a) is 18 stories tall, with 15¼-in.-thick bearing walls. Part of this thickness is attributable to thermal insulation rather than to structural requirements. In Fig. 13-30(b), the floor loads of the 16-story building are car-

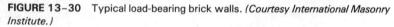

FIGURE 13-30 Typical load-bearing brick walls. *(Courtesy International Masonry Institute.)*

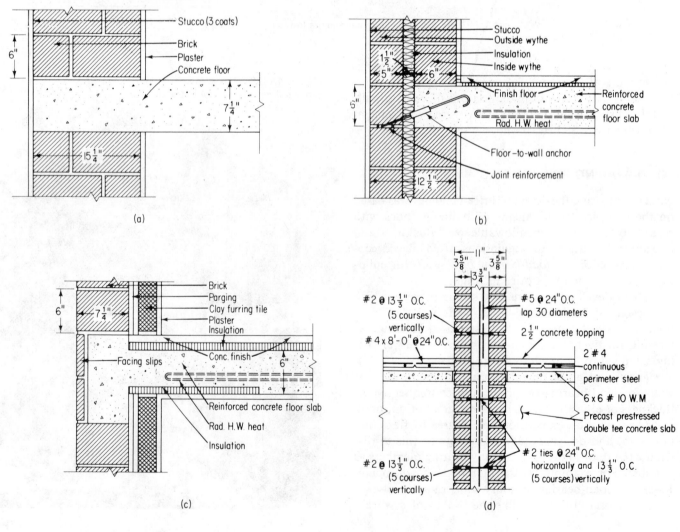

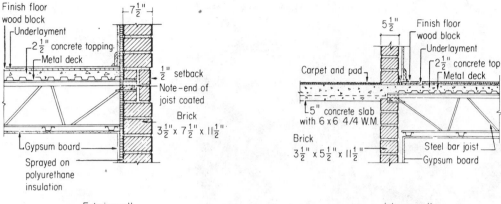

FIGURE 13-31 Open-web steel joist floor frame with load-bearing brick walls. *(Courtesy International Masonry Institute.)*

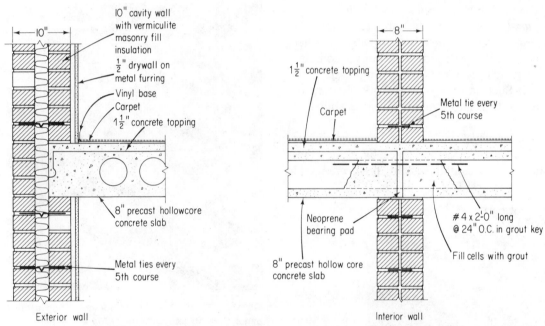

FIGURE 13-32 Brick load-bearing walls with precast concrete floor slabs. *(Courtesy International Masonry Institute.)*

ried on the 6-in. inner wythe of the exterior cavity wall and the 6-in. interior brick bearing partitions. In Fig. 13-30(c), the 14-story building has exterior bearing walls 7¼ in. in thickness. The building represented by Fig. 13-30(d) is 17 stories in height, with 11-in. reinforced brick masonry walls.

Figure 13-31 illustrates the exterior and interior bearing walls of an 8-story apartment building in which the apartment floors are supported by open web steel joists, while the corridor floors are simply 5-in. reinforced concrete slabs, spanning from one load-bearing corridor wall to the other. Figure 13-32 illustrates the use of precast hollow core concrete floor slabs with thin masonry bearing walls.

PREFABRICATED BRICK PANELS

The same techniques used in the prefabrication of concrete block wall panels are being applied to the production of prefabricated brick panels. Some are veneer-type panels, just one brick thick, while others constitute load-bearing walls of combined brick and block, bonded by grout-filled collar joints and wire reinforcement (see Fig. 13-33).

The development of high-bond mortars is a key factor in the increasing use of prefabricated brick panels, since the new mortars will produce 300 to 400 psi of flexural strength, compared to about 150 psi for conventional mortars. However, both kinds are used in manufacturing such units.

FIGURE 13-33 Precast brick facing panels being set in place.

As is the case with block, brick panels may be made in *masonry panel plants* or at the job site, *under all-weather enclosures,* on so-called *launch pads.* In either case, most of the panels are made with mortar containing an additive which produces higher compressive and tensile strength. Such mortar is also more resistant to water penetration, thus improving the weatherability of the panels.

After about 7 days' curing, panels are ready for installation. They are lifted into place by crane (Fig. 13-34) and anchored, usually by welding prelocated weld plates into the structural frame (see Fig. 13-35).

FIGURE 13-34 Curved precast brick section being hoisted by crane.

FIGURE 13-35 Weld plate for anchoring a brick panel to a structural frame. *(Courtesy International Masonry Institute.)*

BUILDING WITH STONE

Although stone has been a major structural material in the past, modern builders use it almost always as a veneer or curtain wall material. As a veneer, stone is laid up to a masonry backup wall in *ashlar* or *rubble* patterns. Ashlar patterns feature cut stone laid with well-defined course lines, while rubble patterns use uncut or semi-cut stone laid with few or no course lines. Stones for ashlar work are produced in thicknesses of from 2 to 8 in., in heights of from 1 to 48 in., and in lengths of from 1 to 8 ft.

Methods of bonding stone veneer to backup walls vary somewhat, depending on the size of stones used. For small stones, ties similar to those used for brick and tile facing work are commonly used. For larger stones, and particularly when the backup wall is nonbearing, a variety of anchor types is available (Fig. 13-36).

Two methods are used to support stone veneer on nonbearing walls. One method involves the use of shelf angles at the spandrel beams, as in Fig. 13-37(a). The stone is held in place on the shelf by a ½-in. rod welded to the horizontal leg of the angle or by dowels protruding through the horizontal leg. In the first instance, the rod lies in a groove cut in the bottom edge of the stone slab, and in the second, the dowel fits into a hole in the edge of the stone.

In the second method, *bond stones* rest on the spandrel beam and provide support for the rest of the veneer. Anchors are not required in the bond stones themselves, but are used between bond stones and regular veneer slabs and between veneer and the backup wall. If the stones are not more than 30 in. in height, two anchors per stone

should be supplied in both the top and bottom beds. When stones are under 24 in. in width, they require only one anchor per stone.

An alternative method of applying stone veneer, shown in Fig. 13-37(d), is particularly applicable if insulation is used in the wall.

Stone is also used to face soffits and columns or to trim brick and tile walls. Trim stones include sills, lintels, quoins, belts, and copings. Stone soffits are hung from the frame by strap hangers which fit into slots in the edges of the soffit slabs (Fig. 13-38). Several of the methods used to face columns stone are illustrated in Fig. 13-39.

Belts and copings are secured by strap anchors or dowels, while sills, lintels, and quoins are built into and supported by the masonry surrounding them. Figure 13-39 shows one method of anchoring coping stones to a parapet wall.

FIGURE 13-36 Stone anchors.

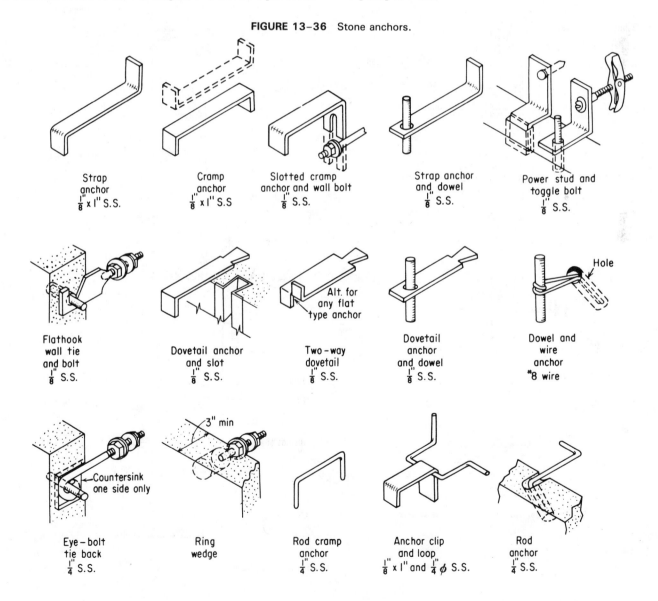

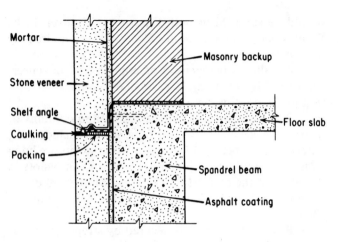

Stone veneer supported on shelf angle

(a)

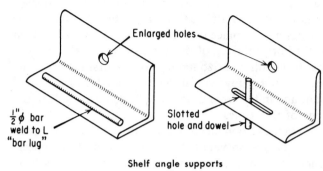

Shelf angle supports

(b)

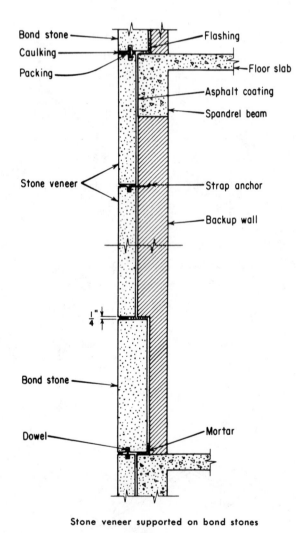

Stone veneer supported on bond stones

(c)

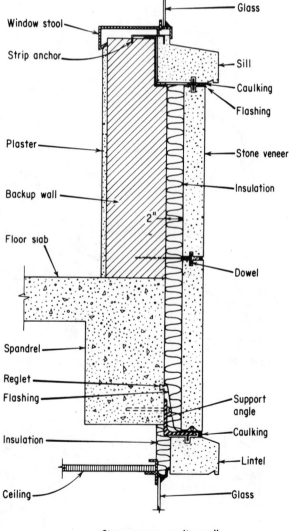

Stone veneer cavity wall

(d)

FIGURE 13-37   Stone veneer support systems.

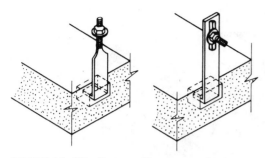

FIGURE 13-38 Stone soffit strap anchors.

Stone ashlar is anchored to solid backup walls with noncorroding corrugated wall ties, and with bond stones where the backup material will allow it. If stone course lines do not coincide with those of the backup material, *toggle* bolts are used to secure the wall ties to the block or tile (Fig. 13–40).

Mortar used for setting stone or for pointing joints should be made with a nonstaining cement. Pointing consists of raking out joints to a depth of ¾ in. and filling them with pointing mortar, packed into place and rubbed smooth to the indicated level. Joints should be wet before

FIGURE 13-39 Stone trim.

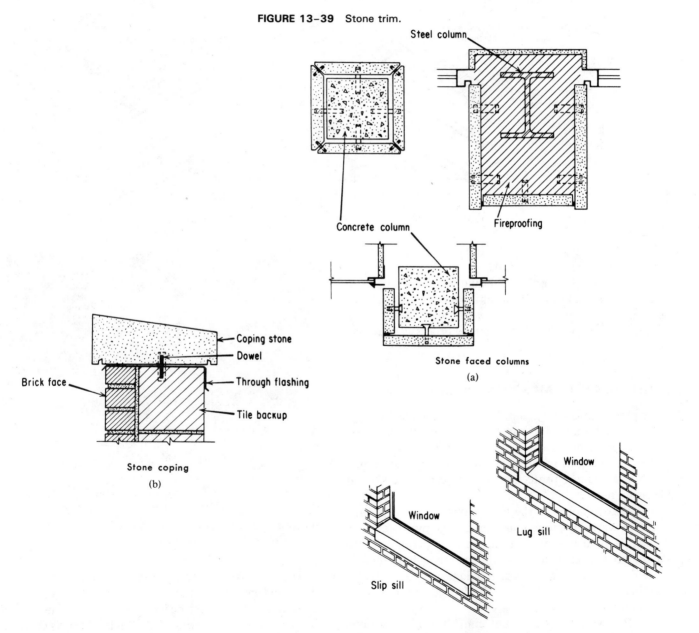

Stone faced columns
(a)

Stone coping
(b)

Stone window sills
(c)

Coping stone — Dowel — Through flashing — Tile backup — Brick face

Steel column — Concrete column — Fireproofing

Window — Lug sill — Slip sill

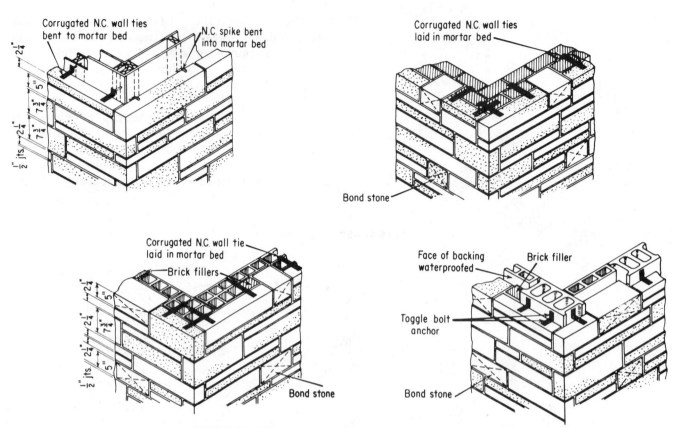

FIGURE 13-40 Anchoring stone to various wall backings.

the pointing mortar is applied. Pointing mortar is omitted if joints are to be caulked.

Wire brushes and acids should not be used to clean stones. As soon as possible after the mortar has set, they should be cleaned with water, soap powder, and fiber brushes. Approved machine-cleaning processes may also be used.

COLD-WEATHER MASONRY WORK

The precautions which must be taken when placing concrete in cold weather also apply to masonry work. Unless protection is provided, the mortar bond may be very poor, resulting in inadequate strength and high permeability. Adequate protection should be provided for at least 48 hr.

Warm mortar is the first requirement. Sand may be kept warm by piling it around a large pipe heated by a gas jet, and it should be turned over periodically to maintain an even distribution of heat throughout the pile (see Fig. 13-41). Warm water should be used also but the temperature of the water will depend on the sand temperature.

But warm mortar is not enough—masonry units should also be warm to prevent the sudden cooling of warm mortar as it comes in contact with cold masonry.

FIGURE 13-41 Keeping mortar sand warm. (*Courtesy Portland Cement Association.*)

It is often difficult to store quantities of bricks or blocks in heated enclosures, and one alternative is to heat smaller quantity of units just before they are to be used. This may be done by covering a small pile of bricks or blocks with a heavy tarpaulin (see Fig. 13-42). With warm mortar and warm masonry, work can go on within a shelter

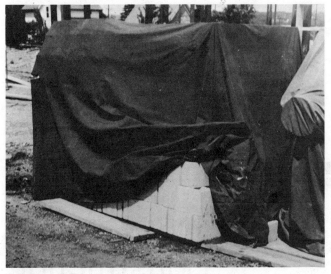

FIGURE 13-42 Heating blocks just prior to use. (*Courtesy Portland Cement Association*.)

FIGURE 13-43 Laying block behind a protecting tarpaulin. (*Courtesy Portland Cement Association*.)

and full bond strength can be developed (see Fig. 13-43). For best results, heat should be maintained for as long as possible.

REVIEW QUESTIONS

1. Explain briefly **(a)** the purpose of a control joint in a block wall, **(b)** the reason for caulking a control joint on the exterior face of the wall, **(c)** the purpose of horizontal joint reinforcement in a block wall, and **(d)** the purpose of a bond beam.

2. By means of neat diagrams, indicate three methods of providing a *lintel* over an opening in a block wall.

3. Outline concisely what you understand by *customized concrete block masonry*.

4. Explain clearly why *tying systems* are so important when prefabricated block panels are being used in a building.

5. Explain the difference between the terms *pattern bond, structural bond,* and *mortar bond,* as used in connection with brickwork.

6. Explain what is meant by **(a)** a brick *wythe,* **(b)** a *weep hole* in a brick wall, **(c)** a *collar joint,* **(d)** *common brick,* and **(e)** *brick veneer.*

7. Describe *why* and *where* you would use *expansion joints* in brick construction.

8. Write a brief summary of *load-bearing brick wall construction,* indicating **(a)** why the system has generated interest, **(b)** the inherent advantages to using such a system, **(c)** the type of building for which it is best suited, and **(d)** how the cost should compare to that of a similar building using curtain wall construction.

9. Explain briefly **(a)** the difference between *ashlar* and *rubble* patterns in stone veneer, **(b)** the purpose of a *bond stone,* **(c)** the reasons for using a *stone belt course* in a masonry wall, and **(d)** how a *stone soffit* may be anchored when using stone for interior finish.

Curtain Wall Construction

The development of steel or concrete structural frames for multistorey buildings has led to the adoption of non-bearing exterior walls known as *curtain walls*. Their primary functions are to protect the interior of the building from the elements and to enhance its appearance.

Curtain walls are not new. They appeared in the nineteenth century, when non-load-bearing walls of clay masonry were first used in tall buildings. Unit masonry is still used as curtain wall material, but there is a trend toward panelizing these products for use in curtain walls. In either case, masonry has to contend, in this field, with large units of lighter, colorful, easily installed panelized materials, where it is shown that they are durable, serviceable, and reasonably priced (see Fig. 14–1).

FIGURE 14–1 Typical curtain walls.

FIGURE 14–2 Exterior bearing walls.

There are a number of advantages resulting from the use of curtain walls in large buildings: (1) There is a reduction in dead load (see Fig. 14–2). Since the curtain wall supports only its own weight, wall thickness may be reduced, thus allowing a reduction in the size of beams, girders, columns, and foundations. (2) Erection time is shortened because of the relatively large size of many curtain wall components. (3) Better insulation is often possible because of the incorporation of insulating materials in the curtain wall panels. (4) More floor space becomes available because walls require less space.

MASONRY CURTAIN WALLS

Masonry curtain walls may be composed of *unit masonry* (see Fig. 14–3) (brick, tile, block, etc.), of *panelized*

FIGURE 14–3 Combination of bearing and curtain walls.

masonry, of *stone panels*, of *precast concrete panels*, or of a *combination of these*.

Unit Masonry

There are two methods of incorporating unit masonry into a building as a curtain wall. One is to build the units into the wall openings formed by columns and beams so that the frame may provide both vertical and lateral support (see Fig. 14–4). The other is to erect the exterior walls independently of any vertical support. The walls carry their own dead load to the foundation (Fig. 14–5), but the structural frame provides the wall with lateral support and carries all other vertical loads. This system allows considerable flexibility between walls and frame. Lateral support is provided by flexible anchors called *looped rods* (Fig. 14–6).

FIGURE 14–4 Brick walls independent of columns. (*Courtesy Morgen Manufacturing Co.*)

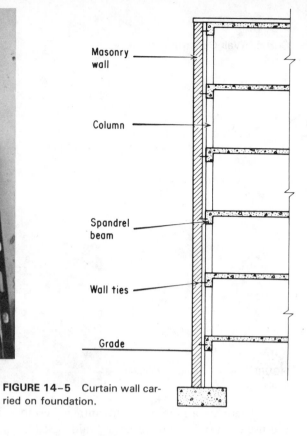

FIGURE 14–5 Curtain wall carried on foundation.

FIGURE 14–6 Lateral supports to beams and columns.

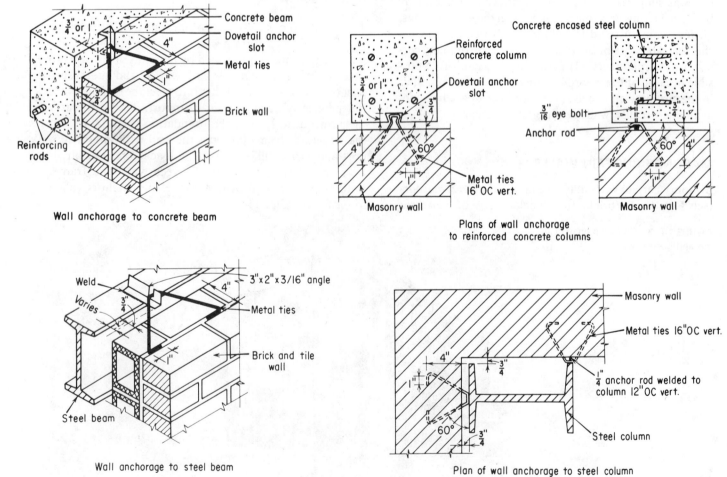

Wall anchorage to concrete beam

Plans of wall anchorage to reinforced concrete columns

Wall anchorage to steel beam

Plan of wall anchorage to steel column

Curtain walls supported entirely by the building frame may be laid and bonded to the spandrel beams with mortar or may be keyed to the beams as in Fig. 14–7. Notice that lateral support is provided here by flat metal ties cast into the column and laid in the mortar joints of the masonry. Also note that the concrete structural frame has been left exposed, as part of the overall design. The faces of a steel structural frame may likewise be left ex-

posed. Figure 14–8 shows a two-wythe brick curtain wall erected in a steel frame so as to leave the column flanges exposed. In Fig. 14–9, a brick cavity wall is used as a curtain wall while leaving the steel frame exposed. Notice the *channels* used to receive the brick and the *cover plates* used to hold the channels. Figure 14–10 illustrates the same type of construction at a beam.

The structural frame is usually concealed by a

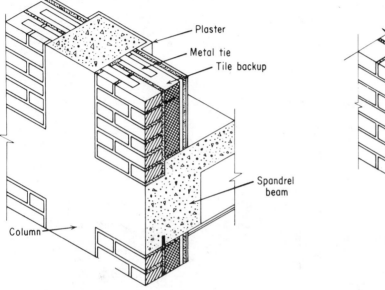

FIGURE 14–7 Curtain wall keyed to a spandral beam.

FIGURE 14–8 Exposed steel column.

FIGURE 14–9 Cavity curtain wall with exposed frame.

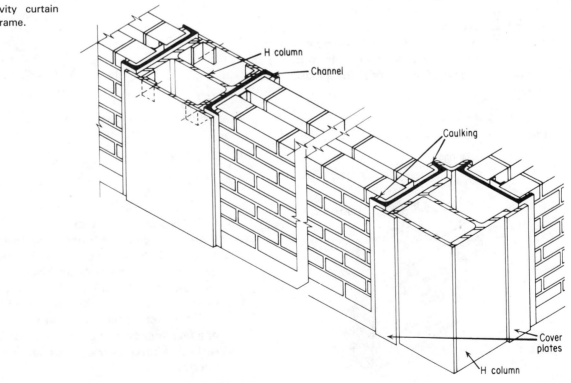

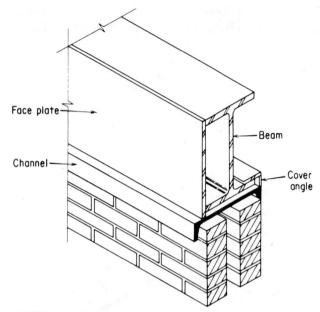

FIGURE 14-10 Cavity wall under beam.

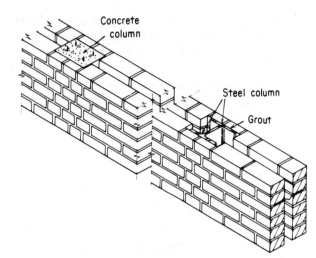

FIGURE 14-12 Structural frame concealed by cavity wall.

masonry curtain wall. In Fig. 14–11, a two-wythe brick curtain wall conceals the exterior of the steel frame. The inner flange is sheathed in furring tile. In this type of construction, the face wythe is carried over the spandrel beams on lintel angles riveted or welded to them. Figure 13–25 illustrates two ways of carrying brick over the face of a beam. A structural frame may be concealed by a cavity curtain wall, as illustrated in Fig. 14–12.

Figure 14–13 illustrates another type of masonry curtain wall, in which a structural tile backup wall is faced with architectural terra-cotta. In this case, the structural frame is steel and fireproofed with concrete. Concrete blocks (standard, lightweight, or cellular concrete) are also widely used as backup material for a masonry curtain wall.

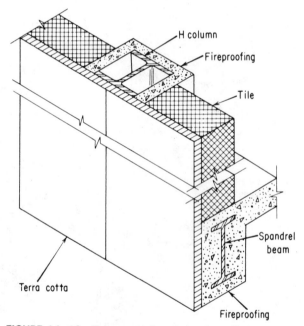

FIGURE 14-13 Terra-cotta-faced curtain wall.

FIGURE 14-11 Steel frame concealed.

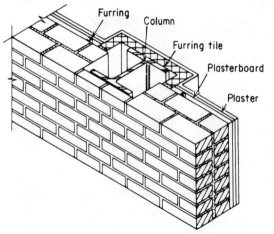

Standard and lightweight concrete blocks are often themselves used as curtain wall material, particularly with a structural concrete frame. Smooth-faced, textured, or sculptured blocks are available for this purpose. Blocks may be laid in running or stack bond, tied to the columns with metal ties (as in Fig. 14–14). Insulation is provided by filling the cells of the blocks with loose fill, such as expanded vermiculite. It is advisable to use either silicone or a *breathing* type of paint waterproofing on the exterior surface. A *vapor-barrier*-type paint should be used on the interior.

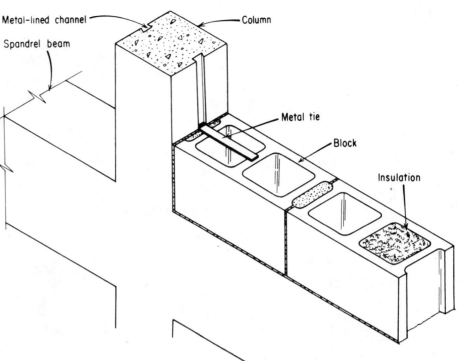

FIGURE 14-14 Concrete block curtain wall.

FIGURE 14-15 Stone panel curtain wall.

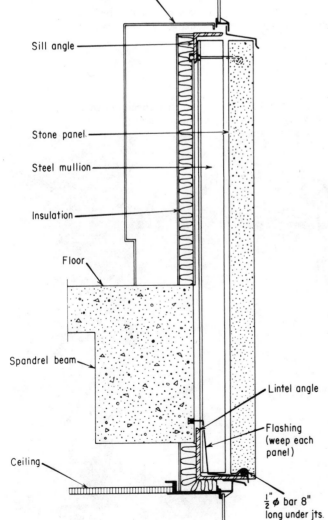

Window stool

Sill angle

Stone panel

Steel mullion

Insulation

Floor

Spandrel beam

Ceiling

Lintel angle

Flashing (weep each panel)

$\frac{1}{2}$" ⌀ bar 8" long under jts.

STONE PANELS

Several types of stone panels are available to be used as curtain walls. One is a 3- or 4-in. thickness of stone dimensioned to fit the particular job. One method of attaching this type of panel is illustrated in Fig. 14-15. In this case, the panels are held in position at the bottom by rods ½ in. in diameter and about 8 in. long, welded to the lintel angle at joints in the stone. Two alternative methods of support are shown in Fig. 14-16.

Another commonly used type of stone panel is a *sandwich* panel, consisting of a stone face, a layer of rigid insulation, and a sheet metal or hardboard back—all cemented together. This type of panel is relatively light and is often supported by a metal subframe or *grid*, which is in turn connected to the structural frame (Fig. 14-17). A great variety of metal grids is available for use with stone or other types of curtain wall panels. Figure 14-17 shows a typical aluminum grid being used with stone sandwich panels.

FIGURE 14-16 Alternative stone panel supports.

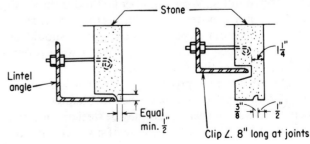

Stone

Lintel angle

Equal min. $\frac{1}{2}$"

$1\frac{1}{4}$"

$\frac{3}{8}$" $\frac{1}{2}$"

Clip ∟ 8" long at joints

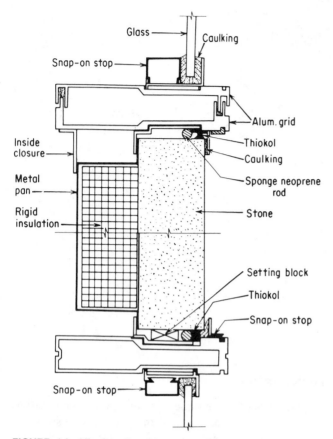

FIGURE 14-17 Details of stone sandwich panel subframe.

FIGURE 14-18 Factory-made precast concrete wall panels. (*Courtesy Portland Cement Association.*)

FIGURE 14-19 Erecting precast concrete panels. (*Courtesy Portland Cement Association.*)

PRECAST CONCRETE CURTAIN WALL

Two types of precast concrete units are used in curtain wall construction. One is a panel which is cast, cured, and finished in a precasting plant (see Fig. 14–18) and brought to the job site for erection in much the same way that stone or masonry panels would be used. Such precast panels are custom-made to the size and with the surface texture specified for a particular building. See Chapter 9 for details of architectural precast concrete wall panels.

Figure 14–19 shows typical precast concrete panels being erected over a structural steel frame, and Fig. 14–20 indicates how those panels are anchored to the frame. Many other anchoring systems are in use, some of which are illustrated in Fig. 9–56. Another method is shown in Fig. 14–21, in which steel dowels project into recesses in the top and bottom edges of panels. Joints should be sealed as described in Chapter 9, using a waterproof, elastic caulking at the exterior face.

The other type of precast concrete unit is produced by the *tilt-up* method, also described in Chapter 9. The structural frame for such panels may be steel or concrete, cast-in-place or precast. In the case of a structural steel

FIGURE 14-20 Panels anchored to frame. (*Courtesy Portland Cement Association*.)

FIGURE 14-22 Panels erected before frame. (*Courtesy Portland Cement Association*.)

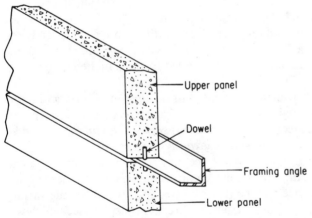

FIGURE 14-21 Dowel anchor for precast concrete panel.

Upper panel

Dowel

Framing angle

Lower panel

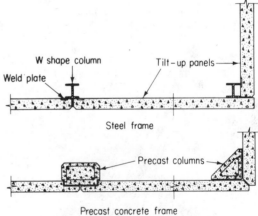

W shape column — Tilt-up panels

Weld plate

Steel frame

Precast columns

Precast concrete frame

FIGURE 14-23 Flush panel tilt-up wall.

FIGURE 14-24 Exposed columns with tilt-up panels.

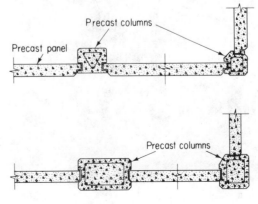

Precast columns

Precast panel

Precast columns

frame, it will be erected prior to the panels being set up. A precast concrete frame may be erected before or at the same times as the panels, while, in many cases, a cast-in-place concrete frame may be placed *after* the panels are erected (see Fig. 14-22).

Panels may be designed to produce a *flush* tilt-up wall, usually the case with a steel frame (see Fig. 14-23) or columns may be exposed (see Fig. 14-24).

FIGURE 14–25   Placing concrete for tilt-up panels. (*Courtesy Portland Cement Association*.)

Forms for the panels are built on the job site, often on the sloor slab; the necessary reinforcing is installed, and the concrete is placed (see Fig. 14–25). Finishing produces the specified surface, and the concrete is properly cured. When the panels have gained sufficient strength, they are lifted or *tilted* into position (see Fig. 14–26). They are held in position at the bottom by dowels projecting from the foundation or by welding plates which have been

FIGURE 14–26   Raising tilt-up panel into place. (*Courtesy Portland Cement Association*.)

cast into the bottom edge of the panel and the top of the foundation. When panels are to be set before the columns, they will be cast with reinforcing bars projecting from their edges to be cast into the columns (see Fig. 14–22). After plumbing and bracing the panels, simple wood forms are built around the openings and concrete is placed. Columns may be made the width of the opening or wide enough that the edges of the panels will be cast into them.

Sandwich wall panels are specified for many buildings where curtain wall construction is being used. Sandwich panels consist of a core of rigid, low-density insulation, covered on both sides with a thin layer of reinforced, high-strength concrete. The panel may be cast with solid outer edges so that the core is completely enclosed, or if adequate *shear connectors* are provided, the insulation may be extended to the panel edges (see Fig. 14–27).

Surface treatment of tilt-up panels is a very important part of the whole operation. Basically, six types of surfaces may be produced: *smooth, textured, patterned, colored, ribbed,* or *exposed aggregate.*

A smooth surface may require only screeding and floating, or if a more uniform appearance is needed, the surface may be troweled (see Fig. 14–28). If the surface is to be painted, a bond-breaker must be used which will not stick to the surface of the concrete and prevent paint from adhering.

Colored surfaces are achieved by the use of white cement or color pigments. A white surface will usually

FIGURE 14–27   Sandwich panel.

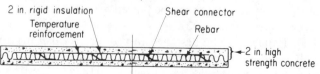

2 in. rigid insulation

Temperature reinforcement

Shear connector

Rebar

2 in. high strength concrete

FIGURE 14-28 Finishing tilt-up panel. (*Courtesy Portland Cement Association.*)

be produced by adding a thin layer of concrete made with *white cement* and *white sand* to the face of the normally cast panel. Color pigments are troweled into a normal concrete surface at specified rates in pounds per square foot to produce the color required.

A textured (roughened) surface may be produced by *brooming, sandblasting, using a cork float,* or *lining the form with rubber matting* or some type of *foamed plastic.*

A patterned surface may be made by forming the pattern with *strips or blocks of foamed plastic* cemented to the casting surface or by using *wood strips* in the same manner. More complicated patterns may be made by producing the shapes desired in a *bed of damp, hard-packed sand* laid over the casting surface.

Ribbed surfaces are formed by placing *ribbed, corrugated,* or *striated liners* on the casting surface as a bottom form. In some cases, wedge-shaped wood strips may be used if there is some way to fasten them down. Such wood strips should be well oiled to prevent them from swelling and coated with bond-breaker for easy stripping.

Exposed aggregate surfaces are produced in several ways. One method is to cast the panel with the face side up and *seed* the aggregate evenly over the fresh concrete surface by shovel or by hand. It is then embedded in the surface by hand float or straightedge until all aggregates are entirely embedded and mortar surrounds and slightly covers all the particles. When the concrete has almost attained its final set, the surface is washed with water and a stiff brush to remove the cement paste and expose the aggregate.

When larger, uniformly sized aggregate is to be used, a commonly used method is to cover the casting surface with a uniform layer of sand to the depth of about one-third the diameter of the aggregate. Pieces of aggregate are then pushed into the sand as close together as possible and wet down with a water spray to settle the sand around each piece. The reinforcement is then set into the form on chairs and the concrete placed.

Still another method is used with flat, irregular shapes and sizes of stone. They are laid face down in the form and the spaces between them filled with dry sand to a depth of about one-third to one-half the thickness of the stone. The stone is moistened and the remainder of the spaces filled with mortar, after which the regular reinforcing and placing of concrete follows. Excess sand may be removed after the panels are raised by brushing or air blasting.

LIGHTWEIGHT CURTAIN WALL

In addition to those previously mentioned, curtain wall panels are available in a wide variety of materials, including plastic, steel, porcelain enamel, asbestos-cement, stainless or galvanized steel, aluminum, copper, bronze, and Galbestos. They are made in single-skin form and in open or closed sandwich panels in a wide range of dimensions. These panels are comparatively lightweight and are often supported on a subframe made of vertical or horizontal bars or on a grid (Fig. 14-29). Some types

FIGURE 14-29 Curtain wall subframe systems.

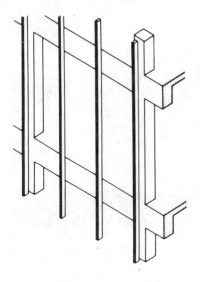
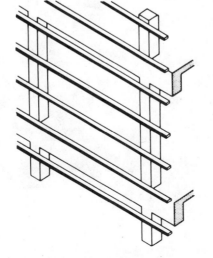
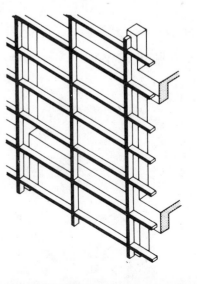

(such as the one shown in Fig. 14–30) require no subframe and are attached to the regular framework, as illustrated in Fig. 14–31. There are many methods for attaching the subframe to the structural frame, four of which are shown in Fig. 14–32.

Figures 14–33 and 14–34 illustrate typical applications of curtain walls. In Fig. 14–33 two sizes of panels are mounted together to an aluminum subframe. Notice the swing stage scaffold used in erecting the panels. In Fig. 14–34, panels and wide mullion strips are mounted

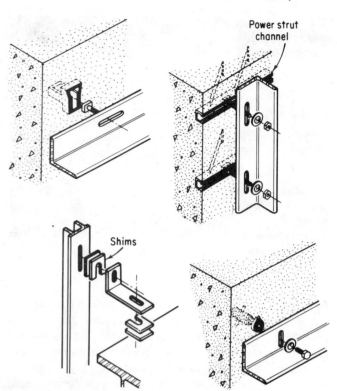

FIGURE 14–32 Methods of attaching a subframe to a concrete or steel structural frame.

FIGURE 14–33 Curtain wall with a subframe.

FIGURE 14–30 Steel sandwich panels being installed.

FIGURE 14–31 Curtain wall panels clipped to a structural frame.

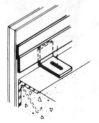
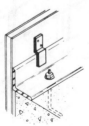
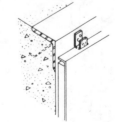

FIGURE 14–34 Curtain wall clipped to a concrete block backup wall.

on the backup wall without the use of a subframe, and a complete scaffolding system is being used during the erection of the panels.

GLASS CURTAIN WALL

Many styles of subframes may be used to mount glass in curtain walls of this kind. Styles depend on whether the wall is single- or double-glazed, on whether the units are fixed or movable, on the thickness of the glass, and on the size of each unit. Figure 14–35 illustrates a typical aluminum subframe for a double-glazed plate glass curtain wall.

The glass curtain wall in the building shown in Fig. 14–36 is transparent from the inside but reflective from the outside during daylight hours.

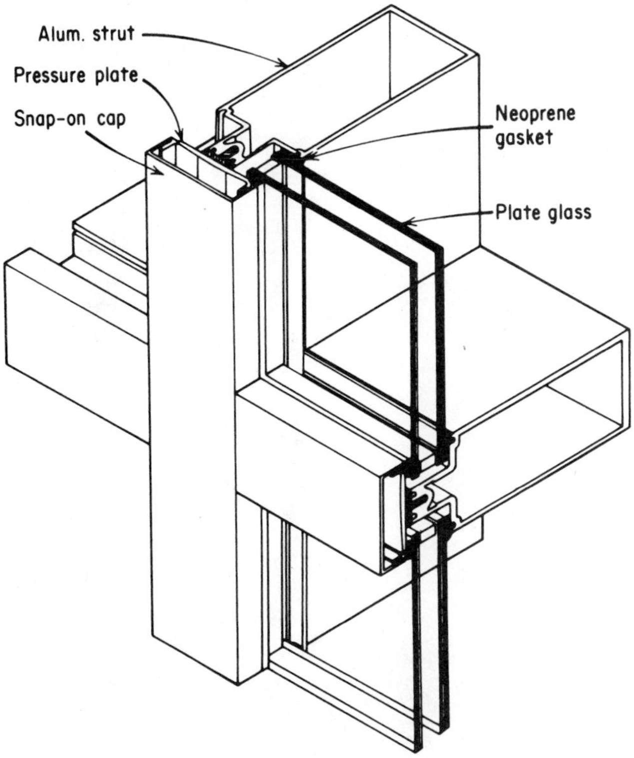

FIGURE 14–36 Glass curtain wall. (*Courtesy Libbey-Owens-Ford Company.*)

FIGURE 14–35 Glass curtain wall subframe details.

Alum. strut

Pressure plate

Snap-on cap

Neoprene gasket

Plate glass

REVIEW QUESTIONS

1. Explain the meaning of each of the following terms: **(a)** curtain wall, **(b)** panelized masonry, **(c)** sandwich panel, and **(d)** spandrel beam.

2. Outline *four* advantages of the use of curtain walls in large buildings.

3. Explain the purpose of each of the following: **(a)** a *backup wall,* **(b)** *looped rods* in curtain wall-to-frame connections, (c) a *cavity wall* in curtain wall construction, and **(d)** a *setting block* used in stone panel erection.

4. By means of neat diagrams illustrate one practical method of providing **(a)** a *textured surface,* **(b)** a *ribbed surface,* and **(c)** a patterned surface to a tilt-up concrete panel.

5. Outline two basic reasons for using a subframe to carry curtain walls over a structural frame.

6. (a) Illustrate by a horizontal section how you think the steel sandwich panels shown in Fig. 14–33 are constructed. **(b)** Outline, with the aid of diagrams, a practical method for attaching those panels to the building.

Building Insulation

Building insulation is generally considered to be any material which has a particularly low coefficient of thermal conductivity. This is illustrated in Fig. 15–1, where a slab of cellular glass successfully resists the passage of

FIGURE 15–1 Good thermal insulator. (*Courtesy Pittsburgh-Corning Corp.*)

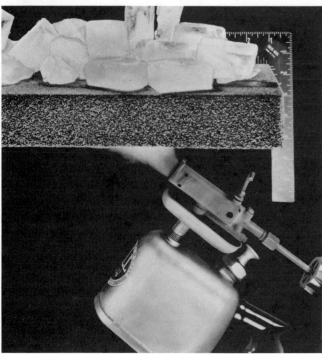

heat. In fact, the term also includes those materials which prevent or reduce the transfer of sound and those which restrict or prevent the migration of moisture and moisture vapor.

THERMAL INSULATION

The ability of a material to resist the passage of heat serves many purposes in building construction. One specific use is the protection of building components against deterioration or failure caused by severe overheating. Materials used for this purpose are generally referred to as *fireproofing*.

Fireproofing Materials

Fireproofing is most commonly used to protect the structural frame of a building, particularly a steel frame (although it may be used for the protection of wood or concrete frames as well). Materials in general use for the protection of a steel frame include regular and lightweight concrete, cellular concrete, gypsum tile, terra-cotta tile, plaster made with lightweight aggregates, sprayed-on asbestos fiber, mineral wool, and cellulose fiber.

Methods of Fireproofing

Fireproofing steel beams and columns with concrete is accomplished by encasing the members in cast-in-place concrete. This involves building a form around the member or using a prefabricated form. When concrete floor slabs are used with a steel frame, fireproofing is usually cast as part of the slab.

    Many systems of forming are available, but three common ones are illustrated in Figs. 15-2, 15-3, and 15-4. In Fig. 15-2, the forms are supported from below by posts. Notice that the fireproofing form is built in such a way that the sides may be removed without interfering

with slab form supports or beam bottom supports. This is done by using *kickers* near the top edges of the sides and ledger strips along their bottom edges. The form shown in Fig. 15-3 is supported by the steel frame. This is accomplished by hangers which are placed over the beams. The formwork is suspended from the legs of these hangers and is in this particular case held in place by standard wedges. Fixed spreaders eliminate the need for soffit spacers.

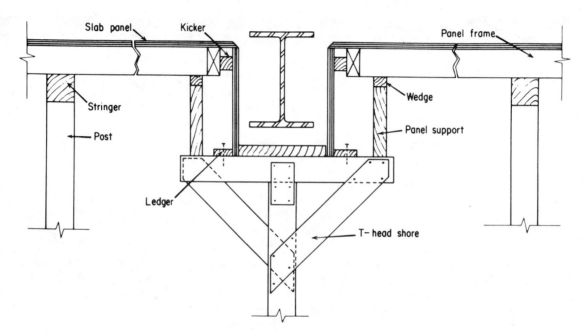

FIGURE 15-2   Fireproofing form supported form below.

FIGURE 15-3   Fireproofing form supported by hangers.

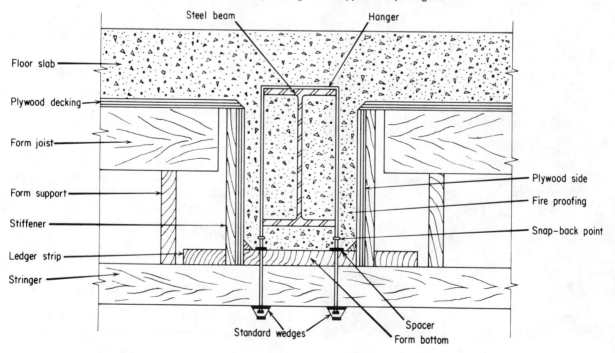

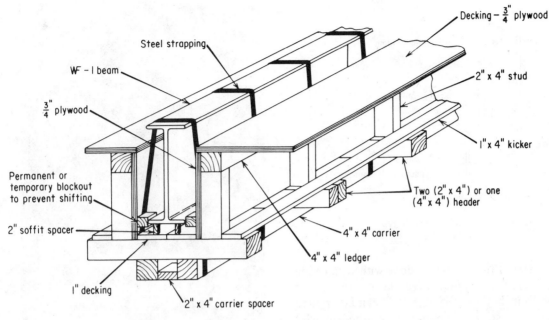

FIGURE 15-4 Fireproofing form supported by steel strapping.

The formwork shown in Fig. 15-4 is also hung from the steel frame, but heavy-duty steel strapping is used in this case to support the form and concrete load. Note that soffit spacers and *blockouts* are used here to hold the form in its proper position.

Columns are fireproofed with concrete by first building or fitting a form around them and then filling it with concrete. It is sometimes advantageous to build the form in two sections—upper and lower. Place and fill the bottom section first, and then the top section.

Cellular concrete makes an excellent fireproofing material because of its high insulation value and its light weight. It is used in 2-, 3-, or 4-in. slabs which may be

fabricated to suit any particular situation. Figure 15-5 illustrates a typical application of cellular concrete fireproofing to a steel girder, while Fig. 15-6 shows the same material used to fireproof a column.

Gypsum wallboard is one of the most common materials used as a covering over framing members to provide fire protection. A $\frac{5}{8}$-in.-thick fire-rated wallboard can provide up to a 1-hr fire rating when properly applied. Figure 15-7 illustrates a steel column covered with two layers of $\frac{5}{8}$-in. fire-rated gypsum wallboard. An assembly of this kind can provide a 2-hr fire rating if the steel column is of sufficient thickness. In addition to being a good fire retarder, gypsum wallboard is economical, light in weight, easy to install, and provides a smooth surface that is easily finished. It is used as a covering on partitions, walls, and ceilings and can be used with wood framing, metal stud framing, concrete, or masonry.

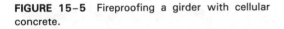

FIGURE 15-5 Fireproofing a girder with cellular concrete.

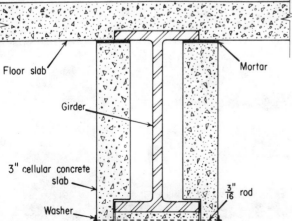

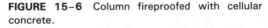

FIGURE 15-6 Column fireproofed with cellular concrete.

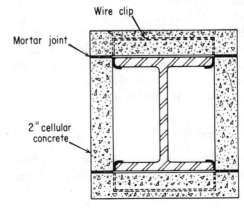

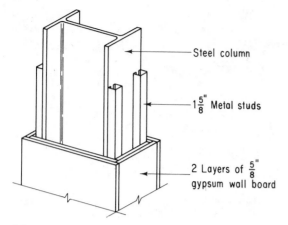

FIGURE 15-7 Steel column protected with gypsum wallboard.

A great deal of fireproofing is done with vermiculite or perlite plaster. Not only do these plasters resist heat transmission, but they are very light and can be applied quite rapidly. The steel members are sheathed in wire mesh or metal lath, stiffened by the addition of light rods if necessary, and the plaster is sprayed over the wire to the required depth.

Asbestos fiber mixed with adhesive may be used the same way or may be applied directly to the steel member. In the latter case, a primer coat must first be applied to the steel in order to bond the fiber layer to its surface.

Asbestos-free sprayed-on insulations such as mineral wool and cellulose fiber are also used as fire retardants. Applied with special spraying equipment, they are often used where irregular surfaces require protection, such as structural steel members and open-web joists. Being asbestos-free, these insulations can be left exposed, as in the case of ceilings where they provide a pleasing textured finish.

Insulation Against Temperature Changes

Another use of insulating materials is to prevent significant temperature changes from taking place in structural members. Such changes would cause movement through thermal expansion. This applies particularly to tall buildings where expansion of long columns could be considerable. Figure 15-8 shows a concrete column being insulated against temperature changes. Cellular glass slabs are held in place by mastic and pins which have been cast into the column. Clips are driven over the ends of the pins projecting through the insulation.

Insulation of Heat Conveyors

An important use of insulation is to prevent heat loss from hot water pipes and hot air ducts. Rigid slabs,

FIGURE 15-8 Insulating a column against temperature changes. (*Courtesy Pittsburgh-Corning Corp.*)

mineral wool blankets, and asbestos insulation are all used for this purpose.

Rigid insulation is shaped around pipes (as shown in Fig. 15-9, and is held in place by fabric or asbestos wrapping. Large pipes may be covered by a heavy layer of plaster made from asbestos fiber mixed with adhesive and water and wrapped in cloth or fireproof paper.

Air ducts maybe sheathed in rigid slab insulation,

FIGURE 15-9 Insulating a pipe with cellular glass. (*Courtesy Pittsburgh-Corning Corp.*)

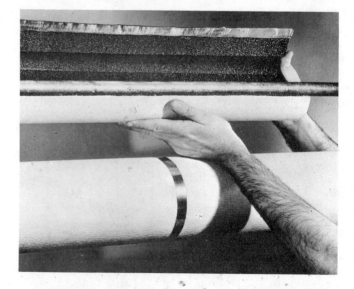

asbestos fiber plaster, rigid boards of corrugated asbestos paper, or mineral wool blankets.

Building Insulation

Heat always seeks lower temperature levels. As a result, whenever there is a difference in temperature between two surfaces of a material, heat will flow to the colder side. This means that when the outside temperature is low, heat from a building will flow to the outside. Conversely, during hot weather, heat tends to flow inward and warm the inside of a building.

Any material will retard the flow of heat to some extent—but in order to effectively control this flow, insulating materials are used in the foundations, floors, walls, ceilings, and roofs of buildings. Heat is transferred by three means: conduction, convection, and radiation; insulation must control all three if it is to be effective. Most insulating materials are effective in controlling heat transfer by conduction and convection, but a special kind of insulation, one with a reflective surface, is necessary to control heat transfer by radiation.

The *thermal conductance, C,* of a material is defined as the amount of heat in Btu (British thermal units) transmitted in 1 hr from surface to surface over 1 sq ft of area of a material or combination of materials producing a temperature difference of 1 °F between the two surfaces. The resistance to heat flow through a material, or *R*-factor, is the inverse of the conductance, (i.e., $1/C$). Each material has its own particular R value, and the greater the R-factor the better are the insulating qualities of the material. Typical R-factors are listed in Table 15–1 for some of the more common building materials.

The overall *heat transmission coefficient,* or *U*-factor, is the amount of heat in Btu transmitted in 1 hr through 1 sq ft of a building section (roof, wall, or floor) for each degree F of temperature difference between air on the cold side and air on the warm side. The U-factor

is calculated by summing the individual R-factors of the various components of the building section and taking the inverse of that total. In equation form,

$$U\text{-factor} = \frac{1}{R_{\text{total}}}$$

Example

A building wall section is composed of the following materials:

- 4 in. face brick
- 2 in. air space
- 8 in. hollow concrete block
- ½ in. drywall

Calculate the overall heat transmission factor, U, for the wall.

Solution

First, determine the total resistance using the values from Table 15–1.

Outside air	= 0.17
4 in. face brick	= 0.44
2 in. air space	= 0.97
8 in. concrete block	= 1.19
Inside air	= 0.68
R_{total}	= 3.90

$$U = \frac{1}{R_{\text{total}}} = \frac{1}{3.90}$$

$$= 0.256$$

If the 2-in. air space is filled with 2 in. of polystyrene insulation the R-factor is increased from 0.97 to 10, giving a total R-factor of 12.93 and a resulting U-factor of 0.077, making the wall about 3⅓ times more resistant to heat loss.

Insulation Application

1. *Perimeter Insulation.* Buildings constructed on surface foundations are subject to perimeter heat loss and footing movements. Perimeter insulation helps control that heat loss and protects the footings from frost action. Figure 15–10 illustrates two methods using rigid insulation to prevent frost penetration under footings of *shallow* foundations. Polystyrene foam is preferred for use as perimeter insulation because of its high thermal resistance, low water absorption, and high compressive strength.

Insulation around the perimeter of a building may be placed on the inside or the outside face of the founda-

TABLE 15–1 Material *R*-factors

Material	R-Factor
1 in. rigid glass fiber	4.00
1 in. polystyrene	5.00
1 in. polyurethane	6.25
4 in. fiberglass batts	12.00
½ in. gypsum wallboard	0.45
8 in. concrete block (normal density)	1.19
4 in. brick	0.44
Stucco	0.20
Outside air surface at 15-mph wind	0.17
Inside air	0.68

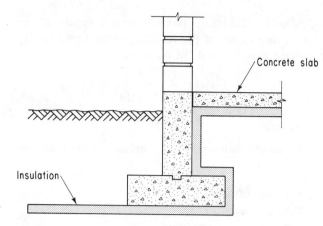

Unheated building

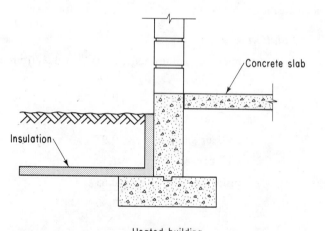

Heated building

FIGURE 15-10 Shallow footings protected against frost action.

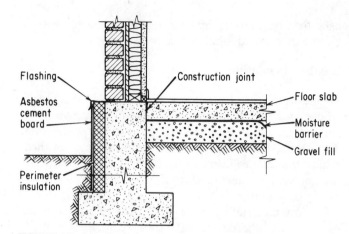

FIGURE 15-11 Perimeter insulation outside.

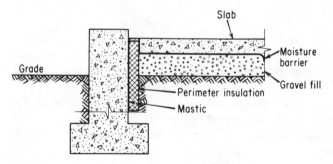

FIGURE 15-12 Perimeter insulation with slab on grade.

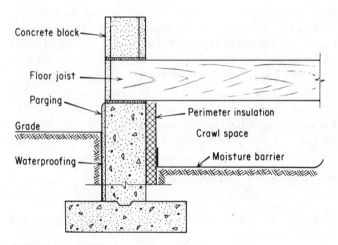

FIGURE 15-13 Perimeter insulation in crawl space.

tion. When applied to the outside (Fig. 15–11), the portion above grade must be protected from physical damage with a covering of treated wood, stucco on metal lath, or asbestos-cement board. If polystyrene foam is used, it must be protected from the sun's rays to prevent deterioration. Many codes recommend that the insulation be carried a minimum of 24 in. below finished grade but better results can be obtained by carrying the insulation down to the top of the footings. When *slab-on-grade* construction is involved, the perimeter insulation should be installed prior to backfilling of the foundation trench. The insulation should be carried down to the frostline or down to the top of the footing (Figure 15–12). Notice that the insulation reaches the top of the floor slab. If the area under the floor slab is to be used as a crawl space, several steps are necessary to install perimeter insulation. First, the ground surface must be covered with a moisture barrier which is lapped at least 2 in. up the wall (see Fig. 15–13). Wall ventilators which will not allow water to enter the foundation must be provided. The insulation must

cover the entire crawl space walls down to ground level, or preferably to the frost line. If the floor slab is to contain heating or air conditioning ducts, insulation should be carried along the ground under them (Fig. 15–14). Rigid insulation can also be used under the slab-on-grade to further reduce heat loss. For this application, a vapor barrier should be used over the insulation and good drainage must be provided under the insulation.

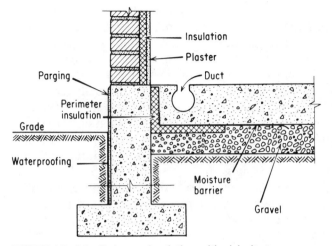

FIGURE 15-14 Perimeter insulation with slab ducts.

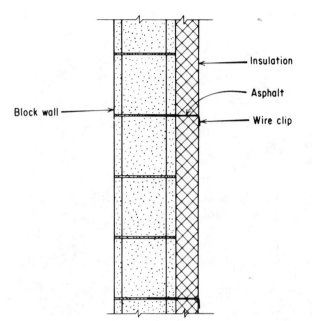

FIGURE 15-15 Insulation held by wire clips.

2. *Foundation Wall Insulation.* Foundation walls are usually insulated with some type of rigid insulation, including insulating boards and rigid slabs or blocks. The insulating boards include wood and cane fiber boards, straw board, and laminated fiber board. Rigid slab insulations include cellular glass (see Fig. 15-8), foamed plastic rigid insulation, and cellular concrete. Additional insulation may be introduced into concrete block walls by pouring loose fill into the block cores during construction. Vermiculite is commonly used for this purpose.

Rigid slab insulations are applied to the wall surface with hot asphalt, portland cement mortar, special synthetic adhesives, or wire clips. Walls must be clean and smooth before insulation is applied to ensure a good bond. It may be necessary to back-plaster the wall to achieve the necessary finish.

If asphalt is used as the adhesive, the surface must first be coated with asphalt primer and allowed to dry. Adhesive is then applied to the insulation by dipping the back face and two adjacent edges of each block in hot asphalt. Blocks are applied to the primed surface while the asphalt is still molten. Cellular glass, foamed plastic, and cellular concrete may all be applied by the hot asphalt method.

When synthetic adhesive is specified, the back face and two adjacent edges of each slab are given a trowel coating of adhesive approximately ⅛ in. thick. Sufficient pressure must be applied to the blocks to assure tight joints and good contact with the wall. Both cellular glass and foamed plastic may be applied with synthetic adhesives.

Rigid slab insulation may be applied to block or brick walls by either of the methods described above or may be held in place by wire clips, one end of which has been embedded in a mortar joint (Fig. 15-15). The edge joints should still be sealed first with hot asphalt or some other adhesive.

Portland cement mortar is frequently employed in the application of cellular concrete and foamed plastic insulations to masonry walls. When using the latter, mortar is often applied by means of a *push box*—a simple device illustrated in Fig. 15-16. To use the push box, place a slab of insulation in the frame with one end directly below the front end of the box (which may be adjusted to regulate the depth of coating), fill the box with mortar, and push the slab through with another directly behind it. Keep the hopper full for best results.

Rigid boards may be applied to foundation walls in several ways. One method is to strap the wall, as in Fig. 15-17, and attach the board insulation to the straps with nails. If the foundation wall is cast-in-place, insulation may be attached inside the inner form face. This may

FIGURE 15-16 Simple push box.

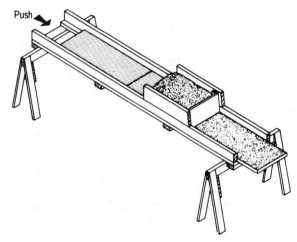

FIGURE 15–17 Foundation wall strapped for insulation.

be done with nails, or, if the boards are thick, with wires through the form sheathing and insulation. The concrete adheres to the face of the insulation, and the insulation remains permanently attached to the wall when the forms are removed. The hot asphalt or the synthetic adhesive method may also be used to secure board insulation to foundation walls.

3. *Curtain Wall Insulation.* Masonry curtain walls may be insulated by any of the methods applicable to foundation walls. Brick or block curtain walls are often built as cavity walls, in which case insulation may be installed within the cavity. It may be loose fill, such as vermiculite, perlite, cellulose fiber, or gypsum fiber, or it may be rigid slab insulation thick enough to fill the cavity. Care should be taken that mortar does not protrude from joints so as to restrict the flow of insulation being poured or blown into place. When using slabs, insulation must be installed as construction proceeds, since the metal ties between the inner and outer wythes of the wall should occur at joints in the insulation.

Another insulating technique for masonry walls involves the use of cellular concrete blocks. These are used to build a backup wall which is faced with a veneer of brick, tile, terra cotta, or stone.

Curtain walls having a stud frame may be insulated with batts placed between the studs, with rigid insulation applied over plywood or other sheathing, or by *foamed-in-place* insulation. Insulation is foamed-in-place by placing a specified amount of plastic resin and activating agent in each space to be insulated. The two materials

react to form a thermosetting, insulating foam which will fill each space to a predetermined depth. The operation is repeated until each space is completely full.

Prefabricated panels for curtain wall construction are very often manufactured with the insulation built in as the core of a sandwich. Such walls usually require no further insulation. If single-skin panels are used, provision must be made for applying the insulation to the back of the panels or to a rigid sheath erected behind them.

Insulation for glass curtain walls is provided by the use of sealed units. In the case of glass blocks or tiles, the two halves or each unit are fused together, enclosing a dry, low-pressure air space. Sheets or plates are hermetically sealed around the edges, in pairs, leaving a space between them.

4. *Ceiling Insulation.* Insulation may be applied directly to the underside of floors, or a suspended ceiling system may be employed. In either case, many kinds of insulation may be used, including insulating boards, rigid slabs, or sprayed-on insulation consisting of vermiculite plaster or asbestos fiber.

When insulation is to be applied under concrete floors, nailing strips should be cast into the bottom of the slab on appropriate centers for the type of insulation being used (Fig. 15–18). The surface must be primed, and the first layer of insulation is applied at right angles to the nailing strips, using hot asphalt or an approved adhesive on the contact face and edges. This layer is further secured with nails and washers (see Fig. 15–18). The second layer is then applied at right angles to the first, using the appropriate adhesive.

Suspended ceilings are usually supported by hanger rods from the roof, floor frame, or the slab above. They support a system of T-bars or metal pans, which in turn support the rigid insulation (Fig. 15–19) or the material which will act as a base for sprayed-on insulation. The space above the insulation must be thoroughly ventilated by free or forced circulation of air.

T-bars of steel or aluminum are suspended on 18- or 24-in. centers by rod hangers which are placed close enough together to prevent any deflection of the completed ceiling. The first layer of rigid insulation is laid between the T-bars after they have been coated with asphalt priming paint. The second layer is applied on top

FIGURE 15–18 Insulation on solid ceiling.

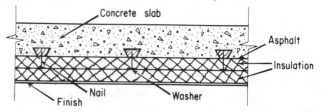

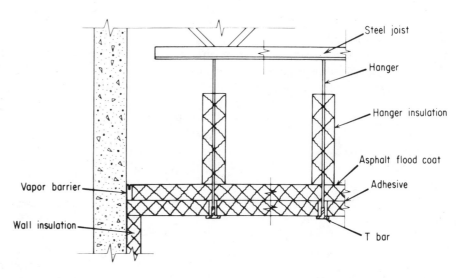

FIGURE 15-19 Suspended T-bar ceiling.

of the first, using a flood coat of hot asphalt or a coating of approved adhesive on the contact surface and edges of each slab. The insulation should be laid in parallel courses, staggering joints between courses and between layers. Finally, the upper surface of the insulation should receive a flood coat of hot asphalt or a ⅛-in. coating of asphalt emulsion. The hanger rods should also be insulated to a distance of about 18 in. from the surface of the insulation (see Fig. 15-19) with at least half the thickness of insulation used on the ceiling.

When sprayed-on insulation is specified, the T-bar or metal pan support system may be used to carry some type of rigid board or gypsum lath (see Fig. 15-20). Another alternative is to suspend a system of *runner channels* and *furring channels* and to wire metal lath to the underside of this framework (see Fig. 15-21). Sprayed-on insulation may be applied to either of these two bases.

Figure 15-22 illustrates the application of sprayed-on insulation on a concrete ceiling.

Framed ceilings may be insulated with batt insulation between the framing members, and any rigid ceiling with accessible space above it may be insulated with loose fill poured or blown into place.

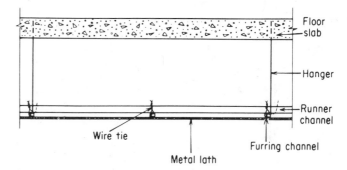

FIGURE 15-21 Suspended ceiling frame.

FIGURE 15-20 Suspended metal pan ceiling.

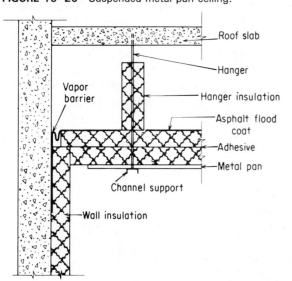

FIGURE 15-22 Sprayed-on insulation on concrete ceiling.

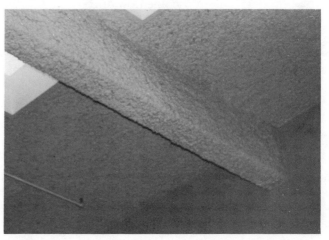

5. *Roof Insulation.* Roof decks may be insulated with any type of rigid insulation, including cellular glass, rigid foamed plastic, straw board, fiber boards (usually double- or triple-laminated), and cellular concrete. With concrete, gypsum, precast slab, or steel decks, no undercoating is normally required, but sheathing paper or roofing felt is nailed over a wood deck (see Fig. 15–23).

To obtain a proper bond between insulation and decking, all surfaces must be dry before and during the application of the insulation. It is applied to the prepared surface by embedding it in hot asphalt, used at the rate of about 25 lb/100 sq ft. On steel decks, the long dimension of the insulation boards should be parallel to the ridges of the deck, with the edges of the boards resting on the ridges.

Use of Reflective Insulation

Reflective insulation may be produced from any metallic substance which has the ability to reflect infrared rays. Aluminum foil has proved to be the most practical and

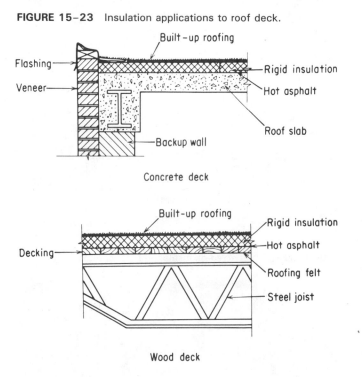

FIGURE 15–23   Insulation applications to roof deck.

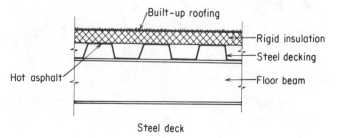

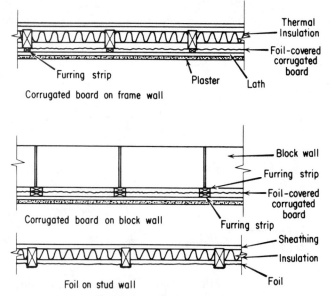

FIGURE 15–24   Typical applications of reflective insulation.

economical material for this purpose. It is produced in several forms, including foil-backed gypsum lath, rigid insulation with foil back, foil laminated to kraft paper, corrugated paper with foil on both sides, and plain sheet foil in rolls.

To be effective, reflective insulation must have an air space of at least ¾ in. in front of it. Figure 15–24 illustrates a few typical applications of foil insulation in which the air space is provided.

SOUND INSULATION AND CONTROL

There are two main problem areas concerning sound and its control to be found in buildings old and new. One is the *improvement of hearing conditions* and the *reduction of unwanted noise* in any given room, and the other is the control of sound transmission from one room to another through walls, floors, and ceilings.

Sound waves travel through the air in the form of small pressure changes alternately above and below the normal atmospheric pressure (see Fig. 15–25). The aver-

FIGURE 15–25   Diagram of a sound wave.

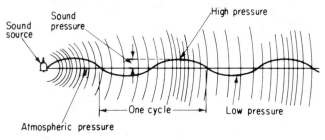

age variation in pressure above and below the normal is called *sound pressure,* which is related to the loudness of a sound. A sound wave is one complete cycle of pressure variation, as illustrated in Fig. 15–25.

The number of times this cycle occurs in 1 sec is the *frequency* of the wave, and the unit of measurement of frequency is called a *hertz,* abbreviated *Hz,* which represents 1 cycle per second.

The improvement of hearing conditions and reduction of unwanted noise is accomplished by the proper *design of inner walls and ceilings* and *control of* the amount of *sound reflected* from walls and ceilings. Reflected sound causes *reverberations* or echoes, which often distort the original sound.

The control of reverberations is accomplished by the use of products which have a much greater ability to absorb sound waves than most building materials. The amount of sound absorbed is measured in *sabins,* one sabin being equal to the sound absorption of 1 sq ft of perfectly absorptive surface (in practice no such surface exists). The fraction of sound energy absorbed, at a specific frequency, during each reflection, is called the *sound absorption coefficient* of that surface. Some surfaces are primarily reflective (for example, glass, concrete, masonry), absorbing perhaps 5% or less of the sound energy. Such materials have sound absorption coefficients of 0.05 or less. On the other hand, some acoustical materials can absorb 90% or more of the sound energy and so have a coefficient of 0.90 or better.

Since most sounds contain a range of frequencies, it is necessary to use an *average* of the absorption coefficients when considering sound absorption. A means of comparing reductions for noise which is mainly in the middle frequencies has been developed. It is called the *noise reduction coefficient* (NRC) and represents the average amount of sound energy absorbed over a range of frequencies between 250 and 2000 Hz.

The exact amount of sound absorption required for various rooms is difficult to state because of individual differences in noise interpretation, but three general rules of thumb may be used as a guide:

1. In larger rooms with low ceilings and widely spaced, medium-density noise sources, such as general offices, the *minimum average absorption coefficient* after acoustical treatment of all surfaces should be 0.20.
2. In smaller rooms containing closely spaced or high-level noise sources, such as business machines, the maximum average absorption coefficient after treatment should be 0.50.
3. To create a definite, noticeable improvement in noise reduction, the average absorption coefficient before treatment or the *total room absorption* should be increased at least three times. However, the application of this rule is subject to the minimum and maximum values of the first two rules.

The ceiling shown in Fig. 15–26 is one example of how part of a room may be designed to affect acoustics.

FIGURE 15–26 Room with effective sound control. *(Courtesy Columbia Acoustics and Fireproofing Co.)*

Sound-Absorbing Materials

Sound-absorbing substances are called *acoustic materials* and can generally be classified in three groups: *acoustic tiles, assembled acoustic units,* and *sprayed-on acoustic material.*

Acoustic tile is intended primarily for ceilings that may be used on upper wall surfaces as well. When used over a solid surface, such as plaster, plywood, hardboard, etc., the surface must be smooth, even, and clean. Tiles may be secured with nails or adhesive, depending on the type of backing used. In any case, the two center lines of the ceiling should be drawn first, and tiling should begin at the intersection of these lines (see Fig. 15–27). If adhesive is to be used, some should first be rubbed on the corners of each tile to act as a prime. A walnut-sized dab of adhesive is then applied to each corner. The tile should be placed close to its final location and forced to slide into place in order to spread the adhesive more completely and ensure a good bond with the backing.

Tiles may be nailed to furring strips attached to the undersides of solid decks or framed ceilings (Fig. 15–28). As usual, tiling should begin along a center line snapped across the furring strips.

Acoustic tile may also be installed in a suspended ceiling framework. Some tiles are made with square edges which rest on T-bars, while others fit over H-bars (Fig. 15–29).

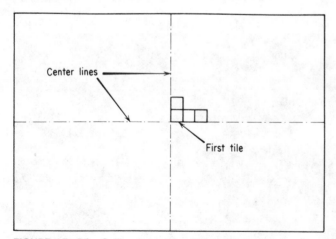

FIGURE 15–27 Ceiling layout for tile application.

FIGURE 15–28 Acoustical ceiling tile on furring strips.

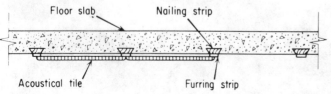

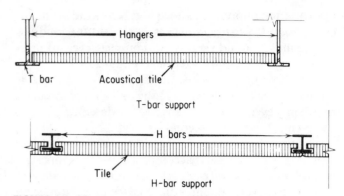

FIGURE 15–29 Acoustical tile in a suspended ceiling.

Assembled acoustic units consist of sound-absorbing material such as mineral wool and fiberglass insulation fastened to hardboard, asbestos board, or sheet metal facing which has been perforated to allow sound waves to penetrate. These units may be fastened to the wall face on furring strips, suspended in front of the wall or from the ceiling, or incorporated into the wall paneling.

Sprayed-on acoustic materials are widely used for sound control because of the relative ease of application on surfaces of almost any shape (see Fig. 15–26). Vermiculite, perlite, and asbestos fiber mixed with adhesive are all used for this purpose, as well as new, sprayed-on types of insulation which do not have an asbestos fiber base. This is a result of safety requirements which place restrictions on the use of such materials as asbestos fiber containing toxic dusts.

Vermiculite plastic is a ready-mixed product requiring only the addition of water. It can be applied over any firm, clean surface such as plaster base coats, masonry, galvanized metal, etc. It is applied in two coats if no more than ½ in. is required or in three coats if a greater thickness is specified. The first coat should be about ⅜ in. thick, straightened with a darby, and allowed to dry. The second coat is then sprayed on to the desired thickness and texture. The temperature should not be less than 55 °F for a week prior to the application of the plastic, during its application, and for long enough after its application to allow it to set thoroughly. Adequate ventilation must be provided.

Asbestos fiber insulation is a prepared material containing an adhesive mixed with the fiber. It is sprayed on any solid surface or on metal lath backing in several thin coats to a total depth of from ½ to ¾ in. The surface to be coated must first be primed with an adhesive (except in the case of metal lath). Prime only as much area as can be sprayed with fiber while the primer is still tacky. The ceiling in Fig. 15–26 has been sprayed with asbestos acoustic insulation. Notice that additional sound absorption is provided by the carpet and the drapes.

Sound Transmission Control

In nearly all types of construction, sound transmission from room to room takes place as a result of diaphragmatic vibration through walls, floors, or ceilings. The vibrations may be initiated by impact, such as a footstep, or by the action of sound waves striking the surface. One is known as *impact transmission* and the other as *airborne transmission*.

Impact transmission normally takes place through floors and ceilings, and there are several ways of reducing it. One is to cover the floor with a resilient material, such as cork tile or a heavy carpet, to absorb the impact. Another is to use a suspended ceiling between floors. This is particularly effective if resilient hangers are used.

Good results are also obtained by using a *floating* floor. The floating floor may be a separate 2-in. concrete slab, isolated from the structural slab by a 1-in. foamed plastic or paper-covered fiberglass quilt, it may be a wood floor nailed to sleepers which are supported on flat steel springs (Fig. 15–30), or it may be a wood floor laid on sheets of resilient material such as fiberboard or foamed plastic (see Fig. 15–31).

The sound insulating efficiency of a wall or floor is known as its *transmission loss* and is measured in *decibels* (dB). It is usually desirable to provide walls and floors with transmission loss ratings of from 35 to 50 dB or better, depending on the purposes for which the building is to be used.

A high transmission loss may be achieved by using either a heavy wall or one composed of two or more relatively independent layers. In either case it is essential that the wall be as airtight as possible.

Several types of construction will produce walls with transmission loss ratings of 50 dB or better. Among them are

1. A single masonry wall weighing at least 80 psf (e.g., 8 in. of brick or 7 in. of concrete), including plaster, if any.

2. A masonry cavity wall consisting of two wythes weighing at least 20 psf each, held together with metal ties, and enclosing a 2-in. cavity.

3. A composite wall, consisting of a basic masonry wall weighing at least 22 psf (e.g., 4 in. of hollow clay tile or 3 in. of solid gypsum tile), with a sheath made of ½-in. gypsum lath mounted with resilient clips (see Fig. 15–32) and plastered with ¾ in. sanded gypsum plaster.

4. A stud wall consisting of 2″ × 4″ studs with ½-in. gypsum lath mounted on resilient clips (see Fig. 15–32) on each side and plastered with ½-in. gypsum plaster. Mineral wool or fiberglass batts are placed between the studs.

5. A staggered stud wall made of 2″ × 3″ studs set at 16-in. centers on a common 2″ × 6″ plate; ½-in. gypsum lath is nailed on both sides and plastered with ½ in. sanded gypsum plaster. Paper-backed mineral wool or fiberglass batts are hung between studs on one side only.

Wall constructions which will produce transmission loss ratings of from 45 to 49 dB include

1. A single masonry wall weighing at least 36 psf (e.g., 4-in. solid gypsum tile or 4-in. brick wall) including plaster, if any.

2. A composite wall similar to #3 above but with furring strips to support the gypsum laths.

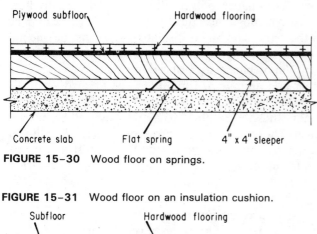

FIGURE 15–30 Wood floor on springs.

FIGURE 15–31 Wood floor on an insulation cushion.

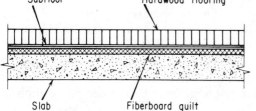

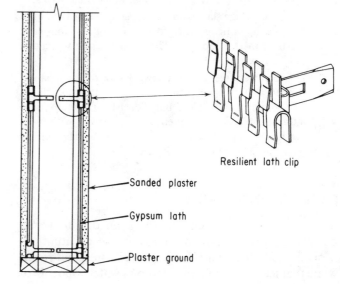

FIGURE 15–32 Sound insulation with lath clips.

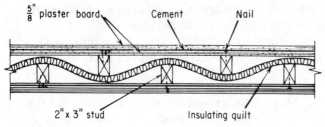

FIGURE 15-33 Sound insulation with staggered studs.

3. A staggered stud dry wall, consisting of two sets of 2″ × 3″ studs at 16-in. centers on a common 2″ × 4″ plate. Two layers of ⅝-in. gypsum wallboard are applied to each face, the first nailed and the second cemented. The joints must be staggered and both surfaces must be sealed. A mineral or glass wool blanket is placed between the boards (Fig. 15-33).

A single masonry wall weighing at least 22 psf including plaster, if any, should have a transmission loss rating of from 40 to 44 dB.

Walls with ratings of from 35 to 40 dB include

1. A 2″ × 3″ or 2″ × 4″ stud wall with ⅜-in. gypsum lath and ½-in. sanded gypsum plaster on both sides.
2. A 2″ × 3″ or 2″ × 4″ stud wall with two layers of ⅜-in. plasterboard on each side, the first one nailed, the second cemented, with staggered joints.

Floors which will have an airborne transmission loss rating of 50 dB or better include

1. A 4-in. solid concrete slab or its equivalent, weighing at least 50 psf, plastered directly on the underside and covered with wood flooring on wood furring strips. However, its impact rating will not be more than 30 dB.
2. A 4-in. concrete slab as above, covered with a 1-in. quilt of foamed plastic or paper-covered fiberglass supporting a 2-in. concrete top slab. The impact rating will again not be more than 30 dB.
3. An open web steel joist framework covered by a paper-backed fiberglass or foamed plastic quilt supporting a 2-in. concrete slab. The ceiling consists of ½-in. gypsum lath mounted on resilient clips and plastered with ½-in. sanded gypsum plaster. This floor will not have an impact rating of more than 30 dB.

MOISTURE INSULATION

Moisture may migrate either into or out of a building, but it is a different type of moisture in each situation. Moisture entering a building is normally in liquid form, while that leaving a building does so as water vapor. Pro-

tection against both is available: *moisture barriers* prevent the passage of water, while *vapor barriers* prevent the passage of water vapor.

Moisture barriers are applied to the outside of a wall or roof to prevent water from entering the structure. On a roof, the roofing material is the moisture barrier. The cladding or exterior finish on a wall is also a moisture barrier, though not perfect in every case. Therefore it is sometimes desirable to apply a moisture barrier (usually an asphalt-impregnated paper) behind the finish. This is generally done in the case of siding, shingles, or other materials which do not have watertight joints.

Masonry walls are protected below ground by the application of a coating of asphalt. Above ground level, masonry walls may be given a coating of transparent sealer to prevent the penetration of water.

Concrete slabs on the ground should have a moisture barrier between them and the earth to prevent the migration of water from the soil into and through the concrete. The best material for this purpose is 6-mil polyethylene film. It is available in wide sheets, but there should be a lap of at least 6 in. where joints are required. Care must be taken that the film is not punctured during the installation of the reinforcement or while placing the concrete.

Vapor barriers are applied inside the structure to prevent moisture vapor from reaching a dew point area *within the wall* where it could condense and collect. This could result in deterioration of the building material or saturation of the thermal insulation, causing a reduction or complete loss of its insulating value.

Materials commonly used as vapor barriers include waxed kraft paper, polyethylene film, foils of aluminum or other metals, and latex paint. They must be applied on the warm side of the insulation and must be one unbroken surface over the entire area.

Vapor barriers produced in strips, such as waxed paper and aluminum foil, should be applied vertically over stud walls or furring strips with their edges lapped on the studs so that they may later be sealed by interior finish. The ends of these strips should extend 6 in. under the ceiling finish and under the flooring. If it is necessary to cut the vapor barrier at openings such as electrical outlets, etc., care should be taken not to make the hole larger than is absolutely necessary.

Polyethylene film is usually available in sheets wide enough to reach from floor to ceiling, and this type of barrier is desirable if there is no practical method of sealing the joints in narrower strips.

Latex paint usually serves a dual purpose—it is decorative and it is a vapor barrier.

Whether or not a vapor barrier should be applied to a ceiling depends on the type of construction. If the space above the ceiling is sealed and cold, a vapor barrier should be used, but if it is ventilated, a vapor barrier is not needed since moisture vapor can escape through the ceiling and then be carried to the outside by cross ventilation.

REVIEW QUESTIONS

1. Define clearly **(a)** *thermal insulation,* **(b)** thermal *bridge,* **(c)** cellular concrete, **(d)** *shoe tile,* and **(e)** *perlite.*

2. Explain clearly why **(a)** cellular concrete is a good fire-proofing material, **(b)** long columns must be insulated, **(c)** hot air ducts must be insulated, and **(d)** vermiculite is a good insulator.

3. Describe an *ideal* thermal insulator.

4. Outline three important steps which must be taken when applying perimeter insulation to the exterior of a foundation wall and explain why they are important.

5. Outline the steps which must be taken to ensure the effectiveness of reflective insulation and explain why they are important.

6. Differentiate between **(a)** *airborne* transmission and *impact* transmission of sound, and **(b)** a *moisture barrier* and a *vapor barrier.*

7. Outline three methods of reducing impact transmission of sound through floors.

8. Determine the *U*-factor of a wall having the following components: stucco, 1 in. polystyrene insulation, 8 in. concrete block, 1 in. polystyrene insulation, and ½ in. drywall.

Finishing

During the construction of the structural frame a number of other elements in the building must be considered. Among these are the heating and ventilating, sewer and water, fire protection systems, and the electrical requirements. In high-rise and retail structures, elevators and escalators must also be incorporated into the structural frame. Many good references are available on the various systems currently in use and it is beyond the scope of this text to deal with the many details.

To ensure that all these various elements are dealt with at the appropriate time, some method of job planning and control must be available. Job planning and scheduling can be accomplished in a number of ways, depending on the size and nature of the project. On small projects the forepersons may plan the activities from day to day using only a simple bar chart and their own experience. On a large construction site a whole group of people may be responsible for the planning and control of the job. Sophisticated methods such as the critical path method of scheduling used in conjunction with a computer allow for very close monitoring of every aspect of the job.

The completion of the building frame and the structural floors and the erection of curtain wall, where applicable, marks the end of one phase in the construction of a large building. The second phase includes the application of exterior veneer facings, installation of outside doors and glass entranceways, window installation, completion of stairs, application of interior wall and ceiling finishes, installation of finish floors, hanging of inside doors, and, in some cases, the installation of partitions.

EXTERIOR FACINGS

Exterior facings include a host of materials of all kinds. Some of them, such as ceramic veneer, stone veneer, or split block, are attached by a *mortar bond* to the backup wall (see Figs. 16–1 and 16–2). Others, such as facing tile or brick veneer (see Fig. 16–3), are tied to the wall by some type of wire tie. Still others—sheet facings such as asbestos-cement, plastic, or sheet steel (see Fig. 16–4)— are attached by means of pins, clips, or snap-on assemblies.

OUTSIDE DOORS

Outside doors will vary widely in material and style, depending on the type of building. They may be made of wood, glass, metal, or combinations of them. They will usually be set in steel or aluminum frames and swing on hinges or pivots (see Fig. 16–4). Steel and aluminum frames are provided with appropriate jamb anchors and often a bottom anchor to be fastened to the floor.

FIGURE 16–1 Split block exterior facing. (*Courtesy National Concrete Masonry Association.*)

FIGURE 16–3 Applying a brick veneer facing. (*Courtesy Morgen Manufacturing Co.*)

FIGURE 16–2 Applying a stone veneer exterior facing. (*Courtesy Morgen Manufacturing Co.*)

FIGURE 16–4 Sheet steel exterior facing. (*Courtesy Stran-Steel Corp.*)

It is common practice for the manufacturer to fit the doors into their frames, so that they are brought to the job site ready for installation. The doors must be removed from their frames and stored until building has progressed to the finishing stage. The frames can be placed in their openings as required.

When wood doors have to be hung on the job, it is important that hinges be properly installed to ensure easy operation. Figure 16–5 indicates hinge clearance details for various thicknesses of door and size of hinge.

Steel swinging doors are hung and operated in similar fashion to wood doors. Hinge recesses and lock holes are formed into the door to match hinge and strike recesses in the frame.

The hardware for glass swinging doors is installed in the shop, and the door is delivered to the job ready to be set in place. Figure 16–6 shows the typical hardware on a plate glass door hung in an aluminum frame.

Various items of hardware may be specified for any particular door. Figure 16–7 shows a number of them and indicates their proper location.

Installation of glass panels flanking outside doors in entranceways will be similar to that of other glass cur-

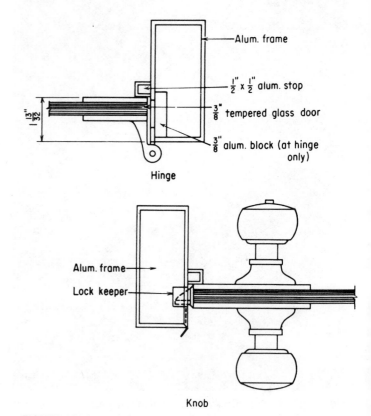

FIGURE 16–6   Hardware on a glass door.

FIGURE 16–5   Hinge clearance details.

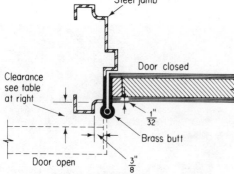

Clearance of stock size butts

Thickness of door	Size of butt	Maximum clearance
$1\frac{3}{8}''$	$3'' \times 3''$	$\frac{3}{4}''$
	$3\frac{1}{2}'' \times 3\frac{1}{2}''$	$\frac{7}{8}''$
	$4'' \times 4''$	$1\frac{5}{8}''$
$1\frac{3}{4}''$	$4'' \times 4''$	$1''$
	$4\frac{1}{2}'' \times 4\frac{1}{2}''$	$1\frac{3}{8}''$
	$5'' \times 5''$	$2''$
$2''$	$4\frac{1}{2}'' \times 4\frac{1}{2}''$	$1\frac{1}{4}''$
	$5'' \times 5''$	$1\frac{3}{4}''$
	$6'' \times 6''$	$2\frac{3}{4}''$
$2\frac{1}{4}''$	$5'' \times 5''$	$1\frac{1}{4}''$
	$6'' \times 6''$	$2\frac{1}{4}''$
	$6'' \times 7''$	$3\frac{1}{4}''$
	$6'' \times 8''$	$4\frac{1}{4}''$

FIGURE 16–7   Location of door hardware.

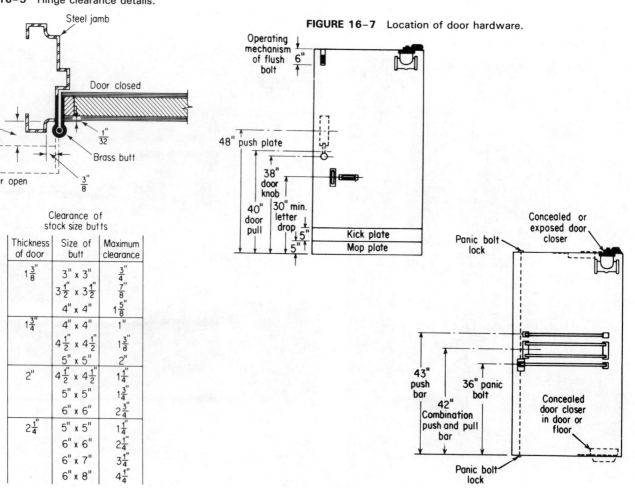

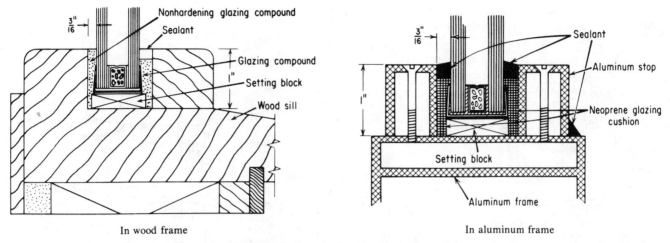

In wood frame

In aluminum frame

FIGURE 16-8 Installation of sealed glass unit.

tain walls. Figure 16-8 illustrates the installation of a glass panel in both wood and aluminum frames.

WINDOWS

Window styles for large buildings include *conventional wood sash* (see Fig. 16-9), similarly styled *windows with aluminum sash, sashless sliding glass panels, window walls* with wood or aluminum sash and large areas of glass curtain wall (see Fig. 16-10), the style depending to some extent on the use for which the building is intended. The important thing with all of them, from the standpoint of construction, is the size of the *rough opening—*

the opening into which the complete window unit will fit. Window manufacturers normally produce data on their products which include the size of the opening required for each unit. Figures 16-11 and 16-12 illustrate typical details on wood and aluminum sash, produced by manufacturers.

FIGURE 16-10 Aluminum frame glass curtain wall.

FIGURE 16-9 Typical wood sash, apartment window.

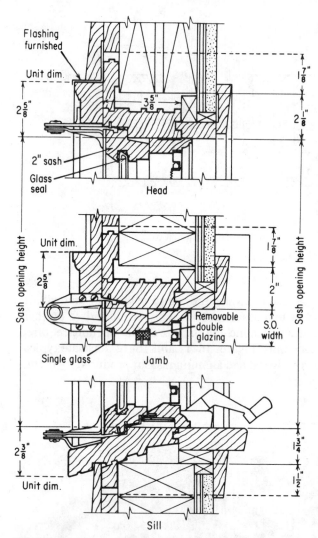

FIGURE 16-11 Typical details on a wood sash window. (*Courtesy Andersen Corp.*)

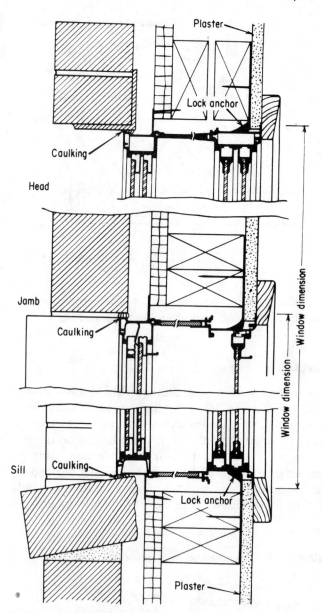

FIGURE 16-12 Typical details on an aluminum sash. (*Courtesy Humphrey Products.*)

PLASTERING

Plaster is well known and widely used for finishing walls and ceilings in many types of buildings. Either portland cement or gypsum plaster may be used, depending on the composition of the wall and the conditions to which the surface will be subjected. Portland cement plaster may be applied to concrete, masonry, and metal lath bases and is used where walls, ceilings, and partitions are subject to rough use or extreme moisture conditions. Gypsum plaster bonds well to gypsum lath, metal lath, fiberboard lath, and gypsum or clay tile. Plaster is usually applied in three coats: the first, or *scratch* coat; the second, or *brown* coat; and the finish coat.

Portland Cement Plaster

Portland cement base plaster is made by mixing 1 part of normal portland or masonry cement with 3 to 4 parts of plaster sand. Sand grading should be as follows:

- *Passing #4   sieve:*   100%
- *Passing #8   sieve:*   80% to 98%
- *Passing #16   sieve:*   60% to 90%
- *Passing #30   sieve:*   35% to 70%
- *Passing #50   sieve:*   10% to 30%
- *Passing #100 sieve:*   not more than 10%

Larger proportions of aggregate to cement may be used when the aggregate is well graded, with a good proportion of coarse particles. When normal portland cement is used, a plasticity agent may be used to increase the workability of the mortar.

The scratch coat should be approximately ⅜ in. thick. It should be *dashed* on concrete walls unless the surface is sufficiently rough to ensure an adequate bond for a troweled coat. The scratch coat must be troweled on clean, dry masonry walls. Before plastering begins, they should be evenly dampened to control suction. Hair or fiber (at the rate of about 1 lb/bag of cement) should be used in a scratch coat applied to metal lath. The scratch coat must be crosshatched before it hardens to provide a mechanical bond for the brown coat. It should be kept damp for at least 2 days immediately following its application and should then be allowed to dry thoroughly.

The surface of the scratch coat should be dampened evenly before beginning the application of the brown coat, which is about ⅜ in. thick, brought to a true, even surface, and then either roughened with a wood float or crosshatched lightly. The brown coat should be damp-cured for at least 2 days and then allowed to dry.

Finish coats may require a slightly finer aggregate, but excessive fineness should be avoided. Color may be added in the form of mineral pigments, at the rate of not more than 6% of the cement, by weight. The usual finish on interior portland cement plaster is a sand float or low-relief texture finish, but plaster may be troweled smooth, if specified. Moist curing should take place for at least 7 days. During plastering operations, the temperature should be maintained at or above 50°F.

Gypsum Plaster

Several types of gypsum plaster are manufactured, each for a specific use. They include *hardwall* plaster, for scratch and brown coats; *cement bond* plaster, for base plaster (to be applied to a concrete surface); and *finish* plasters, for the finish coat over a gypsum plaster base.

Either sand or vermiculite may be used as aggregate with gypsum hardwall plaster for scratch and brown coats. Use 2 parts of dry sand by weight to 1 part of plaster for the scratch coat over gypsum lath, wood lath, or metal lath. For the scratch coat over gypsum tile, brick, or clay tile, use 3 parts of sand by weight to 1 part of plaster. The same general specifications apply for sand used in gypsum plaster as for that used in cement plaster.

The scratch coat is approximately ⅜ in. thick and is crosshatched to receive the brown coat. Curing is carried out in the same manner as for cement plaster.

For the brown coat, use 3 parts of dry sand by weight to 1 part of plaster. This coat is also approximately ⅜ in. thick, floated, and broomed to a straight, even surface, ready for the finish coat.

If vermiculite aggregate is specified, use 1 cu ft of aggregate to 1 sack of hardwall plaster over gypsum lath. If the plaster is to be applied over masonry, 1½ cu ft of aggregate may be used per sack of hardwall. For the brown coat, use approximately 1½ cu ft of vermiculite per sack of plaster.

Cement bond plaster requires only the addition of water. It is applied in two coats to a total thickness of approximately ⅜ in. on ceilings and ⅝ in. on walls. Before it has set, the surface must be broomed to receive the finish coat.

Several types of gypsum plaster are available for the finish coat. One is the widely used gypsum finish plaster. It is mixed with lime putty to make the putty coat. Lime putty is produced either by slaking quick lime or by soaking hydrated lime for about 12 hr. The two ingredients are mixed in the following ratios: 1 part finish plaster to 2 parts dry hydrated lime by weight or 1 part plaster to 3 parts lime putty by volume. Water is added to the mixture to attain the desired degree of plasticity.

This putty finish is applied in two coats over a fairly dry brown coat base. A thin first coat should be ground into the base thoroughly, and a second coat should then be used to fill in the imperfections. The surface is troweled to a smooth finish, sprinkling water on with a brush to provide workability.

A sand float finish is produced by using screened (hairs or fibers have been removed) hardwall plaster and sand in the proportion of 1 part plaster to not more than 2 parts of sand by weight. The maximum size of sand particles will determine the coarseness of the finish. This plaster is applied in two coats to a total thickness of not more than ⅛ in. over a set but not fully cured hardwall base coat. After leveling, the surface is floated with a carpet or cork float to produce the desired texture.

A prepared finish plaster is available which requires only the addition of water. It is mixed at the rate of approximately 2 parts plaster to 1 part water by volume. The plaster should be allowed to soak for at least 30 min before it is used. Application and finishing are essentially the same as for a lime putty coat. This plaster does not produce as white a surface as lime putty, but since it contains no lime, it may be decorated as soon as it is dry.

Plaster of paris and Keene's cement are two other types of gypsum finish plaster with special applications. Plaster of paris is used where a very rapid set is required, and Keene's cement is used where sanitary conditions or high humidity necessitate a hard, impervious surface.

Vermiculite Finish Plaster

A finish plaster in which vermiculite is used as the aggregate is made by mixing vermiculite and screened hardwall at the rate of 100 lb of gypsum to 1 cu ft of vermiculite finish aggregate. The plaster is first mixed with water and the aggregate is added to the putty. More water may be added to make the plaster workable. This

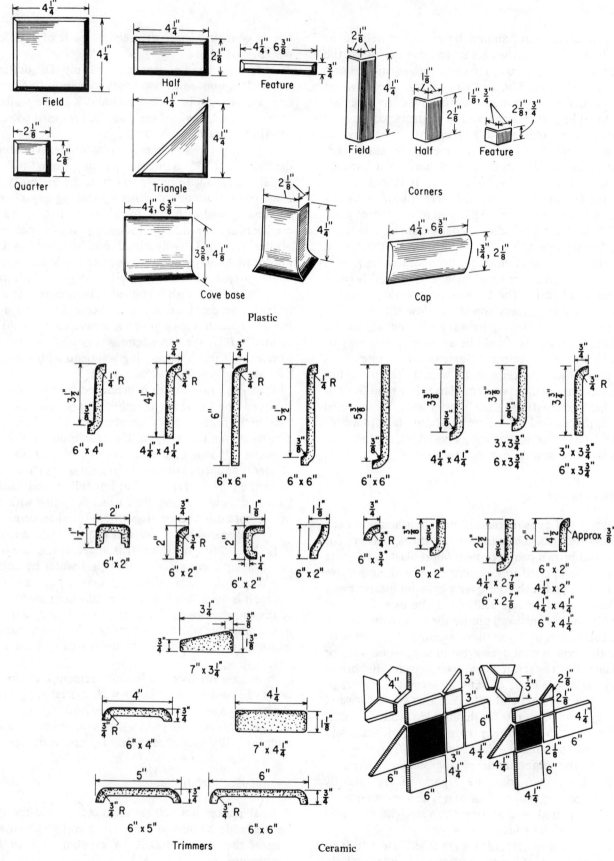

FIGURE 16-13 Typical wall tile.

finish may be applied over any gypsum base plaster to a depth of ⅛ in. The procedure for applying, leveling, and troweling is the same as that used for other finish plasters.

WALL TILE

Several kinds of wall tile are produced in various sizes for interior finishing—including ceramic, glass, steel, copper, and plastic tile. Accessories—such as base, cap, triangles, and corners—are available for most of them. Figure 16-13 illustrates the various shapes available in plastic and ceramic wall tile.

Wall tile may be applied with adhesives to most hard, smooth surfaces such as concrete, hardboard, plywood, gypsum board, or plaster. Ceramic tile may also be set in mortar.

Tile laying should begin at established level and vertical lines on the wall. If tile is to cover the entire wall from floor to ceiling, a level line should be drawn at a convenient height from the floor, measured to accommodate the base and a specified number of tiles. If the tiled surface is to be a *dado,* draw the top line of the tile and a plumb line down the center of the wall. Tile laying should begin at the intersection of these lines. All materials should be at room temperature, 60 to 70 °F. Apply the adhesive to the wall with a notched trowel, using a wavy motion. The coating should be of sufficient thickness so that when tile is pressed against it, the ridges of cement will flatten and contact at least 60% of the back surface of the tile. Do not force the edges of tiles tightly against one another; grouting compound, made for that purpose, is applied to the joints by tube or by narrow spatula after the tiles are in place. Joints are then wiped out to the desired depth.

HARDBOARD

Hardboard with various face designs—such as embossed leather or wood grain, striated, grooved, plastic-coated, either plain or cut into squares like tile—is used for interior finish.

Panels should be *conditioned* before being applied to the walls by allowing each to stand separately on its long edge for at least 24 hr. Do not try to install hardboard panels in abnormally damp areas.

Hardboard may be attached to the wall with nails, with batten strips, or with adhesives. If nails are used, drill a shallow hole at nail locations with a drill slightly smaller than the shank of the nail. Nails should be spaced

about 8 in. apart at intermediate supports and 4 in. apart around the edges. A solid backing must be provided when the board is to be applied with adhesive. Spread adhesive over the entire back surface with a saw-toothed applicator and brace the panels in place until the adhesive sets.

GYPSUM BOARD

Gypsum board is often used instead of plaster as interior finish for walls and ceilings. This is known as *drywall* construction. It may be a single application of ½- or ⅝-in. or a double thickness of ⅜-in. board. The single layer is secured with special nails or by gluing to studs or furring strips. When a double thickness is used, the inner layer is nailed vertically and the outer layer is cemented to it horizontally with a gypsum cement. The outer layer is held in place with double-headed nails which are removed once the cement has set. Nail holes are filled with gypsum joint filler at the same time as the joints are taped and filled. The joints are sized after sanding, and the surface is ready to be painted.

Gypsum board is produced with printed wood grain patterns and with a plastic-fabric-coated surface. The former is applied with nails whose heads have the same color as the pattern, while the fabric-surfaced board is usually held by aluminum battens, producing a paneled effect. The battens are placed at studs, held in place by screws to the stud and are capped with a plastic batten strip (Fig. 16-14).

FIGURE 16-14 Gypsum board over metal studs.

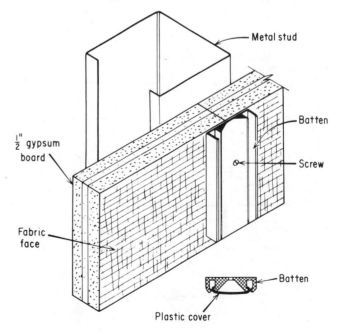

PLASTIC LAMINATES

Plastic laminates in thickness ranging from $\frac{1}{16}$ to $1\frac{1}{2}$ in. are used in various ways for interior finishing. The thin sheets, $\frac{1}{16}$, $\frac{1}{10}$, $\frac{1}{8}$, etc., are used by themselves or laminated to plywood backing to line interior walls. Thin sheets are applied directly by the use of contact adhesives. Edges may be butted together, but aluminum edge moldings are generally used (Fig. 16–15). These moldings may be used with other types of wall paneling as well as plastic laminates.

The application of a thin plastic laminate sheet to plywood backing is usually done in a factory or shop. The plastic laminate may be applied to a flat sheet of plywood or may be *postformed* over a curved surface. Faced plywood sheets or strips are used as paneling, base, or ceiling cove. Figure 16–16 illustrates two methods which may be used to attach panels to walls for a flush joint fit. Paneled effects can be produced in several ways, some of which are illustrated in Fig. 16–16.

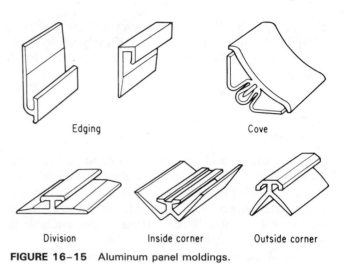

Edging   Cove

Division   Inside corner   Outside corner

FIGURE 16–15 Aluminum panel moldings.

Thick plastic laminates are used for partitions, sliding doors, baseboards, window sills, table tops, etc., and are manufactured to specifications.

FIGURE 16–16 Plastic laminate paneling.

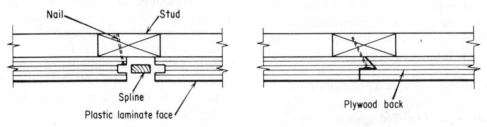

Flush joint panel assemblies

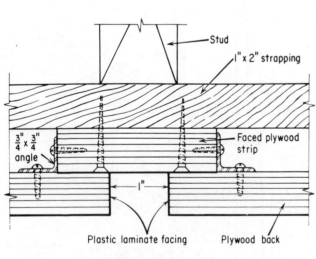

Divided panel assembly

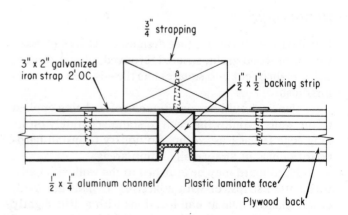

Paneling with aluminum channels

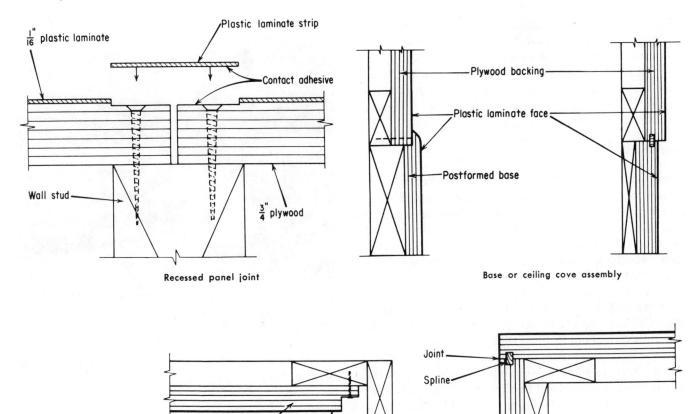

$\frac{1}{16}''$ plastic laminate

Plastic laminate strip

Contact adhesive

Wall stud

$\frac{3}{4}''$ plywood

Recessed panel joint

Plywood backing

Plastic laminate face

Postformed base

Base or ceiling cove assembly

Plywood

Plastic laminate face

Internal

Joint

Spline

External

Panel assembly at corners

FIGURE 16–16 (Continued).

MASONRY FINISHES

Facing with Brick

Interior walls may be faced with standard 4-in.-wide brick or with thin brick veneer units. Standard brick may be either common or face brick, depending on use and appearance. For example, different-colored bricks arranged in varying patterns can produce a pleasing finish to an interior wall (Fig. 16–17).

The procedure for applying face brick to interiors is similar to that used for exterior facing brick except that stack bond is more commonly employed. The brick may be bonded to the backup wall with metal ties or by a mortar coat between backup and face brick. Particular attention should be paid to the mortar joints. Sand should be

FIGURE 16–17 Varying brick patterns.

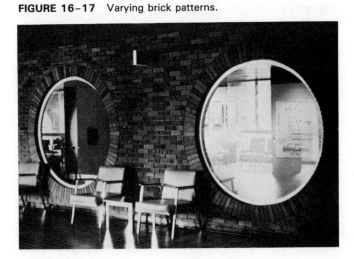

fine enough to pass a #16 sieve. Where nonstaining mortar is specified, ammonium or calcium stearate is added at the rate of 3% of the weight of cement used.

Interior curtain walls or load-bearing walls faced with brick provide good sound insulation as well as a high degree of fire protection in addition to providing a durable wall finish (Fig. 16–18).

Facing Tile

Facing tiles are available in thicknesses of from 2 to 8 in. and may be laid with either horizontal or vertical cells. Tile laid with horizontal cells should have divided bed and head joints. Enough mortar should be used so that the excess will ooze from the joints as the unit is pressed into place. Excess mortar is then struck off and the joints are tooled to a flush or concave surface.

When facing tile is laid with vertical cells, the mortar bed should not carry through the wall on the connecting webs and end shells except at corners and at ends. The cells should not be filled with mortar but should instead be left open to allow drainage of any moisture penetrating from the surface. Weep holes should be provided at the bottom of the wall to allow moisture to escape.

Mortar for tile is similar to that used for face brick. If pointing mortar is specified, regular mortar is raked back, and pointing mortar is applied with a tuck pointer's trowel. Pointing mortar should consist of 1 part portland cement and ⅛ part of hydrated lime to 2 parts of fine sand (50 mesh or finer). Ammonium or calcium stearate is added at the rate of 2% of the weight of cement used.

Brick and tile surfaces should be cleaned with burlap as work progresses and later scrubbed with a stiff brush and water when the walls are completed. Mortar stains on unglazed brick or tile may be removed by using a wash consisting of 1 part hydrochloric acid to 9 parts

FIGURE 16–18 Interior brick curtain wall.

FIGURE 16–19 Ceramic wall finish.

of water by volume. Do not use this solution on glazed brick or tile.

Ceramic Mosaic

Ceramic mosaics are small, flat tile units which have been assembled into sheets in the plant and held by cement on a polyethylene-coated backing material or by a paper sheet over the tile faces. Back-mounted units are applied to the wall with tile cement, while face-mounted ones are usually applied by pressing the sheets of tile into a wall coating of mortar. When the mortar has set, the paper mounting is scrubbed from the face of the tile.

Joints are grouted with either fine mortar or prepared grouting compound once the tiles are in place (see Fig. 16–19).

STONE FACING

Stone used for interiors is normally applied as a veneer over walls and around columns. Cut stones in thicknesses of from 1 to 4 in. and of various face dimensions are employed. The procedure for laying stone in interior work is essentially the same as that used for stone exteriors. Small stone units are laid in mortar beds and anchored to the wall with corrugated metal ties (see Fig. 13–40). Larger slabs may be secured by any of the stone anchors illustrated in Fig. 13–36. Because of the stones' weight, two wedges are placed under the bottom edge of each stone in a course and are allowed to remain there until the mortar bed hardens. The wedges are then removed and the holes are filled with mortar (see Fig. 16–20).

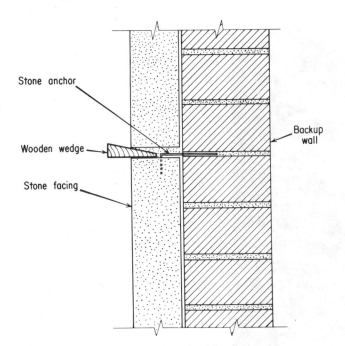

FIGURE 16–20 Stone panel wedged in place.

STEEL FACING

Interior facing material of sheet steel is available in several forms. Steel wall tile has already been mentioned. A somewhat similar product is made to fit over standard concrete blocks (Fig. 16–21). Joints between blocks must be raked out to allow the facing unit to fit around the edges. Adhesive is applied to the back of the panel, and it is then pressed into place over the face of the block.

CEILINGS

The type of ceiling to be used will depend on a number of factors, including the type of *structural floor* in the building, the location of the *mechanical services,* the *intended use* of the building, and whether *acoustical treatment* is required.

Some designers will specify that, for a particular use, the underside of prestressed concrete floor slabs, such as single or double Ts, ribbed slabs, or waffle slabs, may be painted and left exposed, as illustrated in Fig. 16–22.

Where the building's mechanical services are located below the floor slabs (see Fig. 16–23), they will have to be hidden by *suspending* the ceiling below them. Heavy wires or small rods, attached to the underside of the floor slab, support a system of T-bars. Careful attention must be paid to this operation in order to have all the bars hanging perfectly level. The T-bars, in turn, support some type of ceiling panels. In Fig. 16–23, for example, workmen are installing *steel-faced* ceiling panels.

FIGURE 16–21 Steel facing units for concrete block. (*Courtesy Steel Co. of Canada.*)

FIGURE 16–22 Underside of waffle slab as finished ceiling. (*Courtesy Portland Cement Association.*)

FIGURE 16–23 Suspended ceiling system. (*Courtesy Steel Co. of Canada.*)

FIGURE 16–24 Acoustical tile in suspended ceiling.

The acoustical treatment of the ceiling may involve plastering with acoustical plaster, if there is solid support for the plaster base. If suspended ceilings are required, acoustical panels, such as those shown in Fig. 16–24, may be used.

STAIRS

Stairs of one kind or another are a necessary requirement in any building of more than one floor. Plans sometimes call for prefabricated stairs, and temporary stairs are then required until the permanent ones have been installed. These will usually be made of wood. In other cases, stairs will be "roughed in" as construction progresses from floor to floor and are later finished as part of the interior finishing schedule.

Stairs may be of wood, concrete, stone, or steel, but the same general principles apply to their planning and design:

1. The stairwell opening must be long enough to allow ample headroom under the opening header. The usual minimum is 6 ft 4 in.

2. The stair must be at a slope that makes climbing as easy as possible. The angle of the line of flight should be from 25° to 35°.

3. The height of risers must be kept within reasonable limits. In many public buildings the maximum rise is 6 in., and in most cases risers should not be allowed to exceed 7 in.

4. The tread must be wide enough to provide safe footing (10 in. is the minimum width that should be considered).

5. All risers must be exactly the same height.

6. Long, straight flights of stairs without a break should be avoided. Most codes provide a maximum allowable length of stair without a landing.

7. Stairs should be at least 36 in. wide.

8. Nonslip treads should always be provided.

9. Handrails must be provided at the proper height (36 to 42 in.)

10. Stairs with treads of varying width from one end to the other (*winders*) should be avoided whenever possible.

Wood Stairs

Wood stairs may be made with an *open* or *cutout* stringer, a *semihoused* stringer, a *housed* stringer, or a *built-up* stringer (Fig. 16–25).

The exact riser height is determined by measuring the vertical distance from one finished floor to another and dividing that distance by the desired riser height. The result, to the next whole number above or below that figure, will provide two alternative numbers of risers which may be used. Those two numbers divided into the total rise will provide a choice in the exact height of riser to be used. The width of treads is determined by the total run available for the stairs (keeping in mind that there will normally be one less tread than riser in the complete flight) and by the requirement for headroom (if any).

It must be remembered that with all types of wood stairs except those with housed stringers, the height of the bottom riser must be reduced by an amount equal to that of the thickness of tread being used (see Fig.

16–25). If stairs are resting on a subfloor, the thickness of the finished floor must be added to the height of the bottom riser.

Stringers for housed stairs may be laid out from either the back edge or the front, and edges must be straight and parallel. The top and bottom ends of housed and semihoused stringers can be extended to provide the *easement* required (see Fig. 16–25).

Steel Stairs

Steel stairs may be made in several ways. One simple method is to use two structural channels, back to back, and weld or bolt steel treads between them.

Another method of building a steel stair is illustrated in Fig. 16–26. Two or more structural steel members (depending on the width of the stair) are used as stringers. They may be channels, I beams, or heavy plates. Tread brackets are welded to the top edge or to the inside face of the stringers, and steel treads or *subtreads*

FIGURE 16–25 Wooden stair stringers.

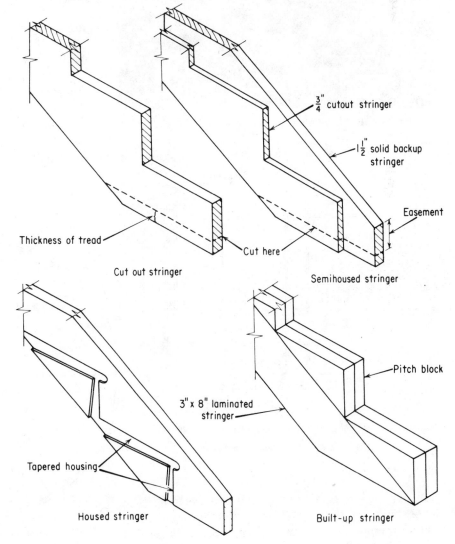

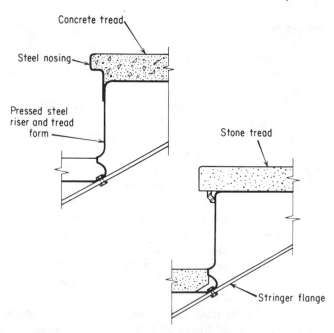

FIGURE 16–27   Stone or concrete treads on a steel stair.

are then welded to the brackets. In the stair shown in Fig. 16–26, the subtreads are hollow steel pans which will later be filled with concrete, terrazzo, or magnesite.

A somewhat similar type of stair is made by bolting pressed steel risers and treads to steel stringers (Fig. 16–27). The tread mold may receive a stone slab or may be a pan to be filled with poured-in-place tread material.

FIGURE 16–26   Built-up steel stair.

FIGURE 16–28   Spiral stair.

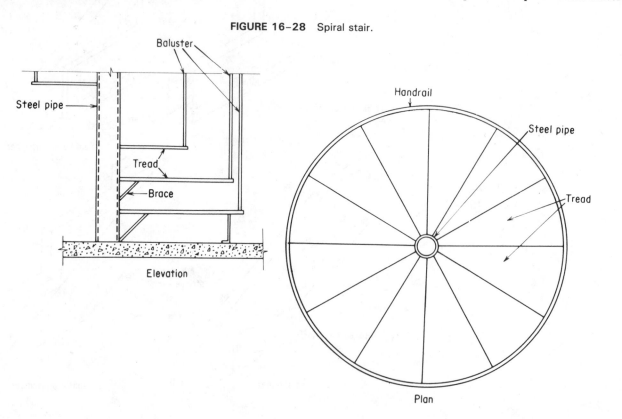

A spiral steel stair is made by welding nonslip treads or subtreads in rising succession around a central axis such as a steel pipe. This is one type of stair in which the tread is wider at one end than the other, and every effort should be made to build them as wide as possible in the normal line of travel.

Spiral stairs are gaining increasing attention for buildings such as apartment blocks because they take up less space than conventional stairs and because, properly designed and finished, they can be very attractive (see Fig. 16–28).

Concrete Stairs

Reinforced concrete stairs may be precast or cast-in-place. Precast units must be tied into the structure, which may be done by casting the topping slab around the lower end,

as in Fig. 16–29. Anchor plates cast into the end of the stair may be welded to a matching plate in the lower floor. The top end of the stair usually rests on a beam ledge.

The method used to form a cast-in-place stair will depend on the type of stair to be constructed. A *closed* string or flight (between enclosing walls) will require forms that are different from those used for an *open* flight. If the stair is cast monolithically with the landing and upper floor, the method must again be different from that used to form a stair cast between two existing floors.

A stair to be cast between existing walls requires a soffit form (Fig. 16–30) and riser forms held in place by inverted stringers. The stringers must be supported independently of the soffit form, and this may be done by wedging across the stair, from one stringer to the other.

An open flight stair also requires the soffit form, but in this case the stringers rest on the framework which

FIGURE 16–29 Concrete stairs.

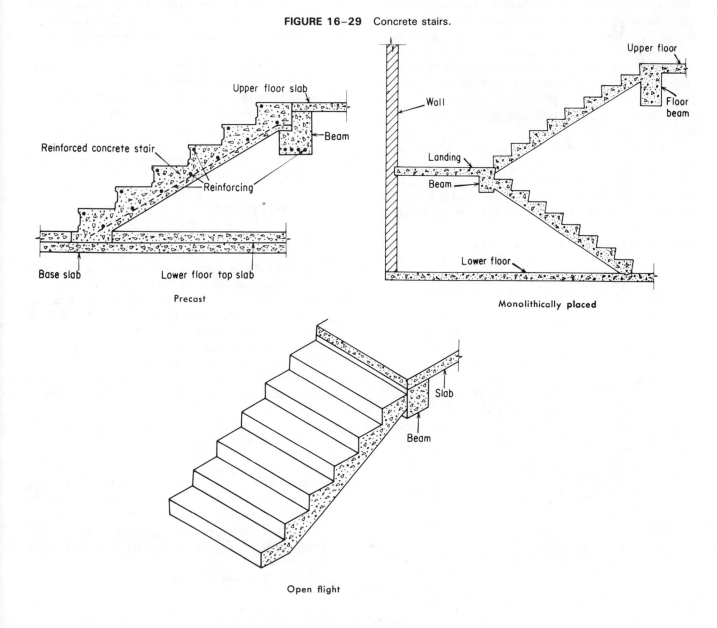

Precast

Monolithically placed

Open flight

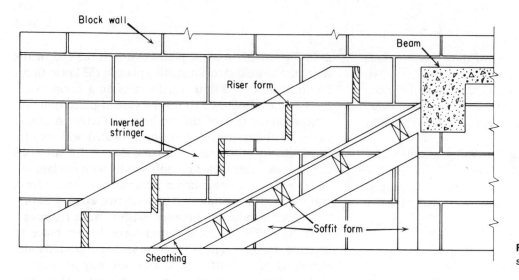

FIGURE 16-30 Form for a closed stair.

supports the soffit form (see Fig. 16–31). The bottom edge of the riser forms should be beveled so that the surface of the treads may be troweled back to their junction with the next riser. A wide stair may require extra support at the center of the risers to prevent them from bowing under the pressure of the concrete. A 2″ × 6″ cut out as a stringer and inverted may be placed against the face of the risers and braced at the bottom to provide that support.

Figure 16–32 illustrates one method of forming a stair being cast monolithically with floor or landing.

Circular cast-in-place concrete stairs present special forming problems. In the first place, the stringers must be flexible enough to be bent. Second, although the total rise of both inside and outside stringers is the same, the total run of each varies because of the difference in radius

FIGURE 16-31 Form for an open stair.

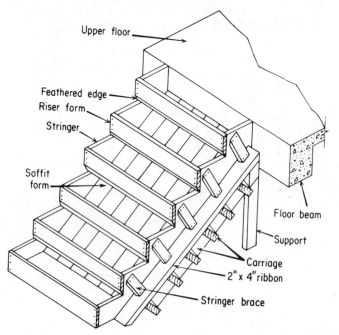

of the circles circumscribing the inside and outside edges of the stair. Finally, the soffit must be made in sections of relatively thin material if it is to fit the warped surface of the underside of the stair. To lay out and build the forms for a circular concrete stair, proceed as follows:

1. Draw circles representing the inside and outside circumferences of the stair to scale (Fig. 16–33).
2. Measure the total rise and also determine the number and height of risers required. This will be the same for both inside and outside stringers.
3. Determine the number of treads and calculate the width of each on the outside circumference. This is the tread run for the outside stringer.
4. Step off these tread widths around the outside circle and join each point to the center of the circles. The distance between points at which the inside circle is cut represents the tread width at the inside of the stair and is the tread run for the inside stringer.
5. Calculate the total run for both stringers and lay both out against a common total rise (Fig. 16–33).
6. Make the vertical distance from the junction of the rise and the run to the bottom edge of the stringer (distance x, Fig. 16–33) the same for both stringers.
7. Build inside and outside circular walls to carry the stair form. Make the necessary allowance for the thickness of the stringer material, as in Fig. 16–33.
8. Bend the stringers around the face of the walls and fasten them in place.
9. Make a template for one section of soffit. Lay out a full-size plan view of one tread and draw in the center line. On either side of this center line, lay out half the length of distance a at one end and half the length of distance b at the other, at right angles to the center line. Join the ends of these lines. The figure LMNO represents the approximate shape of the soffit section.
10. Cut this template out of thin material and shape its ends until the piece fits snugly against the form walls under the stringers. Cut as many soffit sections of this shape as there are treads in the stair.

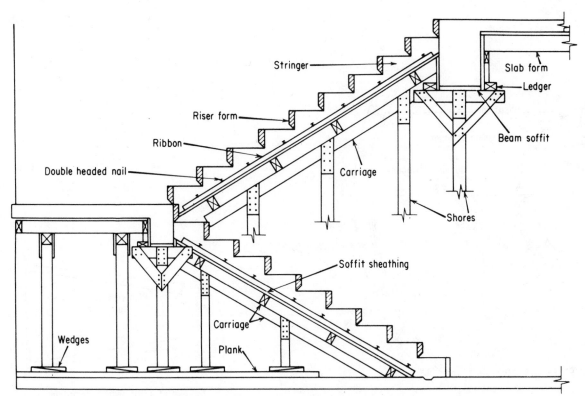

FIGURE 16-32 Form for a monolithically cast stair.

FIGURE 16-33 Layout for half-circle stair.

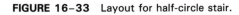

Note: AB = AB
AE = 2 AB
CD = CD
CE = 2 CD

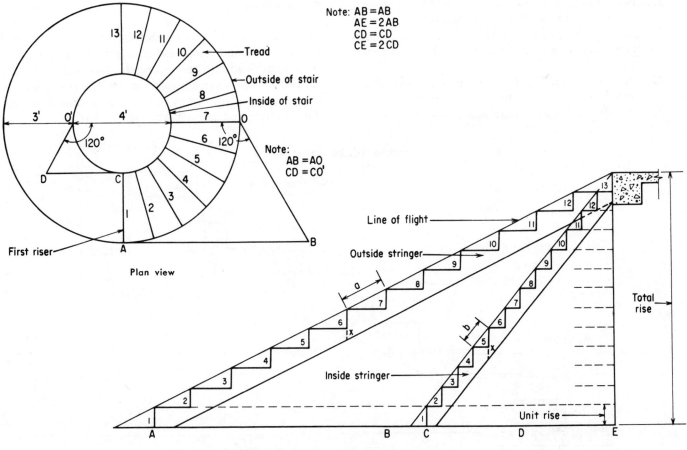

Note:
AB = AO
CD = CO'

Plan view

Layout of stringers

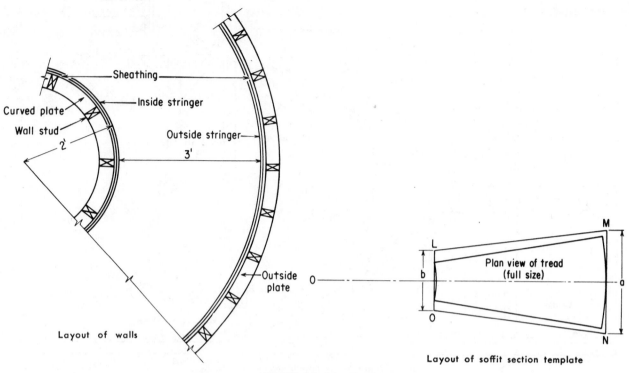

FIGURE 16-33 (Continued).

11. Nail the soffit sections to the bottom edge of the stringers, nailing a ribbon under them for support.

12. Nail riser forms with tapered bottom edges to the stringer risers, and nail blocking in front of each to provide further support.

13. Place the necessary shoring under the center of the soffit to keep the soffit sections in line.

14. Place the specified reinforcement in the stair form.

Stone Stairs

Stone stairs are usually a combination of either stone and steel or stone and reinforced concrete. In the stone and steel combination, steel stringers and subtreads are capped with stone treads (see Fig. 16-27). In the second case, the stair is formed by concrete, and stone slabs provide the finished surface (see Fig. 16-34).

FIGURE 16-34 Stone stair.

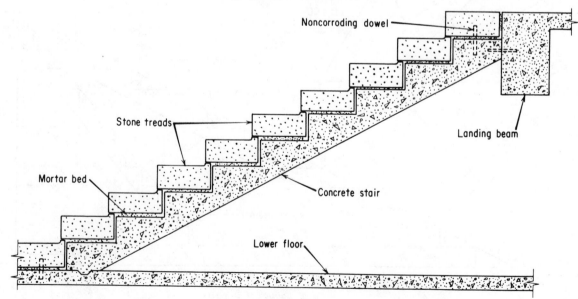

REVIEW QUESTIONS

1. Explain: **(a)** under what conditions portland cement plaster is preferred to gypsum plaster, **(b)** why masonry walls should be dampened before applying a plaster *scratch coat,* **(c)** what purpose *hair* or *fiber* serves in a scratch coat, **(d)** what is meant by *crosshatching* a scratch coat, and **(e)** why more *sand* or *lightweight aggregate* may be used in mixing the *brown coat* than the scratch coat.

2. Define the following terms used in stair design and construction: **(a)** line of flight, **(b)** headroom, **(c)** tread run, **(d)** total rise, **(e)** baluster, **(f)** newel post, **(g)** winder, **(h)** easement, and **(i)** soffit.

3. Outline the reason why **(a)** the vertical edges of wood doors should be dressed to a slight bevel, **(b)** riser forms for cast-in-place concrete stairs should have their bottom edges tapered to the back face, **(c)** hardboard panels should be *conditioned* prior to installation, and **(d)** *setting blocks* are used in the installation of plate glass windows.

Index